THE ELEMENTS

							VIIA	O
			IIIA	IVA	VA	VIA	**1** **H** 1.00797 ± 0.00001	**2** **He** 4.0026 ± 0.00005
			5 **B** 10.811 ± 0.003	**6** **C** 12.01115 ± 0.00005	**7** **N** 14.0067 ± 0.00005	**8** **O** 15.9994 ± 0.0001	**9** **F** 18.9984 ± 0.00005	**10** **Ne** 20.183 ± 0.0005
IB	IIB		**13** **Al** 26.9815 ± 0.00005	**14** **Si** 28.086 ± 0.001	**15** **P** 30.9738 ± 0.00005	**16** **S** 32.064 ± 0.003	**17** **Cl** 35.453 ± 0.001	**18** **Ar** 39.948 ± 0.0005
28 **Ni** 58.71 ± 0.005	**29** **Cu** 63.54 ± 0.005	**30** **Zn** 65.37 ± 0.005	**31** **Ga** 69.72 ± 0.005	**32** **Ge** 72.59 ± 0.005	**33** **As** 74.9216 ± 0.00005	**34** **Se** 78.96 ± 0.005	**35** **Br** 79.909 ± 0.002	**36** **Kr** 83.80 ± 0.005
46 **Pd** 106.4 ± 0.05	**47** **Ag** 107.870 ± 0.003	**48** **Cd** 112.40 ± 0.005	**49** **In** 114.82 ± 0.005	**50** **Sn** 118.69 ± 0.005	**51** **Sb** 121.75 ± 0.005	**52** **Te** 127.60 ± 0.005	**53** **I** 126.9044 ± 0.00005	**54** **Xe** 131.30 ± 0.005
78 **Pt** 195.09 ± 0.005	**79** **Au** 196.967 ± 0.0005	**80** **Hg** 200.59 ± 0.005	**81** **Tl** 204.37 ± 0.005	**82** **Pb** 207.19 ± 0.005	**83** **Bi** 208.980 ± 0.0005	**84** **Po**	**85** **At**	**86** **Rn** (222)

63 **Eu** 151.96 ± 0.005	**64** **Gd** 157.25 ± 0.005	**65** **Tb** 158.924 ± 0.0005	**66** **Dy** 162.50 ± 0.005	**67** **Ho** 164.930 ± 0.0005	**68** **Er** 167.26 ± 0.005	**69** **Tm** 168.934 ± 0.0005	**70** **Yb** 173.04 + 0.005	**71** **Lu** 174.97 ± 0.005

95 **Am** (243)	**96** **Cm** (247)	**97** **Bk** (247)	**98** **Cf** (249)	**99** **Es** (254)	**100** **Fm** (253)	**101** **Md** (256)	**102** **No** (253)	**103** **Lw** (257)

Atomic Weights are based on C¹²—12.0000 and Conform to the 1961 Values

Printed in U.S.A.

CHEMISTRY FOR LABORATORY TECHNICIANS

STANLEY M. CHERIM

Department of Chemistry,
Community College of Delaware County
Media, Pennsylvania

SAUNDERS GOLDEN SUNBURST SERIES

W. B. Saunders Company
PHILADELPHIA • LONDON • TORONTO

W. B. Saunders Company: West Washington Square
Philadelphia, PA 19105

1 St. Anne's Road
Eastbourne, East Sussex BN21 3UN, England

1 Goldthorne Avenue
Toronto, Ontario M8Z 5T9, Canada

Listed here is the latest translated edition of this book together
with the language of the translation and the publisher.

Spanish (1st Edition) — NEISA, Mexico 4 D.F., Mexico

Chemistry for Laboratory Technicians   ISBN 0-7216-2515-0

Print No.:   9   8   7   6

With love to my wife,
Solveig,
and our children,
Greg and Lise

Preface

This book has been written to satisfy the needs of junior colleges, community colleges, and technical institutes which are offering courses in chemical technology or programs leading to an Associate Degree in this field. It is hoped, also, that this text may be useful for industrial research laboratories that provide in-house training for their technicians. *Chemistry for Laboratory Technicians*, therefore, is not merely a simplified general chemistry text, but it is rather a practical, laboratory-oriented introduction to those topics and skills most relevant to the laboratory technician. However, the author does not mean to imply that the topics and skills relevant to the laboratory technician are irrelevant to other people. A one-semester chemistry course for non-science majors may find this text meeting its requirements if the instructor shares the author's opinion that a book dealing with the activities of a chemist in the laboratory is more exciting to the non-science student than a watered down, high school level textbook of general chemistry.

Just as theory without practice misses the point with regard to what chemistry is all about, so does a practical approach devoid of theory miss by failing to satisfy basic human curiosity. Few technicians could really be content with pushing buttons and flipping switches on instruments whose operation and purposes are mysterious. It is the author's contention, growing out of his currently offered courses in Chemical Technology and Principles of Chemistry (for non-science majors), that essential chemical and instrumental theory can be presented in a lucid, readable, and understandable way. The text presumes no prior chemistry training, although some previous experience might permit the omission of selected fundamental topics.

Examination of this text will reveal a biochemical bias. While there is a critical need for technicians in the research laboratories of universities, pharmaceutical companies, and hospital clinical laboratories, the author believes that the topics and skills discussed in his text are fundamental to most types of chemical industry. The field of chemical technology is too broad for any one text to be ideally suited to the whole spectrum of industrial activity in which the technician may participate. The reasoning behind the author's choice of discussing a topic in basic biochemical research instead of an analytical procedure in metallurgy, for example, is based not on a belief that biochemistry is more important but rather on a strongly held feeling that a better service is performed if a teacher operates within the limits of his own primary area of competence.

The order of topics in the text proceeds from the scientific essentials in mathematics and measurement, through two chapters on basic chemistry, to the great importance of the science library and the kind of scientific writing and record keeping expected of a laboratory technician. There is a constant emphasis on the important role of the technician in the research laboratory.

The chapters continue with an extensive discussion of solutions, oxidation-reduction, electrochemistry, and an introduction to organic chemistry. An introductory chapter on biochemistry and a chapter on radioactivity bring the text to the final chapters on chemical instrumentation. The discussions of electronic instruments focus on what they do, and, with very limited mathematical language, on how they do it. No attempt has been made to include specific operating manuals for selected instruments, since this can most effectively be learned on the job. The problems to be solved at the end of each chapter are, in most cases, the kinds of problems, at the level of mathematics, that a technician may be expected to handle.

The author is indebted to Professor Donald G. Slavin of the Philadelphia Community College and Professor Virginia J. Harris of the Community College of Delaware County for their many helpful suggestions, criticisms, and words of encouragement during the course of the manuscript writing. Thanks also to Professor John E. Stutzman of San Jacinto College (Texas) for his comments.

The fine artwork of John Hackmaster and the cordiality and thoughtful advice of the staff of W. B. Saunders Company, are gratefully acknowledged.

Many thanks, also, to Miss Margaret M. McAuliffe, who performed the remarkable job of interpreting the author's handwriting and typing the greater part of the manuscript, and to Mrs. Helen Whiteway, who helped when time was pressing.

STANLEY M. CHERIM

Wallingford, Pa.

Contents

Scientific Mathematics 1

Many students are needlessly terrified by the mathematics involved in the scientific disciplines. Science undoubtedly fails to attract many imaginative and creative people because of their fear of numbers. This is sad, because so much of the trouble is due to a distorted viewpoint produced by the lack-luster teaching of courses that seem hardly related to reality. It often comes as a shock to students to learn that quadratic equations may be used to solve real problems instead of simply being drill exercises in an algebra textbook.

In this chapter and throughout the text an attempt will be made to present mathematics as an understandable and manageable (perhaps even beautiful) tool for relating scientific concepts to entirely real measurements and calculations. There is no alternative to an immediate involvement in arithmetic, because the often stated axiom that mathematics is the language of science is just as true as ever. A scientist must observe, describe, and hypothesize, and in any kind of precise measuring system he must use numbers. The processing of these numerical data means multiplication, division, and so forth, and this is mathematics.

SCIENTIFIC NOTATION

This is a method of expressing numbers that is used in science because of its efficiency. It is writing with exponents. The exponent indicates how many times a value is multiplied by itself.

The following are some examples:

$$2^2 = 2 \cdot 2 = 4$$

$$2^3 = 2 \cdot 2 \cdot 2 = 8$$

$$10^3 = 10 \cdot 10 \cdot 10 = 1000$$

In this last example, a very useful focal point is provided in that the base number 10 is the heart of the metric system, which will be discussed at length because of its tremendous importance.

Note the relationship between the exponent and the number of digits (zeros in this case) in the number on the right.

$$1 \times 10^{③} = 1{,}000.$$
$$1 \times 10^{②} = 100.$$
$$1 \times 10^{⑤} = 100{,}000.$$

It is immediately seen that the exponent is exactly the same as the number of digits. This, of course, simplifies the process of writing a large number by using the exponential form:

$$1{,}000{,}000. = 10^6$$

$$325{,}000. = 3.25 \times 10^5$$

When a number is less than *one*, the same method applies, but the direction in which the decimal is shifted is reversed. This reversal is indicated by using *negative* exponents.

$$0.1 = 1 \times 10^{-1}$$

$$0.001 = 1 \times 10^{-3}$$

$$0.0002{,}5 = 2.5 \times 10^{-4}$$

Changing the Sign of the Exponent

If it is desirable to change a negative exponent to a positive one, this may be done by writing the **reciprocal**. The reciprocal is really the result of changing a number from the numerator to the denominator of a fraction (or changing the denominator to a numerator) without changing its value.

$$\frac{100}{1} = \frac{1}{0.01}$$

In other words,

$$\frac{10^2}{1} = \frac{1}{10^{-2}}$$

Other examples:

$$\frac{1}{10^{-5}} = 10^5$$

$$\frac{6}{2 \times 10^{-2}} = \frac{6 \times 10^2}{2} = 3 \times 10^2 = 300$$

Proving the last example:

$$\frac{6}{2 \times 10^{-2}} = \frac{6}{0.02} = 300$$

Rules for Exponents in Multiplication and Division

Consider the example $10^2 \times 10^4$ and substitute a parenthesis for the "times" sign ($\times$).

$$10^2(10^4) = 10 \cdot 10(10 \cdot 10 \cdot 10 \cdot 10)$$

You will observe that the multiplication in its expanded form shows the number 10 multiplying itself six times. Therefore, the answer can be written 10^6. This amounts to *adding* the exponents rather than multiplying them. If it was assumed that $10^2(10^4) = 10^8$, it would be incorrect.

The rule derived from this is *When multiplying exponential values, add the exponents*.

Example:

$$10^2(10^5)(10^{-3}) = 10^4$$

because

$$2 + 5 - 3 = 4$$

So far as division is concerned, examine the example $\dfrac{10^5}{10^2}$

$$\frac{10^5}{10^2} = \frac{\cancel{10} \cdot \cancel{10} \cdot 10 \cdot 10 \cdot 10}{\cancel{10} \cdot \cancel{10}} = 10 \cdot 10 \cdot 10 = 10^3$$

This is just the same as *subtracting* the exponents. The rule derived is *When dividing exponential values, subtract the exponents*.

For example:

$$\frac{10^7}{10^5} = 10^2$$

$$(7 - 5 = 2)$$

$$\frac{10^3}{10^8} = 10^{-5}$$

$$(3 - 8 = -5)$$

An alternative method is to use the reciprocals, thereby converting divisions into multiplications:

$$\frac{10^7}{10^5} = \frac{10^7(10^{-5})}{1} = 10^2$$

$$\frac{10^3}{10^8} = \frac{10^3(10^{-8})}{1} = 10^{-5}$$

THE METRIC SYSTEM

This is the measuring system used throughout most of the world in everyday living, and by practically all scientists in the laboratory. Regardless of what is being measured—*lengths, areas, volumes, mass* (or *weight* as a loosely applied synonym) or *temperature*—the system has the advantage of being based on *tens*, both multiples and divisions of ten.

It is a real advantage to use the metric system exclusively, without the time-consuming and unproductive conversion to the familiar English units of inches, quarts and pounds. It is analogous to communicating in a foreign language: until a person can *think* in the other tongue, instead of mentally forming a sentence in English and translating, he finds communication slow and frustrating. Similarly, people working in the field of science have to reach the point where they are thinking in metric units. Perhaps it will be more understandable if you remember that our American monetary system is basically a metric one.

Metric Units of Length and Area

The basic unit of linear measure is the meter (m). Think of this as roughly a large stride, or the distance from your *left* ear to the fingertips of your *right* hand outstretched (Fig. 1-1).

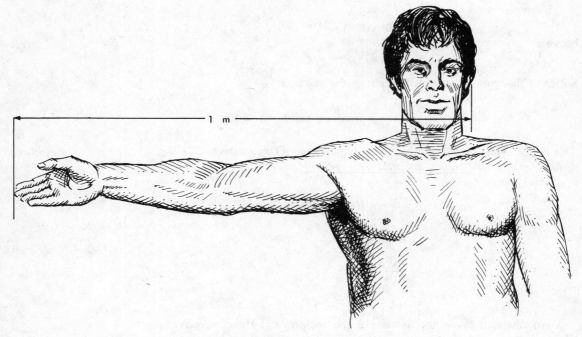

Figure 1-1. An approximation of a meter.

A hundredth of a meter is a *centi*meter (cm), the prefix *centi* always meaning a hundredth part of.

$$1 \text{ cm} = 10^{-2} \text{ m} = 0.01 \text{ m} = \frac{1}{100} \text{ m}$$

The centimeter is about the width of a fingernail (Fig. 1-2).

The large practical unit is equivalent to 1000 meters and is called the kilometer (km). The prefix *kilo* always means a multiple of 1000. For example, a "kiloglop" would mean 1000 "glops." You might imagine 100 km as the distance between Philadelphia and the Atlantic Ocean.

On the other side of the scale, the *millimeter* (mm) is next. This is one-thousandth of a meter, or one-tenth of a centimeter (see Fig. 1-3).

Figure 1-2. An approximation of a centimeter.

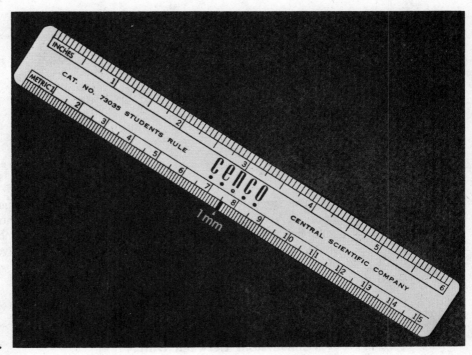

Figure I-3.

The millimeter is about equivalent to the thickness of a dime. The prefix *milli* always means a thousandth part.

$$1 \text{ mm} = 10^{-3} \text{ m} = 0.001 \text{ m} = \frac{1}{1000} \text{ m}$$

$$1 \text{ mm} = 0.1 \text{ cm} = \frac{1}{10} \text{ cm}$$

Most of you are familiar with millimeters in connection with photographic film. You take 8 mm home movies or use a 35 mm camera for color slides. The dimensions, in these instances, refer to the width of the film.

Further subdivisions are more difficult to visualize. The *micron* (μ, the Greek letter mu) is a thousandth of a millimeter. The dimensions of bacteria might be measured in microns. The prefix *micro* always means a millionth part.

$$1 \ \mu = 10^{-6} \text{ m} = 0.000001 \text{ m} = \frac{1}{1,000,000} \text{ m}$$

Subdividing the micron into a thousand parts brings you to the *milli-micron* ($m\mu$).* This tiny unit of measure is most appropriate for measuring the dimensions of large molecules or the lengths of light waves (Fig. 1-4).

This latter use has special significance to the laboratory technician who uses instruments based on the light-absorbing or light-emitting properties of substances.

---

* It should be noted that scientists are gradually shifting to the prefix "NANO" to indicate a billionth part (10^{-9}). The $m\mu$ would then be written "nm", and be called a nanometer.

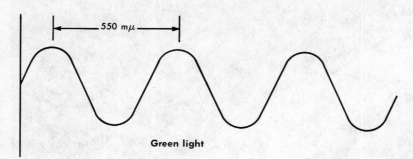

Green light

Figure I-4. Measurement of a green light wave.

The smallest useful unit of linear measure is the **angstrom** (Å), pronounced ANG-strum by most English speaking people, and it is a tenth of a millimicron. Angstroms are used to measure the dimensions of small molecules and individual atoms as well as very short wavelengths of light. A typical atom has a diameter of one to two angstroms.

$$1 \text{ Å} = 10^{-1} \text{ m}\mu = 0.1 \text{ m}\mu = \frac{1}{10} \text{ m}\mu$$

Conversions Between Units

A practically foolproof way of interconverting metric units is found in the use of labels and **conversion factors**. A conversion factor is a numerical value that relates different units of measure and permits an easy switch by multiplication or division.

A conversion factor is obtained by the following steps:

a. Take a stated equivalence:

$$1 \text{ cm} = 10^{-2} \text{ m} \ (1 \text{ cm} = 0.01 \text{ m})$$

b. State it in terms of positive exponents:

$$1 \text{ m} = 10^2 \text{ cm}$$

c. Express this equality as

$$10^2 \text{ cm } per \text{ m}$$

d. The term *per* always signifies a fraction:

$$10^2 \text{ cm/m} \quad \text{or} \quad \frac{10^2 \text{ cm}}{\text{m}}$$

e. Write the reciprocal value for m:

$$10^2 \text{ cm m}^{-1}$$

This technique of assigning an exponent of -1 on a letter may be translated as *per*, therefore,

$$10^2 \text{ cm m}^{-1} \text{ means 100 centimeters per meter.}$$

f. Most importantly, 10^2 cm m^{-1} is a *conversion factor*.

Example 1

How many meters are there in 12,500 cm?

a. 12,500 cm $= 1.25 \times 10^4$ cm

b. Visualizing the change, assume that many centimeters (small units) would equal few meters (comparatively large units). The method most likely to yield a smaller number is *division*. So, divide by the conversion factor:

$$\frac{1.25 \times 10^4 \ \cancel{\text{cm}}}{10^2 \ \cancel{\text{cm}} \ \text{m}^{-1}} = 1.25 \times 10^2 \ \text{m}$$

1. the cm terms cancel out

2. $\dfrac{10^4}{10^2} = 10^2$ by subtracting exponents

3. the reciprocal of $\dfrac{1}{m^{-1}} = \dfrac{m}{1}$

Example 2

Convert 5.50×10^{-5} cm to mμ.

1. Obtain an equivalence:

$$10^7 \ \text{m}\mu = 1 \ \text{cm}$$

2. Write it as a conversion factor:

$$10^7 \ \text{m}\mu \ \text{cm}^{-1} \ (10^7 \ \text{m}\mu \ \text{per cm})$$

3. Try multiplication, since 1 cm equals many (10 million) mμ:

$$5.50 \times 10^{-5} \ \cancel{\text{cm}} \ (10^7 \ \text{m}\mu \ \cancel{\text{cm}^{-1}}) = 5.50 \times 10^2 \ \text{m}\mu$$

a. cm and cm^{-1} cancel out because cm^{+1} (cm^{-1}) $=$ cm^0 $= 1$

b. $10^{-5}(10^7) = 10^2$

4. The label "foolproof" can be illustrated by the *mistake* of dividing instead of multiplying:

$$\frac{5.50 \times 10^{-5} \ \text{cm}}{10^7 \ \text{m}\mu \ \text{cm}^{-1}} = 5.50 \times 10^{-12} \ \text{cm}^2 \ \text{m}\mu^{-1}$$

a. The reciprocal of $\dfrac{1}{10^7 \ \text{m}\mu^{+1} \ \text{cm}^{-1}} = \dfrac{10^{-7} \ \text{cm}^{+1} \ \text{m}\mu^{-1}}{1}$

b. The answer is read as 5.50×10^{12} centimeters squared per millimicron, which is nonsense!

TABLE I-I. CONVERSION FACTORS (LINEAR MEASUREMENT).

Unit	Symbol	Conversion Factors
kilometer	km	10^3 m km^{-1}
meter	m	basic unit
centimeter	cm	10^2 cm m^{-1} 10^5 cm km^{-1}
millimeter	mm	10^1 mm cm^{-1} 10^3 mm m^{-1}
micron	μ	10^6 μ m^{-1} 10^4 μ cm^{-1} 10^3 μ mm^{-1}
millimicron	mμ	10^9 mμ m^{-1} 10^7 mμ cm^{-1} 10^6 mμ mm^{-1} 10^3 mμ μ^{-1}
angstrom	Å	10^{10} Å m^{-1} 10^8 Å cm^{-1} 10^7 Å mm^{-1} 10^4 Å μ^{-1} 10^1 Å mμ^{-1}

Metric Units of Volume

The basic unit of volume is the liter (ℓ). This is close to the volume of the familiar quart of milk in your refrigerator. The liter is also very nearly the same as 1000 cubic centimeters (cm³). For all practical purposes, cm³ and milliliters can be used interchangeably. The situation is similar to the more familiar choice that we have between cubic feet and quarts or gallons as measures of volume. In short, liters are usually applied to liquid and gas volumes, while solids are measured in cm³.

Since 1000 cm³ roughly equals 1 liter, a single cm³ would approximate a *milli*liter (ml) (see Fig. 1-5).

$$1 \text{ ml} = 10^{-3} \text{ liters} = 0.001 \text{ liter} = \frac{1}{1000} \text{ liter}$$

It is extremely useful to note that a ml of cold water weighs about a gram.

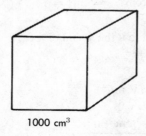

1000 cm³

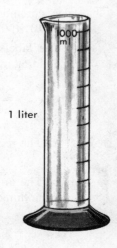

1 liter

Figure I-5.

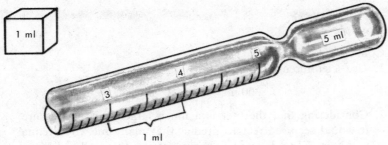

Figure 1-6.

This density of water as 1 g ml⁻¹ varies with the temperature, but for practical purposes this variation is not very significant. Figure 1-6 shows a volume of 1 ml.

The next step down toward smaller units of volume leads to the *micro-liter* (μl or λ [lambda]).

$$1 \ \mu l = 10^{-3} \ ml = 0.001 \ ml = \frac{1}{1000} \ ml$$

TABLE 1-2.   CONVERSION FACTORS (VOLUME).

Unit	Symbol	Conversion Factor
liter	ℓ	basic unit
milliliter	ml	10^3 ml ℓ^{-1}
microliter	μl	10^3 μl ml⁻¹
	λ	10^6 μl ℓ^{-1}
cubic meter	m³	basic unit
cubic centimeter	cm³	10^6 cm³ m⁻³

A microliter of water weighs about a milligram. This rather small unit of volume, measured with small and·expensive micropipets, has wide application in the laboratory. For example, gas chromatography, fluorimetry, spectrophotometry, and other instrumental techniques in analysis require micropipets (Fig. 1-7).

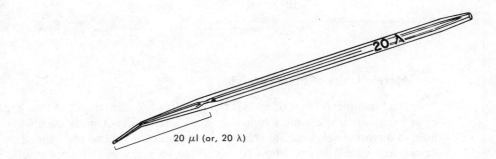

20 μl (or, 20 λ)

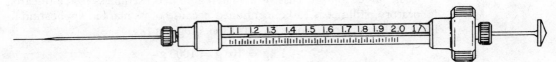

Figure 1-7.   *Below,* A syringe type pipet capable of delivery from 0.1λ to 2.0λ.

Example 3

Convert 300 ml to liters.
1. Use scientific notation:

$$300 \text{ ml} = 3 \times 10^2 \text{ ml}$$

2. Considering that the liter is a much larger unit of measure, it would seem correct to predict that many ml would equal few liters. Try division as the operation that will produce a smaller numerical value:

$$\frac{3 \times 10^2 \text{ ml}}{10^3 \text{ ml } \ell^{-1}} = 3 \times 10^{-1} \ell = 0.3 \ell$$

3. Once again, to support the "foolproof" aspect of this method, suppose you multiply by mistake:

$$3 \times 10^2 \text{ ml } (10^3 \text{ ml } \ell^{-1}) = 3 \times 10^5 \text{ ml}^2 \ell^{-1}$$

The resulting units, milliliters squared per liter, are totally meaningless.

Example 4

Suppose an addition of 0.4 μl of some unknown is injected into a gas chromatograph, but the calculations depend on volume measured in ml units.
1. Adjust 0.4 μl to $4 \times 10^{-1} \mu$l
2. Since a microliter is such a small fraction of a milliliter, a very small numerical value is expected as an answer. Division seems to be indicated:

$$\frac{4 \times 10^{-1} \mu l}{10^3 \mu l \text{ ml}^{-1}} = 4 \times 10^{-4} \text{ ml}$$

Metric Units of Mass (Weight)

The fundamental unit of mass is the gram (g). A gram might be thought of as being about the same weight as a dime. The units of mass are always used to quantitatively describe solids, whereas units of volume are applied to liquids and gases. It's pointless to talk about so many milliliters of a solid because you unavoidably include the air spaces among the solid particles in the measuring process.

The larger unit of mass, while having limited usefulness in the research laboratory, still needs to be mentioned because of its marked application to large scale industrial processes. It is the *kilo*gram (kg).

$$1 \text{ kg} = 10^3 \text{ g} = 1000 \text{ g}$$

An example that might help you to visualize a kg is the five pound bag of potatoes which is roughly equivalent to two kilograms.

High precision work often involves the use of the *milli*gram (mg), which is *three orders of magnitude* lower than the gram. An order of magnitude is a power of an exponent. For example, 10^4 is two orders of magnitude larger than 10^2.

$$1 \text{ mg} = 10^{-3} \text{ g} = 0.001 \text{ g} = \frac{1}{1000} \text{ g}$$

A small "pinch" of some solid reagent would probably weigh in the mg range. If you weighed the ink used in writing your name, it would be about three or four milligrams.

When extremely potent or expensive (or both) materials are required in an experiment, it may be necessary to use an even smaller unit of mass called the *micro*gram (μg). The microgram is sometimes symbolized as γ (gamma). The prefix *micro* indicates one-millionth, as usual.

$$1 \text{ } \mu\text{g} = 10^{-6} \text{ g} = 0.000001 \text{ g}$$

TABLE I-3. CONVERSION FACTORS (MASS).

Unit	Symbol	Conversion Factor
kilogram	kg	10^3 g kg^{-1}
gram	g	basic unit
milligram	mg	10^3 mg g^{-1}
microgram	μg	10^3 μg mg^{-1} 10^6 μg g^{-1}

Example 5

For purposes of calculation convenience, convert 4.2 mg to grams.

1. Select the conversion factor and reason that since milligrams are small compared to grams, many mg will equal few g. The need for a small number as an answer suggests division.

$$\frac{4.2 \text{ mg}}{10^3 \text{ mg g}^{-1}} = 4.2 \times 10^{-3} \text{ g}$$

2. Assume an error in reasoning to support the "foolproof" hypothesis.

$$4.2 \text{ mg} (10^3 \text{ mg g}^{-1}) = 4.2 \times 10^3 \text{ mg}^2 \text{ g}^{-1}$$

The units, milligrams squared per gram, are nonsense, as expected.

Example 6

You are asked to weigh 0.0067 mg of a salt. How many micrograms is this?

1. Convert 0.0067 mg to scientific notation:

$$6.7 \times 10^{-3} \text{ mg}$$

and select the conversion factor:

$$10^3 \ \mu\text{g mg}^{-1}$$

2. Reasoning that micrograms are very small units, you would have to multiply in order to get many of them:

$$6.7 \times 10^{-3} \text{ mg} \ (10^3 \ \mu\text{g mg}^{-1}) = 6.7 \ \mu\text{g}$$

3. Observe that $10^{-3}(10^3) = 10^0 = 1$ and $\text{mg}^{+1} \ (\text{mg}^{-1}) = \text{mg}^0 = 1$. When exponents add up to unity, the effect is what is called **cancellation**.

TEMPERATURE

Temperature is a measure of how vigorously molecules are moving. This movement may be described as vibrating, rotating on an imaginary axis, actually moving from one place to another, or any combination of these three types of motion. Temperature is *not* heat, nor is it a measure of heat. Heat is a form of energy and its unit of measure is the calorie. The technician needs to be familiar with temperature systems first, and then the relationships between the particular units of temperature and heat can more profitably be taken up later.

Fahrenheit and Celsius (Centigrade) Scales

A feature common to both the Fahrenheit and Celsius temperature scales is the arbitrary zero point. An arbitrary zero signifies a starting point selected for some convenience. The difference between the zero points amounts to 32 Fahrenheit units. This must be included as a **correction factor**. whenever you convert from one scale to the other.

An immediately noticeable difference between the two scales is the size of a degree (see Fig. 1-8).

There are 180 F degrees covering the same distance as 100 C degrees. This equivalence 180 F° = 100 C° may be reduced to 1.8 F° = 1 C°, which can be stated as a conversion factor, 1.8 F C^{-1} (or, 1.8 Fahrenheit units per Celsius unit).

Using both the conversion factor and the correction factor, Fahrenheit and Celsius degrees can be easily interconverted.

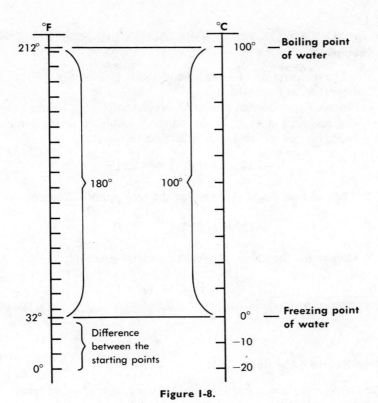

Figure I-8.

Example 7

If the temperature in the laboratory is 72° F, what is this in degrees Celsius?

1. Correct the difference between the zero points by subtracting the extra 32 degrees:

$$72° \text{ F} - 32° \text{ F} = 40 \text{ F}°$$

2. Use the conversion factor in a way that will produce the smaller numerical value expected:

$$\frac{40 \text{ F}°}{1.8 \text{ F}° \text{ C}°^{-1}} = 22.2° \text{ C}$$

3. Combine the two operations to derive a general equation:

$$°\text{C} = \frac{(\text{F} - 32 \text{ F})}{1.8 \text{ F C}^{-1}}$$

correction factor

conversion factor

In more simplified form

$$C = \frac{\text{F} - 32}{1.8}$$

Example 8

The temperature of a dry ice and alcohol mixture is $-70°$ C. Convert this to Fahrenheit.

1. Since Celsius degrees cannot be corrected by 32 Fahrenheit degrees (just as you cannot directly subtract inches from feet), the conversion factor must first be used.

$$-70 \, \cancel{C} \, (1.8 \, F \, \cancel{C}^{-1}) = -126 \, F$$

2. Then add on the $32°$ F to adjust the zero point difference.

$$-126 \, F + 32 \, F = -94° \, F$$

3. Combining the above steps into a general equation:

$$°F = 1.8 \, C + 32$$

The Kelvin (Absolute) Scale

The striking difference between the Kelvin scale and other temperature scales is the fact that the zero point is not an arbitrary starting point, but it is a point that indicates an absence of temperature, i.e., where molecular motion theoretically stops. Basing the definition on gases, it is the point at which a volume of gas would be theoretically reduced to zero. It is primarily because of gases and their relationship to the Kelvin scale that the technician needs to be concerned with this scale. Since the size of a Kelvin degree is the same as a Celsius degree, all that is needed is a correction factor for the differences between the zero points. This difference is about 273 degrees. In other words, the absolute zero point is 273 degrees below the Celsius zero point (Fig. 1-9). The equation easily derived from this relationship is

$$°K = °C + 273$$

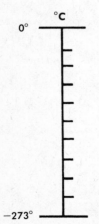 

Figure I-9.

Example 9

If the temperature of a water bath is 27° C, convert this to Kelvin degrees.

$$°K = °C + 273°$$

$$°K = 27° + 273°$$

$$K = 300 \text{ degrees}$$

SIGNIFICANT FIGURES

The Difference Between Precision and Accuracy

Talking about the significance of digits is a matter of accuracy. The question then is what sensible approach should be made toward the handling of numbers so that you neither make a fetish of significant figures (often abbreviated in speech to "sig figs") nor confuse precision with accuracy.

Making a fetish of "sig figs" really amounts to extreme fussiness regarding digits even in calculations that are meant to be crude approximations. A common sense approach is far preferable to any rigid system that a technician might impose upon himself. The confusion between accuracy and precision becomes lamentable when a person deludes himself into believing that a number to eight decimal places represents exquisite accuracy. Precision is a matter of *reproducibility*. The technician must be able to perform operations in the laboratory repeatedly and dependably in the same way. But it is possible to constantly repeat the same error! This is where accuracy comes in.

If a technician mistakenly assumes a flask to hold 200 ml while, in fact, it holds 210 ml, he can beautifully reproduce the error, exactly, countless times. A target serves as a good analogy (Fig. 1-10).

Working with significant figures means that the accuracy (suggested by the number of decimal places) of an answer is limited by the least accurate fact obtained.

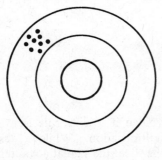
High precision
Poor accuracy

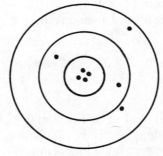
High accuracy
Poor precision

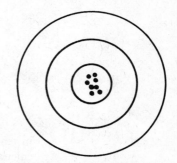
High accuracy
High precision

Figure 1-10.

Example 10

Suppose you are asked to calculate the average of the following weights:

$$12.22 \text{ g}$$
$$11.50 \text{ g}$$
$$11.6 \text{ g}$$
$$12.272 \text{ g}$$

1. Note the third number. It is 11.6, not 11.60.
2. This means the number could be 11.60, 11.61, 11.62, 11.63, or 11.64.
3. Since the second digit after the decimal is unknown, the answer to the problem is limited to three digits:

$$
\begin{array}{l}
12.2 \\
11.5 \\
11.6 \\
12.3 \\
\hline
47.6
\end{array}
\qquad
\frac{47.6}{4} = 11.9
$$

Answer: 11.9

Example 11

Calculate the area of a circle having a diameter of 6.30 cm.
1. The equation is

$$\text{Area of circle} = \pi r^2$$

2. If the diameter is 6.30 cm, the radius is 3.15 cm.
3. π is 3.14, so

$$\text{Area} = 3.14(3.15)^2$$

Using a calculator, we may obtain the answer:

$$\text{Area} = 30.842650 \text{ cm}^2$$

This is absurd! The accuracy of the answer is limited to *two* decimal places because the original diameter had only two decimals.
4. The answer should read:

$$\text{Area} = 30.84 \text{ cm}^2$$

A great deal of self-deception can be avoided if a slide rule is used instead of long methods of division and multiplication. While a desk calculator is very rapid and accurate, it lacks the slide rule's great advantage of portability. Since a slide rule gives numbers to only three digits, you cannot make the mistake of confusing the use of many decimal places for great accuracy.

THE SLIDE RULE

If you were given the two numbers 63213517 and 48193204 and asked whether you preferred to add or subtract them rather than multiply or divide (assuming your response to a dull job would be the typical choice of short and fast), you would choose to add or subtract the two large numbers. The beauty of the slide rule is the fact that it is based on a number system that allows us to replace multiplication with addition and division with subtraction.

Understanding the Slide Rule

Recall the rules for the multiplication and division of exponents which say that (1) the sign of multiplication means the addition of exponents:

$$2^2(2^4)(2^3) = 2^9$$
$$10^4(10^{0.301}) = 10^{4.301}$$

and (2) the sign of division means the subtraction of exponents:

$$\frac{2^7}{2^4} = 2^3, \qquad \frac{10^{0.778}}{10^{0.301}} = 10^{0.477}$$

This is the answer! If you can convert any number to exponential form, you can substitute addition and subtraction for multiplication and division. Because of the convenience of a decimal system, all exponents will be related to the base 10. Furthermore, there is a name given to the exponent of the base 10, which is *logarithm* (abbreviated to log). While this is not a very sophisticated definition, it is a very useful one.

$$10,000 = 10^4 \text{ and the log is } 4$$
$$1000 = 10^3 \text{ and the log is } 3$$
$$100 = 10^2 \text{ and the log is } 2$$
$$10 = 10^1 \text{ and the log is } 1$$
$$1 = 10^0 \text{ and the log is } 0$$
$$0.1 = 10^{-1} \text{ and the log is } -1$$
$$0.01 = 10^{-2} \text{ and the log is } -2$$
$$0.001 = 10^{-3} \text{ and the log is } -3$$

From this sample you can see that

$$100(1000) = 10^2(10^3) = \underbrace{2 + 3}_{\text{logarithms}} = 5 = 10^5 = 100,000.$$

This is fine and simple for nice round numbers, but what about the numbers 2 through 9, 11 through 99, and so on?

Number		Exponent		Log
1	=	10^0	=	0
2	=	$10^?$	=	?
3	=	$10^?$	=	?
⋮		⋮		⋮
10	=	10^1	=	1

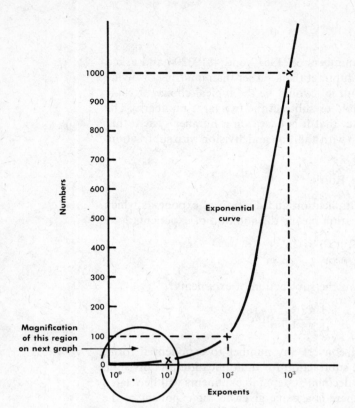

Figure 1-11. Exponential curve.

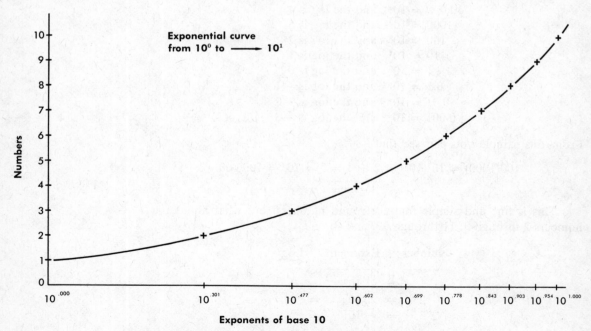

Figure 1-12. Magnification of section of graph in Figure 1-11.

Observe the log of 1 equaling zero and the log of 10 equaling one. The conclusion must be that the exponents of the numbers between 1 and 10 must have fractional values. The easiest way of finding these numbers is to draw a graph and pick the answers from the resultant line (see Figs. 1-11 and 1-12).

Summarizing the points that are taken from the curve,

Number	Exponent	Log
1	$10^{0.000}$	0.000
2	$10^{0.301}$	0.301
3	$10^{0.477}$	0.477
4	$10^{0.602}$	0.602
5	$10^{0.699}$	0.699
6	$10^{0.778}$	0.778
7	$10^{0.843}$	0.843
8	$10^{0.903}$	0.903
9	$10^{0.954}$	0.954
10	$10^{1.000}$	1.000

they may be used in the following way:

$$2(3) = 10^{0.301}(10^{0.477}) = 0.301 + 0.477 = 0.778$$

$$10^{0.778} = 6$$

Another example is

$$\frac{8}{4} = \frac{10^{0.903}}{10^{0.602}} = 0.903 - 0.602 = 0.301$$

$$10^{0.301} = 2$$

Of course, the purpose of the slide rule is to solve problems more difficult than 2 times 3 or 8 divided by 4. The important thing, however, is that the theory and method are the same, regardless of the complexity of the calculation.

The relationship between logarithms and the slide rule is due to the fact that the slide rule gives substance to the fractional exponents. For example, $10^{0.301}$ is equivalent to the number 2. The fraction 0.301 (approximately $\frac{3}{10}$) is called the logarithm. But $\frac{3}{10}$, simply in the form of an unrelated fraction has limited value. A person might ask you if you would like $\frac{3}{10}$. Well, $\frac{3}{10}$ of *what*? If it's $\frac{3}{10}$ of a million dollars—fine! If it's $\frac{3}{10}$ of his personal debts—no thanks! The slide rule gives substance to this fraction by using a length of some real material, such as wood or plastic, and then measuring off $\frac{3}{10}$ of whatever its length is. In other words, the fraction is translated into a distinct length of material. Take a 10 cm line and mark off $\frac{3}{10}$:

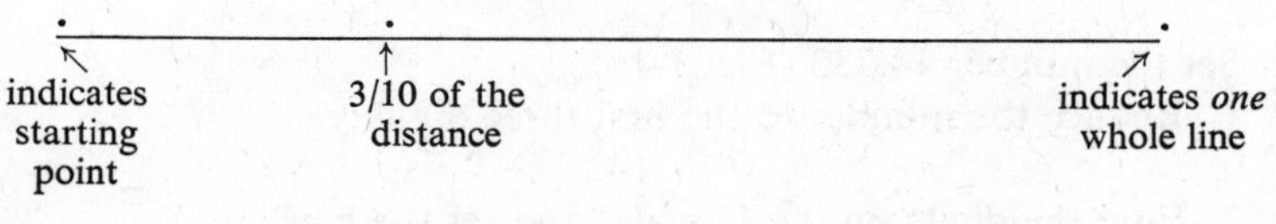

indicates   3/10 of the   indicates *one*
starting   distance   whole line
point

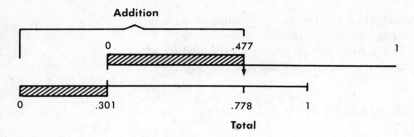

Figure 1-13.

Now, instead of simply adding fractions (logarithms of the numbers between 1 and 10) on paper, you can add lengths of slide rule material. Look again at $3(2) = 6$. Since $3 = 10^{0.477}$, and $2 = 10^{0.301}$, you can add those two fractions of slide rule length (Fig. 1-13). You know that $10^{0.778}$ equals 6, and you see only the number 6 on the slide rule.

The Reading and Setting of Numbers on the Slide Rule

If you look at the C or D scales of your slide rule, you will note that the distance between 1 and 2 (logarithms 0.000 and 0.302) is much larger than the distance between 9 and 10 (logarithms 0.954 and 1.000). For this reason, the subdivisions can be shown in much greater detail on the left hand side, while careful visual estimation is required on the right.

The slide rule is good for only the first three digits of a number, and the operator must always allow some flexibility for the value of the third digit. All numbers having four or more digits must be rounded off. Some examples might be helpful in the following comparison.

the number given	the digits used on slide rule
2.01	201
375621	376
81.508	815
0.00026	26
7.0084	701

A natural question at this point is *What happens to the decimal points in the given numbers that are ignored when digits are set on the slide rule?* These temporarily forgotten decimal points are incorporated in the scientific notation which comes after the slide rule manipulation.

Example 12

Set the number 14.730 (Fig. 1-14).
1. Reduce the number to the first three digits:

147

2. Find the digits on the D scale, and set the hairline:

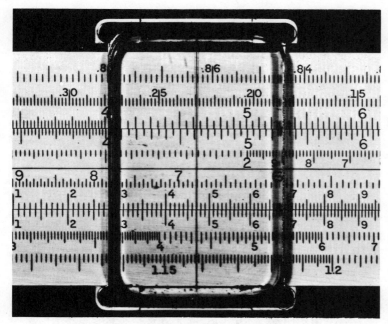

Figure 1-14. The number 147.

Example 13

Set the number 0.03637 (Fig. 1-15).
1. Reduce the number to three digits rounded off:

$$364$$

2. Find 364 on the D scale and set the hairline:

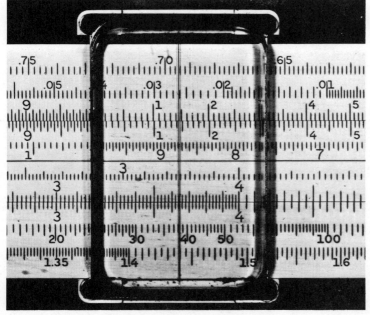

Figure 1-15. The number 364.

Example 14

Set the number 8.23 × 10³ (Fig. 1-16).
1. Use the three digits 823.
2. Find 823 on the D scale and set the hairline:

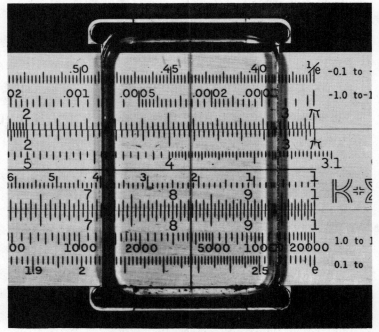

Figure 1-16. The number 823.

The Multiplication Process

It should be pointed out that while the C1 scale (often printed in red in the middle of the slide rule) gives you reciprocals of numbers (for example, the reciprocal of 2 is $\frac{1}{2}$, which is the same as 0.5), it is even more useful in making the multiplication process more efficient (see Fig. 1-17).

Multiplication uses the C1 and D scales. *Both* multipliers are set on the hairline and the answer is read on the D scale under an endpoint (the 1 or 10) of the C scale.

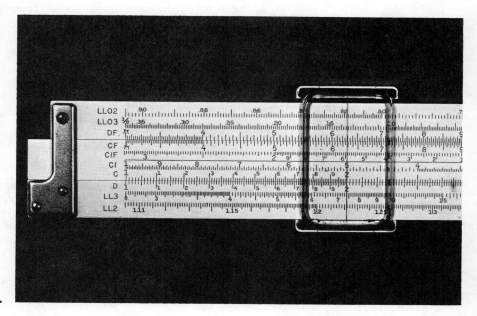

Figure I-17.

Example 15

Multiply 602 (31.87) (Fig. 1-18).

1. Select either number (let's say 602) and set it on the D scale with the hairline.
2. Move the 319 (31.87 rounded off) of the C1 scale to the hairline.
   Caution: don't forget that the numbering of the C1 scale runs in reverse—right to left.
3. Read the answer (just three digits) under the endpoint of the movable section. It is 192.
4. Use scientific notation with numbers drastically rounded off to find the true answer:

$$\text{let } 602 \text{ be } 600 = 6 \times 10^2$$

$$\text{let } 31.87 \text{ be } 30 = 3 \times 10^1$$

Multiply as indicated:

$$(6 \times 10^2)(3 \times 10^1) = 18 \times 10^3$$

5. The answer then is a two digit number $\times 10^3$ or, preferably, a one digit number $\times 10^4$.

$$19.2 \times 10^3 \qquad \text{or} \qquad 1.92 \times 10^4$$

2 digits 1 digit

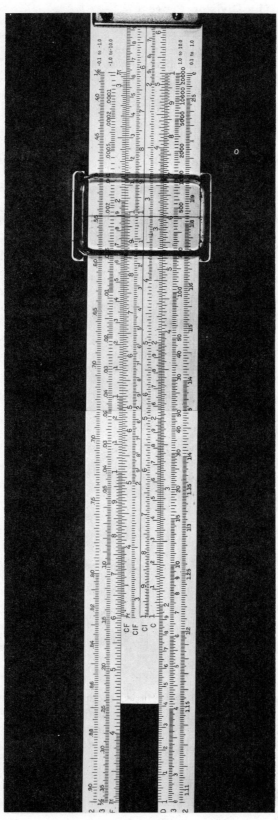

Figure I-18.

The Division Process

Division uses the C and D scales. In this case, the *dividend* must be set on the D scale, the *divisor* on the C scale, and the *quotient* is found under the endpoint of the moveable section of slide rule. If you have

$$\frac{6}{2} \quad \begin{array}{l} \text{dividend} \\ \text{divisor} \end{array} \quad (\text{or } 2\overline{)6}\ )$$

the 6 goes on the D scale and the 2 on the C scale. The number (3) will appear under the endpoint.

Example 16

Divide 471.6 by 23,041.0 (Fig. 1-19).

1. The digits 472 go on the D scale and the hairline is set at that point.
2. The digits 230 of the C scale are moved to the hairline. The three digit answer is 204.
3. Use scientific notation to determine the actual answer:

$$\text{let } 471.6 \text{ be } 500 = 5 \times 10^2$$

$$\text{let } 23,041.0 \text{ be } 20,000 = 2 \times 10^4$$

Divide as indicated:

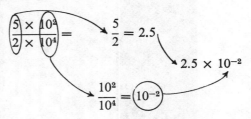

4. The answer is one digit, then the decimal point $\times 10^{-2}$

$$2.04 \times 10^{-2}$$

1 digit

Once you have developed facility in manipulating the slide rule and handling the scientific notation method of determining the decimal point and exponents, you can easily advance to the point of solving more complex calculations. There is no substitute for practice at this time. It is analogous to typing insofar as you can easily learn where to place your fingers in the "touch system," but until you practice at length you cannot gain skill, speed, and confidence.

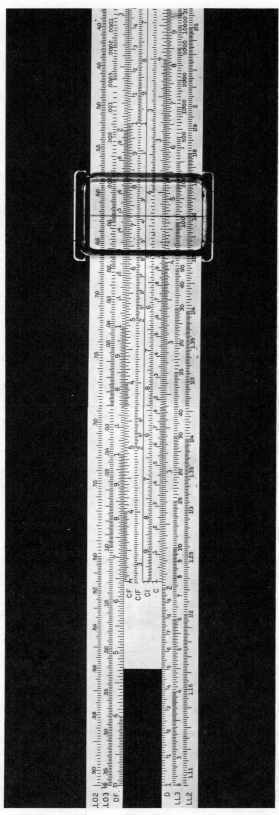

Figure 1-19.

PROBLEMS

Set A

1. Convert the following numbers to exponential form by scientific notation.

a. 320

b. 16,000

c. 9,124

d. 60,700,000

e. 0.05

f. 0.000238

g. 1.92

h. 0.000000149

2. Use the slide rule for the following calculations of three digit answers.

a. $25(731) =$

b. $5917(30610) =$

c. $0.00842(61.28) =$

d. $\dfrac{48147}{113.8} =$

e. $\dfrac{30.57}{0.0081} =$

f. $\dfrac{52.7(0.0336)}{24132} =$

3. Use scientific notation to determine the placement of decimal points and orders of magnitude for the three digit answers in Problem 2.

4. Make the following calculations, with special regard for significant figures:

a. $23.015 \text{ g} + 1.22 \text{ g} + 78.167 \text{ g} =$

b. $\dfrac{16.45}{4.1} =$

c. $1.18 \times 10^{-2}(3.0512 \times 10^{-4}) =$

d. $\dfrac{12(604.1)(303.5)}{760(273)} =$

5. Perform the following metric system conversions:

a. $2.5 \text{ km} = ? \text{ m}$

b. $32 \text{ cm} = ? \text{ mm}$

c. $125 \text{ m}\mu = ? \text{ cm}$

d. $2.8 \times 10^3 \text{ Å} = ? \text{ m}\mu$

e. $25 \mu = ? \text{ mm}$

f. $6.2 \times 10^{-5} \text{ cm} = ? \text{ m}\mu$

6. More metric system conversions:

a. $32 \text{ g} = ? \text{ kg}$

b. $2.2 \text{ g} = ? \text{ mg}$

c. $4 \times 10^{-3} \text{ g} = ? \mu\text{g}$

d. $500 \mu\text{g} = ? \text{ mg}$

e. $0.4 \text{ mg} = ? \text{ g}$

f. $1.5 \text{ mg} = ? \mu\text{g}$

7. Still more metric system conversions:

a. $230 \text{ ml} = ? \ell$

b. $1500 \mu\text{l} = ? \text{ ml}$

c. $0.03 \ell = ? \text{ ml}$

d. $85 \text{ cm}^3 = ? \text{ ml}$

e. $0.09 \text{ ml} = ? \mu\text{l}$

f. $5 \times 10^{-3} \text{ ml} = ? \lambda$

8. Temperature conversions (Fahrenheit and Celsius):

a. $35°\ C = ?\ °F$

b. $80°\ F = ?\ °C$

c. $-4°\ F = ?\ °C$

d. $-200°\ C = ?\ °F$

9. Temperature conversions (Celsius and Kelvin):

a. $5°\ C = ?\ °K$

b. $-20°\ C = ?\ °K$

c. $400°\ K = ?\ °C$

d. $273°\ K = ?\ °C$

10. Using conversion factors: A very important constant in nature is the speed of light, 3×10^{10} centimeters per second. Change this value to meters per second by using the conversion factor 10^2 cm m^{-1}.

Set B

11. Convert the following numbers to exponential form by scientific notation:

a. 1630

b. 23,800

c. 83.16

d. 5,782,000

e. 0.022

f. 0.00017

g. 3.810

h. 0.0000064

12. Use the slide rule for the following calculations of three digit answers:

a. 16(485)

b. 302(5173)

c. 0.0441(3.72)

d. $\dfrac{5913}{81.2}$

e. $\dfrac{75.92}{0.0339}$

f. $\dfrac{19.7(0.086)}{6230}$

13. Use scientific notation to determine the placement of decimal points and orders of magnitude for the three digit answers in Problem 12.

14. Make the following calculations, with special regard for significant figures:

a. $42.137 + 8.05 + 51.351 =$

b. $\dfrac{58.17}{3.2} =$

c. $4.17 \times 10^3(1.0314 \times 10^5) =$

d. $\dfrac{14.1(308.4)(59.17)}{420(6182)} =$

15. Perform the following metric system conversions:

a. $351\ m = ?\ km$

b. $14\ mm = ?\ cm$

c. $650\ m\mu = ?\ cm$

d. $52\ m\mu = ?\ Å$

e. $0.02\ mm = ?\ \mu$

f. $1.3 \times 10^{-6}\ cm = ?\ m\mu$

16. More metric system conversions:

a. $0.4\ kg = ?\ g$

b. $512\ mg = ?\ g$

c. $7 \times 10^4\ \mu g = ?\ g$

d. $4.5\ mg = ?\ \mu g$

e. $2 \times 10^{-3}\ g = ?\ mg$

f. $6.7 \times 10^5\ \mu g = ?\ mg$

17. More metric system conversions:

a. $0.12 \ \ell = ? \ ml$

b. $3.7 \times 10^{-5} \ ml = ? \ \mu l$

c. $185 \ ml = ? \ \ell$

d. $28 \ ml = ? \ cm^3$

e. $0.8 \ \mu l = ? \ ml$

f. $5 \times 10^5 \ \lambda = ? \ ml$

18. Interconvert the following Fahrenheit and Celsius measurements:

a. $5° \ C = ? \ °F$

b. $160° \ F = ? \ °C$

c. $-70° \ C = ? \ °F$

d. $-120° \ F = ? \ °C$

19. Interconvert the following Celsius and Kelvin measurements:

a. $60° \ C = ? \ °K$

b. $-80° \ C = ? \ °K$

c. $3000° \ K = ? \ °C$

d. $230° \ K = ? \ °C$

20. An important constant in nature is known as Planck's constant. It has the value of 6.62×10^{-27} erg sec. Convert this to kcal by using the conversion factor, 4.2×10^{10} erg kcal^{-1}.

SUPPLEMENTARY READING

Kline, M.: *Mathematics in Western Culture.* Oxford University Press, New York, 1953.

A fascinating account of the impact of mathematics on Western culture.

Arnold, J. N.: *The Complete Slide Rule Handbook.* Prentice-Hall, Englewood Cliffs, New Jersey, 1954.

An excellent supplementary text for those who use the slide rule. Problems and answers included.

Masterton, W. L., and Slowinski, E. J.: *Mathematical Preparation for General Chemistry.* (paperback) W. B. Saunders Company, Philadelphia, 1970.

Provides a thorough preparation for solving mathematical problems in chemistry.

Reid, C.: *From Zero to Infinity.* T. Y. Crowell Co., New York, 1966.

Excellent chapter on the explanation of the base e and natural logarithms. Very readable.

Measurement 2

This chapter deals with one of the most important activities of the laboratory technician. The process of making measurements, recording, organizing, and reporting data, and the application of tentative statistical analysis is absolutely critical. No matter how brilliantly an experiment is designed, it may be worthless (or actually harmful) if the technician is not committed to accuracy and precision in his work. Faulty lab technique can be disastrous. There is the physical danger of explosion, production of toxic gases, and the exposure of fellow workers to corrosive chemicals that may cause severe burns. There is also the economic liability involved in the loss of many experimental animals and valuable time when "fouled-up" experiments have to be repeated. When directions in an experimental procedure call for an addition of 1.5 ml of a solution to each of ten test tubes, this means *precisely* 1.5 ml and not approximately 1.5 ml.

The laboratory technician needs to have a real appreciation of his value and status as part of a scientific team. The realization that he is much more than just a small cog in a big wheel goes a long way in establishing the self-esteem needed in doing a job which he can be proud of and doing it with the serious dedication so becoming to a person who is playing an important role.

GRAVIMETRIC METHODS

To perform a gravimetric determination means to find out how much something weighs. This procedure is absolutely critical to the success of most laboratory activities. The technician must decide what type of balance (commonly called a scale) is most appropriate to the required precision (reproducibility), and which can be used with speed. The choice has to be a thoughtful one. There are different kinds of balances, each one best suited to a particular task. A balance that sacrifices precision for speed won't do when precision is needed.

BALANCES

Figure 2-1.

The precision of a balance usually is related to its capacity. A balance designed to weigh kilograms can hardly be expected to have the kind of precision that could give reproducible weighings to the tenth of a milligram. A partial classification of balances is given in the following chart.

Type of Balance	Capacity	Precision	Model Type	Speed
coarse	~ 2 kg	0.1 g	triple beam	moderate
			top loading	fast
			double pan	moderate
analytical	~ 200 g	0.1 mg	top loading	fast
			single pan	fast
			double pan	slow
semi-micro	~ 100 g	0.01 mg	double pan	slow
			single pan	fast
micro	~ 30 g	1 μg	double pan	slow
			single pan	fast

Note: This little sign $\sim$ ("squiggle") is often used by scientists to mean *approximately.*

There are so many types of balances on the market today that any attempt toward a detailed discussion of their manipulation would go far beyond the scope of this text. A few are shown in Figure 2-1. When the technician is introduced to a particular balance in his lab, he can easily master the required procedure. However, there are some steps of gravimetry and gravimetric analysis that have general application regardless of the particular type of scale that may be used.

Standard Operating Procedure

1. Check the cleanliness of the balance. A camel's-hair brush may be put to good use.
2. Check the balance for level position, since it could have been nudged accidentally.
3. Check the zero point. In the case of the newer electronic digital readout balances, there is a tendency for the illuminated scale to drift.
4. Weighing bottles must be clean and dry. It is a good idea to keep them in a desiccator (a drying chamber) (Fig. 2-2).
5. If glassine paper (similar to waxed paper) is used, the balance should be adjusted to zero just as with the weighing bottles.
6. Use forceps, tongs, spatulas, or lint-free paper in the weighing process, since your fingers add moisture and may contribute to contamination.
7. Discard any amounts of spilled chemicals rather than risk contaminating the original container.
8. Weighed materials should be transferred to labeled containers or used immediately.
9. When the weighing is completed, make sure control dials are reset at zero, any glass door is closed, and any electrical switches are turned off.

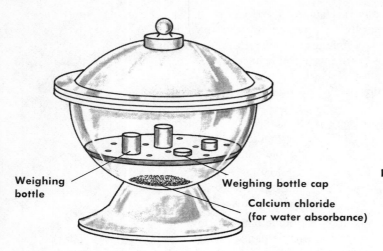

Figure 2-2. A laboratory desiccator.

Weighing
bottle

Weighing bottle cap

Calcium chloride
(for water absorbance)

The newer balances available in laboratories today make the weighing process remarkably easy. It is no longer necessary to make coarse pre-weighings, since this step is incorporated into the system.

Since the single pan, electronically operated, digital readout balance is the one you will most likely use in any modern laboratory, a consideration of *how* it works is appropriate. The Mettler Balance Corporation has supplied the illustration in Figure 2-3 which is also shown diagrammatically in Figure 2-4.

Figure 2-4 illustrates an interesting principle which is the reverse of the centuries-old method of balancing an object on one pan by *adding* weights to a second pan (Fig. 2-5).

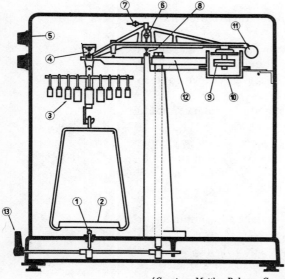

(Courtesy Mettler Balance Corp.)

Figure 2-3. Substitution weighing balance. *1*, Pan arrest. *2*, Sample pan. *3*, Weights. *4*, Knife edge supporting pan holder. *5*, Weight lifter control. *6*, Sensitivity set screw. *7*, Null balance set screw. *8*, Knife edge supporting beam. *9*, Counterweight. *10*, Air damper (dashpot). *11*, Deflection scale. *12*, Beam arrest. *13*, Beam and pan release control.

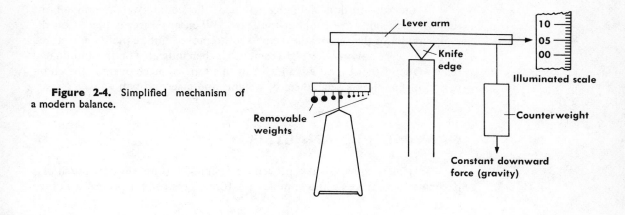

Figure 2-4. Simplified mechanism of a modern balance.

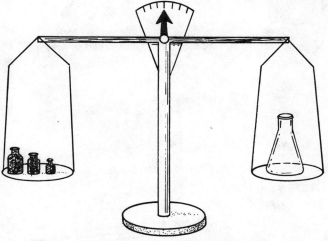

Figure 2-5. Ancient two-pan balance.

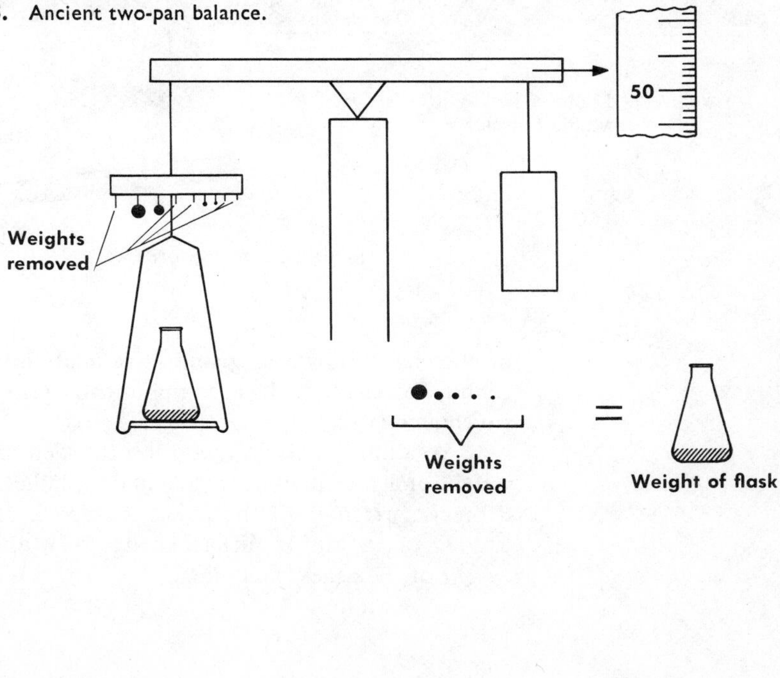

Figure 2-6.

In the substitution weighing balance, the counterweight and all the removable weights are equal at the start. When an object is placed on the pan, enough of the removable weights are removed until equality is restored (Fig. 2-6). The amount of weight removed is then indicated on the illuminated scale. One of the several advantages of this type of mechanism is the elimination of handling a second set of weights and a second pan.

GRAVIMETRIC ANALYSIS

Gravimetric analysis is a procedure designed to find out how much of a particular substance is in a complex mixture. This is an aspect of a branch

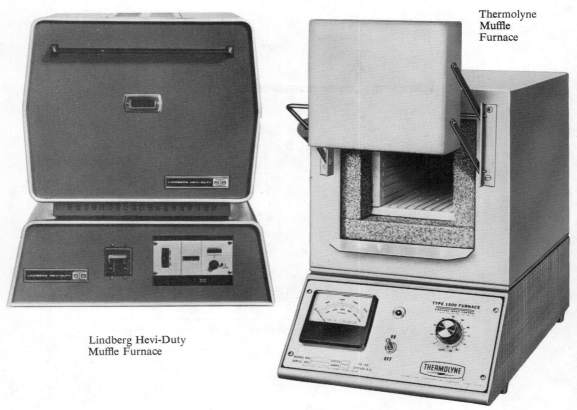

Thermolyne
Muffle
Furnace

Lindberg Hevi-Duty
Muffle Furnace

Figure 2-7. Muffle furnaces.

of chemistry known as **quantitative analysis**. The measurement of the percentage of elements in a compound by gravimetric analysis is often called **stoichiometry**.

A sample analysis might be the determination of the percentage by weight of ash (mineral content) in corn flakes. The procedure would involve a precise *weighing* of the flakes, complete reduction to ashes in a furnace (known as a muffle furnace) (Fig. 2-7), and a final determination of the *weight* of the ashes remaining.

Example 1

Find the percentage of ash in corn flakes.
1. Weight of sample:
$$1.543 \text{ g}$$
2. Weight of ash:
$$0.091 \text{ g}$$
3. Percentage of ash:
$$\frac{0.091 \text{ g}}{1.543 \text{ g}} = 0.0589 = 5.89\%$$

Example 2

Find the percentage of silver in a coin.
1. Dissolve weighed coin (3.50 g) in nitric acid and then add sodium chloride to form solid white flakes (precipitate) of silver chloride.
2. Filter the solid (see Fig. 2-8) and wash it free of all the dissolved material with distilled water.

Two Methods of Folding Filter Paper

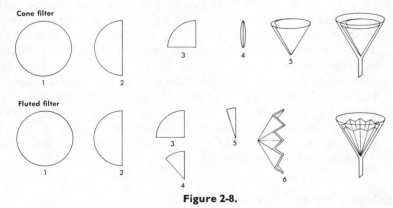

Figure 2-8.

3. Dry the silver chloride and determine the weight (4.20 g).
4. Given the information that silver chloride is 75 per cent silver, take
$$75 \text{ per cent of } 4.20 \text{ g} = 0.75(4.20) = 3.14 \text{ g}$$
5. The final step then is to find out what per cent of the 3.50 g coin is 3.14 g of silver
$$\frac{3.14 \text{ silver}}{3.50 \text{ coin}} = 0.90 = 90 \text{ per cent silver}$$

VOLUMETRIC METHODS

The measurement of volume is equally as important as weight determinations; however, of the two, volumes of liquids and gases are invariably easier to measure and generally more useful.

A distinction should be made between glassware used for measurement and glassware or plastic used for storage. The former, called **volumetric** glassware is usually more costly because of the precision and accuracy required in its manufacture. For this reason, these vessels should not be tied up for storage purposes. Special attention must be given to cleanliness and gentle handling in order to avoid both contamination and breakage.

Bottles used for the storage of solutions must be selected with regard to the volume needed, the resistance to chemical reaction with the solution being stored, the effects of heat and cold, and the sensitivity of the solution to light. Further consideration of these points follows:

1. Determine the volume needed. The prospect of pouring 500 ml of solution into a 400 ml bottle is frustrating, while storing 50 ml of solution in a liter flask is a waste of shelf space.
2. Check the chemical resistance of the container to the solution being stored. There are some organic solvents, such as acetone and ether, that may react with some types of plastic bottles. Reference to a table dealing with these possibilities is included in the Appendix for your use. There are some solutions such as hydrofluoric acid and strong alkalis (sodium and potassium hydroxide) that should not be stored in glass bottles because of chemical interaction.
3. Consider the effects of heat and cold. Some containers become brittle if they are stored under refrigerated conditions, and the effect could be disastrous if they are placed in a $-70°$ C dry ice bath. Heat-sensitive containers should not be sterilized in an **autoclave** (a chamber of steam under pressure used to destroy microorganisms) where temperatures of $140°$ C might reduce them to shapeless globs of plastic. See the Appendix for the table dealing with this matter.
4. Check for sensitivity to light. There are some solutions, such as silver nitrate or acetic acid, that undergo decomposition or some other type of change when exposed to light. Brown glass bottles would be indicated in this case.

Labeling Storage Bottles

When a solution has been prepared and transferred from a measuring vessel to a storage bottle, there are several important items that should be printed on the label (see Fig. 2-9).

1. The *formula* or *name* (whichever is shorter or more familiar).
2. The *strength* (*concentration*) of the solution. This subject is taken up in detail in Chapter Seven.
3. The *date of preparation*, which will avoid the error of using solutions that may have deteriorated with age.
4. *Special information* such as danger to skin, flammability, and pH (see Chapter Eight).
5. The *technician's name* (or initials), so that any questions about the solution can be directed to the preparer.

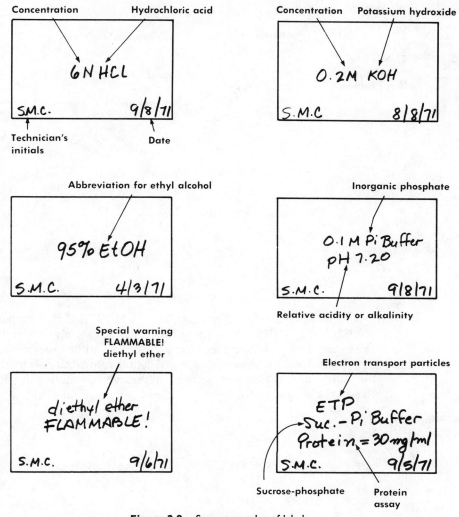

Figure 2-9. Some examples of labels.

Volumetric Glassware

The several types of glassware used for measuring liquids are graduated cylinders, burets, volumetric flasks, and pipets. These items vary in capacity, purpose, and precision. The volumetric flasks and graduated cylinders, for example, are designed to measure what they *contain* more accurately than what they *deliver* when poured. The amount of liquid adhering to the walls of such large vessels prevents accurate measurement of poured volumes. A piece of glassware designed for contained volume measurement is usually marked TC (to contain). A buret or pipet (really a simplified buret) is more likely made to accurately measure the volume delivered, and they are appropriately marked TD (to deliver). If the temperature varies greatly from the 20° C indicated, the precision is reduced because the volume occupied is more at higher temperatures and less at colder temperatures.

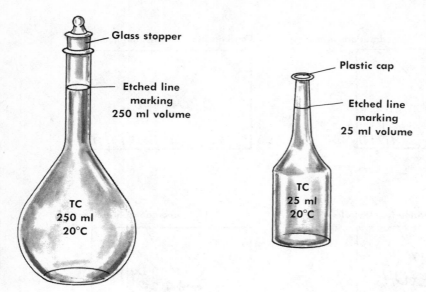

Glass stopper

Etched line
marking
250 ml volume

TC
250 ml
20°C

Plastic cap

Etched line
marking
25 ml volume

TC
25 ml
20°C

Figure 2-10. Volumetric flasks.

Volumetric flasks. These flasks are used for the preparation of solutions (Fig. 2-10). Volumetric flasks vary from about 10 ml to 1 liter or larger. They are quite expensive and should not be used for storage.

Graduated cylinders. Graduated cylinders are routinely used in the lab for quick and reasonably accurate measurement of liquids (Fig. 2-11). Reading any graduated piece of glassware must take into account the

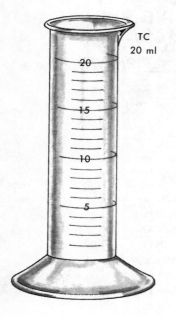

TC
20 ml

20

15

10

5

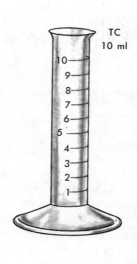

TC
10 ml

10
9
8
7
6
5
4
3
2
1

Figure 2-11. Graduated cylinders.

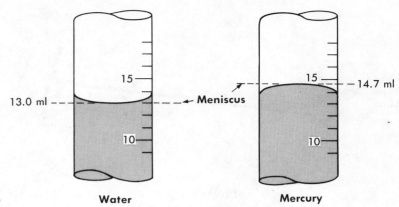

Figure 2-12. Reading the meniscus.   Water   Mercury

tendency of liquids to adhere to the walls of the vessel. The standard procedure is to hold the vessel at eye level and read the lowest point when it sags or the highest point when it bulges upward. The name given to either level is the **meniscus** (Fig. 2-12). Some plastic cylinders eliminate this bulge, because water or water solutions do not adhere to such material.

Burets. The use of burets for the process of titration will be discussed in Chapter Eight. Two are shown in Figure 2-13.

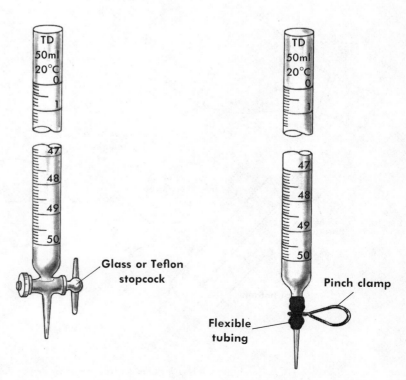

Figure 2-13. Titration burets.

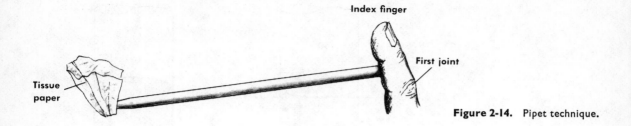

Figure 2-14. Pipet technique.

Pipets. Pipets are designed to be used much like a straw. When poisonous or otherwise dangerous solutions are to be drawn, a rubber bulb or a more sophisticated controlled volume plunger attachment should be used rather than your mouth. The tip of the pipet should be wiped dry of any hanging drops by using a tissue paper before transferring any liquid (Fig. 2-14). The flow of liquid is controlled by a gentle rolling motion of the finger as it is clamped against the mouth section of the pipet after filling. The index finger, just before the first joint, has been found most suitable. This is a skill that definitely requires practice before attempting any serious use. Attachments for use instead of sucking by mouth are shown in Figure 2-15. Figure 2-16 shows various types of pipets, ranging from microliter (λ)

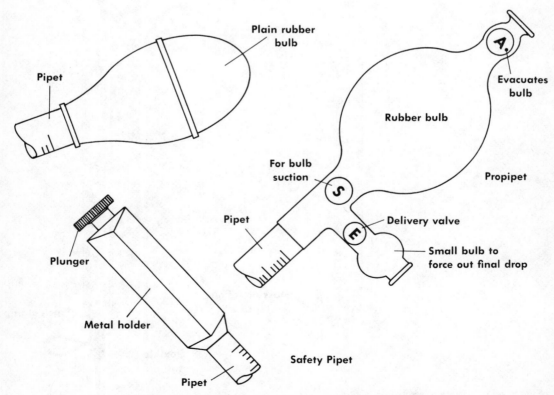

Figure 2-15. Volume control and safety devices for pipets.

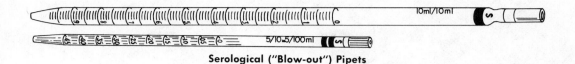

Serological ("Blow-out") Pipets

LANG-LEVY PIPET

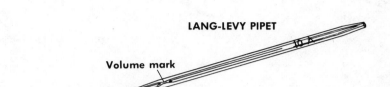

Volume mark

Micro Pipet (Transfer)

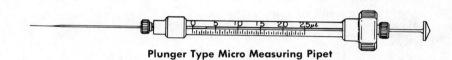

Plunger Type Micro Measuring Pipet

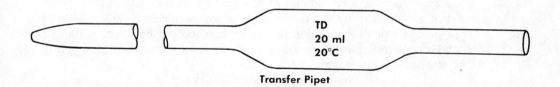

TD
20 ml
20°C

Transfer Pipet

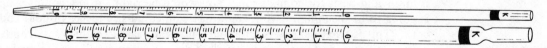

Measuring Pipets

Figure 2-16.

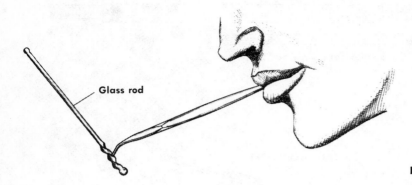

Glass rod

Figure 2-17.   Micro-pipet technique.

volumes to 25 ml volumes. Because of the extremely small volumes (1 to 500λ) delivered by the transfer micro-pipets, the actual transfer is often made by blowing out the tiny drop onto a glass rod and then using the rod as the transfer instrument (Fig. 2-17). This avoids splashing, which could occur if the technician blows a few microliters of solution too forcefully into a small container.

DENSITY AND SPECIFIC GRAVITY

When a label on a bottle of sulfuric acid reads "sp. gr. 1.84," the question may be raised as to what this means and what this information is used for. Sp. gr. is an abbreviation for specific gravity, which is a physical characteristic of matter directly related to density in the metric system. Density is a precise type of description that takes into account both weight and volume. Gravimetric and volumetric methods are both required to find density and thence specific gravity.

If you were asked to compare a room full of cork to a small cube of gold in the palm of your hand, you might have a nagging suspicion that this is an invalid basis for comparison. You would be right. If any comparison is going to be meaningful, you would insist on equal size pieces of gold and cork. This is what density is all about—it is the mass (weight) of a definite volume. The definite volume is usually taken to be one milliliter (or cubic centimeters) for liquids and solids, and one liter for gases. These are known as **unit** volumes. So, by definition, the density of a substance is its mass per unit volume.

In order to develop this relationship mathematically so that a useful equation may be derived, start with the observation that the size (volume) of an object is directly related, or proportional, to its mass. A shorthand expression of this statement is

$$M \quad \propto \quad V$$
$$\text{(mass)} \qquad \text{(volume)}$$

sign of proportionality

Converting the shorthand statement into a mathematical one requires a **proportionality constant**:

$$M = kV$$

proportionality constant
this makes it mathematical

The term **constant** means just what the word suggests—a value that is fixed and unchanging. If the equation is rearranged by simple algebra

1. $\dfrac{M}{V} = \dfrac{k\cancel{V}}{\cancel{V}}$ (divide the equation by V)

2. $k = \dfrac{M}{V}$

it results in the constant being equal to the <u>mass divided by the volume.</u> Consider silver.

Mass	÷	Volume	=	k (g per cm³)
21 g	÷	2 cm³	=	10.5 g cm⁻³
31.5 g	÷	3 cm³	=	10.5 g cm⁻³
42 g	÷	4 cm³	=	10.5 g cm⁻³
52.5 g	÷	5 cm³	=	10.5 g cm⁻³

It is observed that the value of the constant for silver is 10.5 g cm⁻³ (see Fig. 2-18). This constant which is grams per cubic centimeter, obtained by dividing mass by volume, is so important that it is given the distinctive name, **density**. The derived equation, then, is

$$D = \frac{M}{V}$$

With this equation in hand, a whole host of practical problems can be solved.

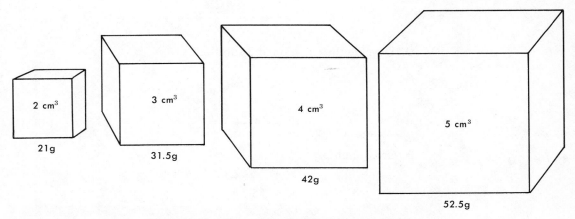

Figure 2-18. Direct proportionality of mass and volume.

Example 3

Find the density of copper if a 57 cm³ block weighs 52 g.
1. Organize the data:

D	?
M	52 g
V	57 cm³

2. Substitute in the equation:

$$D = \frac{M}{V} = \frac{52 \text{ g}}{57 \text{ cm}^3} = 0.91 \text{ g cm}^{-3}$$

Example 4

Find the density of benzene if 214 ml weigh 184 g at 16° C.
1. Organize the data:

$D_{16° \text{ C}}$	?
M	184 g
V	214 ml

2. Substitute in the equation:

$$D_{16° \text{ C}} = \frac{M}{V} = \frac{184 \text{ g}}{214 \text{ ml}} = 0.86 \text{ g ml}^{-1}$$

Example 5

Find the weight of 30 ml of a liquid having a density of 1.84 g ml⁻¹.
1. Organize the data:

D	1.84 g ml⁻¹
M	?
V	30 ml

2. Develop the equation and substitute:

$$D = \frac{M}{V}$$

$$\frac{1.84 \text{ g ml}^{-1}}{1} = \frac{M}{30 \text{ ml}}$$

3. Cross multiply:

$$M = 1.84 \text{ g ml}^{-1} (30 \text{ ml})$$
$$M = 55.2 \text{ g}$$

Example 6

If a metal has a density of 9.71 g cm^{-3} and weighs 205.6 g, what volume will it occupy?

1. Organize the data:

D	9.71 g cm^{-3}
M	205.6 g
V	?

2. Develop the equation and substitute:

$$D = \frac{M}{V}$$

$$\frac{9.71 \text{ g cm}^{-3}}{1} = \frac{205.6 \text{ g}}{V}$$

3. Cross multiply:

$$9.71 \text{ g cm}^{-3}\, V = 205.6 \text{ g}$$

4. Divide the equation by the density:

$$\frac{\cancel{9.71 \text{ g cm}^{-3}}\, V}{\cancel{9.71 \text{ g cm}^{-3}}} = \frac{205.6 \text{ \cancel{g}}}{9.71 \text{ \cancel{g} cm}^{-3}}$$

$$V = 21.2 \text{ cm}^3$$

The fact that density has the dimensions of mass and volume is what distinguishes it, as a physical characteristic of matter, from specific gravity. Specific gravity is a comparison in the form of a fraction, or ratio, between the density of some substance and the density of water. Water, very conveniently, has a density of 1 g per milliliter ($D_{water} = 1$ g ml^{-1}).

The equation for specific gravity (sp. gr.) is

$$\text{Sp. Gr.} = \frac{D_{substance}}{D_{water}}$$

Example 7

What is the sp. gr. of sulfuric acid if its density is 1.84 g ml^{-1}?

$$\text{Sp. Gr.} = \frac{1.84 \text{ \cancel{g} \cancel{ml}}^{-1}}{1 \text{ \cancel{g} \cancel{ml}}^{-1}} = 1.84$$

With this answer you see that the numerical values of density and specific gravity are the same; only the units of mass and volume are missing from specific gravity.

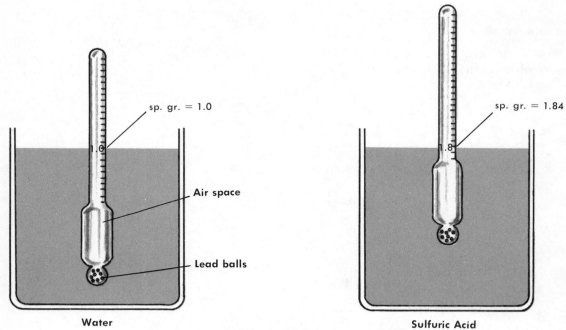

Figure 2-19. Specific gravity determination by hydrometer.

Thus, it is seen that the knowledge of specific gravity, obtained from a label, tells you the weight of definite volume, which is the starting point of other useful calculations. Chapter Seven develops this point.

A **hydrometer** is an instrument used to measure the specific gravity of liquids. A weighted bulb at the bottom causes the hydrometer to sink until it displaces a volume of liquid that has an equal weight. The sp. gr. is determined visually by a calibrated scale that extends above the surface of the liquid (Fig. 2-19).

ELECTRICAL MEASUREMENTS AND INTERRELATIONSHIPS

The modern research laboratory is greatly dependent on electronic instrumentation in the pursuit of knowledge. The acts of plugging an exotic hunk of electronic hardware into a wall socket, building or repairing the instrument, or using it intelligently to perform some critical measurements in an experiment all serve to involve the technician in electronics. Admittedly, there is a gigantic difference in the skill and understanding required to build, use, or merely switch on an instrument. This chapter deals with some of those minimum essentials of vocabulary, fundamental relationships, and simple testing that a technician today must be able to grasp. Discussion of a more specific nature regarding electronic instruments will be taken up in

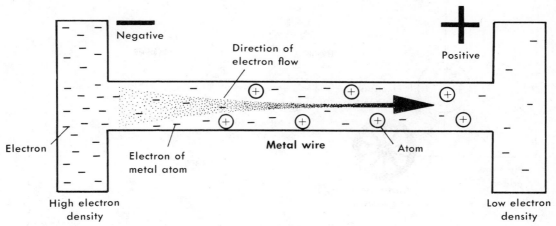

Figure 2-20.

the final chapters. A short course in electronics is far beyond the scope of this text, but a good stride forward can be made.

Electronics is the technology of electron behavior. The electrons, often described as units of negative charge (*negative*, an arbitrary sign used as a contrast to the opposite charge of *positive*), are an integral part of an atom, but they are also found abundantly outside the atom. Electrons can be harvested by running a comb through your hair, scraping shoes across a carpet, or pulling a wool sweater over a nylon blouse. Electrons can be "boiled out" of a glowing wire or lamp filament. They can be "kicked out" of metal by the energy of sunlight, and electrons can be induced to flow through a conducting wire by breaking the invisible lines of force between two magnets. There is also a tendency for electrons to flow between two different metals.

Some of these examples of electron sources suggest the common definition of electricity which is the *flow* of electrons through a conductor (usually a wire). The electrons flow from a region of high density (many negative charges) to a region of low density (comparatively few electrons present and therefore called positive) (Fig. 2-20).

A well known axiom in electricity is that "opposite charges attract and like charges repel." This statement is largely explained by our observations of nature in which we see a tendency of unequal forces to strive toward a balance, or equilibrium. In the case of electrons, this drive toward balance is achieved by the electrons flowing from the high population area to the low. For example, if opposite charges are placed on two metal spheres by a static electricity generator, you can see the force of attraction in operation. The contact that results allows the flow of negative charge (call it the electric current) to move from the negative sphere to the positive. The resulting balance is seen when the two spheres, now having similar charges, begin to repel each other (Fig. 2-21).

A more vivid example of the tendency of electricity to flow from high electron density areas to low is seen when you scrape your shoes over a carpet as a means of gathering free electrons from the fibers and then approach a radiator. The load of electrons on your body leaps to the radiator and produces a visible spark.

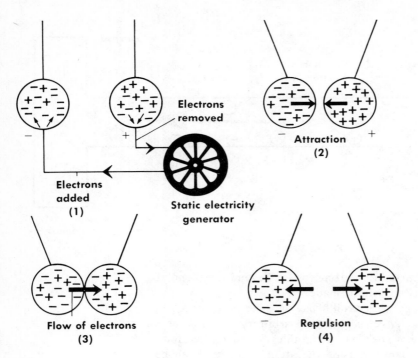

Figure 2-21.

Electrical Units

One commonly hears about units called volts, amps, ohms, watts, and components such as transistors, rectifiers, and vacuum tubes, to mention a few. You may be familiar with the VTVM (vacuum tube voltmeter), oscilloscope, and signal generator. While all of these terms cannot be adequately discussed, a start can be made, and the best place to begin is with the fundamental units of electricity.

The volt. The volt may be described as a unit of electrical *pressure*. A high voltage is not necessarily dangerous; it simply means that there is a difference in the electron density between two separate points so that the one point, having the greater electron population, possesses a kind of potential energy, i.e., stored up electrical energy that will move to the region of low density if a pathway (such as a wire) is provided. **Potential**, or **potential difference**, is a common synonym for voltage.

We are all familiar with 1.5 volt dry cells (batteries), the 110 volt lines in our buildings, or the many thousand volt lines from the electric company generators. A useful analogy in visualizing electric pressure is to think of the water behind a dam. The water pressure (grams per square centimeter or pounds per square foot) in this situation is similar to electrical pressure.

The ampere (amp). The amp is a measure of the *amount* of electricity actually being drawn from a supply line. There is a relationship between amps and volts, but it is not as simple as one might suppose. It is possible to draw a great deal of water from a small dam as it is also conceivable to draw a trickle from a huge dam. The larger dam, however, would be more efficient for obtaining large amounts of water with the least strain.

Familiarity with amps comes about through our use of household appliances and fuse boxes or circuit breakers. You may have experienced annoyance when too many appliances on the same line draw more current than the safety fuse will permit and the fuse is "blown." The water analogy here is like a faucet set in the dam. The amount of water that can be drawn depends on how far the valve is opened to supply the existing need. If the valve is opened too far, it may blow apart ("blow its fuse"). Some electronic instruments with complex circuitry have parts requiring very small quantities of current, measured in milliamps or microamps.

The ohm.   The ohm is a measure of the degree of *resistance* to the flow of current. This is effectively done by using materials that are poor conductors of electricity. A resistor in an electronic circuit is shown in Figure 2-22.

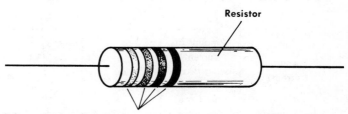

Resistor

Figure 2-22.   Color code describes the precision and the value of resistor in ohms.

The water analogy would consider the dam itself as a fixed resistance to the flow of water and the valve on the faucet as a variable resistance in the way a rheostat or potentiometer in electronics are variable resistors.

The relationship of these three units—volt, amp, and ohm—are mathematically stated in **Ohm's law**.

$$\text{Amount of current} = \frac{\text{Voltage}}{\text{Resistance}}$$

Symbolically:

$$I = \frac{V}{R}$$

The watt.   One other common relationship is described as a measure of *power*, and this unit is the **watt**. Watts are equal to the product of volts and amps.

$$W = IV$$

A high powered water flow would be obtained if the valve of the faucet were opened wide at the base of a high dam. Light bulbs provide a familiarity with watts.

In summary, Figure 2-23 illustrates the principal electrical units in terms of the water analogy.

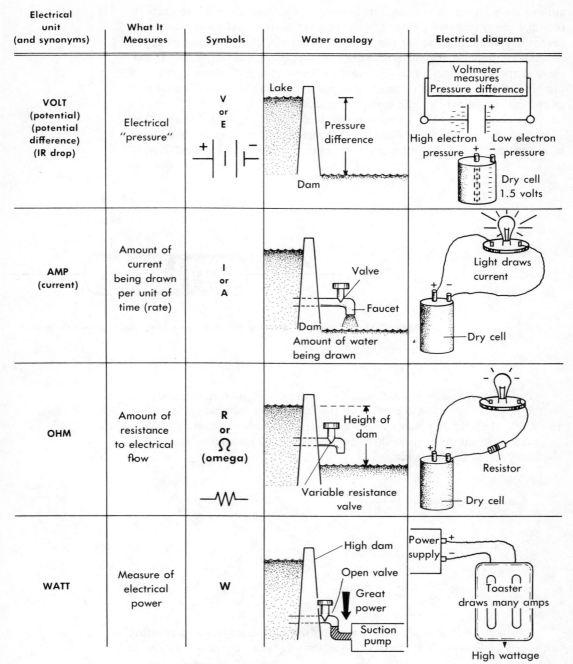

Electrical unit (and synonyms)	What It Measures	Symbols	Water analogy	Electrical diagram

Figure 2-23. Electrical units related to a water analogy.

The Circuit Diagram

Circuit diagrams are extremely useful in the building and maintenance of electronic instruments. However, they require some special instruction before they can be properly appreciated. A look at the schematic (diagram) on the back of your radio will provide some insight into the complexity of electronic diagrams. A little sample (Fig. 2-24) will serve as an illustration.

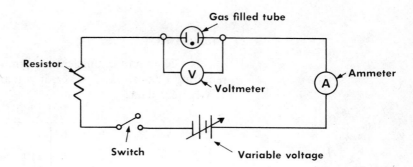

Figure 2-24.

Example 8

What is the amount of current that can be drawn from a 1.5 volt dry cell when a 3000 ohm (3 kΩ) resistance is in the circuit?
1. Organize the data:

V	1.5
R	2×10^3
I	?

2. Develop the equation:

$$I = \frac{V}{R} = \frac{1.5}{3 \times 10^3} = 0.5 \times 10^{-3} = 5 \times 10^{-4} \text{ amps}$$

or

$$0.5 \text{ mA} \quad \text{(milliamps)}$$

Example 9

What resistance value must be in a circuit if you wish to draw 20 mA from a 50 volt battery?
1. Organize the data:

V	50
R	?
I	20 mA = 0.02 amps = 2×10^{-2} A

2. Develop the equation:

$$I = \frac{V}{R} \qquad 2 \times 10^{-2} = \frac{50}{R}$$

$$R = \frac{50}{2 \times 10^{-2}} = 25 \times 10^2 = 2500 \text{ ohms}$$

or

$$2.5 \text{ kΩ} \quad \text{(kilo ohms)}$$

Example 10

Is a 20 amp fuse sufficient to permit use of an air conditioner requiring 1200 watts of power, if you are on a 110 volt line?

1. Organize data:

W	1200
I	?
V	110

2. Develop the equation:

$$W = IV \qquad 1200 = 110I$$

$$I = \frac{1200}{110} = 10.9 \text{ amps}$$

3. The answer is *yes*. Only ~11 amps are drawn and the fuse permits up to 20 amps.

The Meter

While there are a variety of meters for measuring electrical units, their complex circuitry can be simplified to illustrate the basic principles on which meters are built (see Fig. 2-25).

1. A pointer is calibrated to sweep across a scale marked off in some units (volts, amps, ohms).
2. The pointer is attached to a spring-mounted electromagnet that is caused to rotate by the current coming from the object being measured.

A Brief Glossary of Electronic Technology

Alternating current (AC)—This is the common form of current. Call it household electricity if you will. When the positive and negative poles reverse themselves sixty times per second, it is called 60 cycle AC current.

Ammeter—Similar to the galvanometer, but equipped with range selectors permitting the measurement of larger currents in amperes.

Capacitor—A device used to store current in electronic circuits (Fig. 2-26). The amount of current stored can be hazardous. In a TV set, capacitors are usually shielded to prevent tampering.

Direct current (DC)—This is a flow of current with a fixed direction. A dry cell provides direct current since the electron flow is always from one specific pole (electrode).

Galvanometer—A very sensitive instrument used to detect small amounts of current flow.

Impedance—This is the total resistance to current flow in AC circuits. It includes the resistance due to resistors, capacitors, and inductance. The unit is the **ohm**.

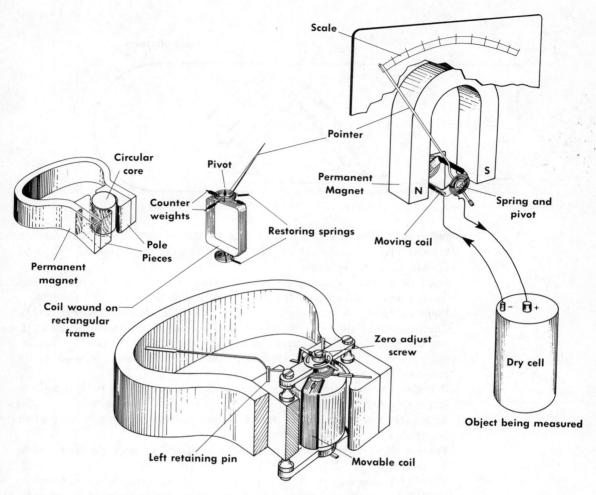

Figure 2-25. Diagrammatic representation of the parts of an electrical meter.

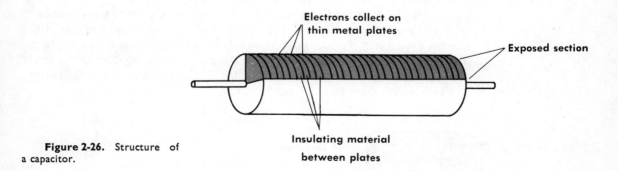

Figure 2-26. Structure of a capacitor.

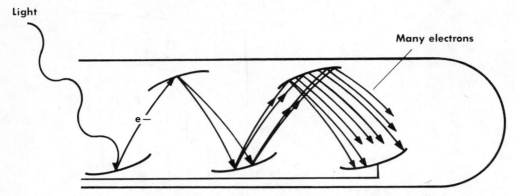

Figure 2-27. Multiplication of the photoelectric effect.

Inductance—The resistance to change in the amount of current flowing through a wire, usually achieved by placing a coil of wire in circuit. The unit of inductance is the **henry**.

Oscilloscope—A measuring instrument that provides a visual image of electronic signals on a screen. It is capable of registering very rapid changes.

Photomultiplier—A very important tube in optical instruments (Fig. 2-27). The transmitted light in colorimeters and spectrophotometers, for example, causes electrons to be ejected (photoelectric effect). The photomultiplier takes the small current of a few electrons and multiplies it to more easily measurable size.

Rectifier—A device used to block the electron flow from one pole, effectively converting AC to DC.

Transistors and vacuum tubes—Devices used primarily to amplify very small currents.

Transformer—A device used to increase (step-up) or decrease (step-down) the voltage (Fig. 2-28).

VOM (volt-ohm meter)—Similar to the VTVM, but operated on batteries.

VTVM (vacuum tube voltmeter)—A common test instrument to measure DC and AC voltages as well as resistance. A little instruction and practice can make you very proficient in the use of this instrument.

This brief glossary contains but a fragment of the many terms common to electronic technology. If your interest is aroused, there are numerous texts available to further your knowledge. An excellent start would be an attempt to build your own electronic instrument. There are many commercial sources of kits. The directions included in the kits are remarkably clear, which makes them ideal for the beginner.

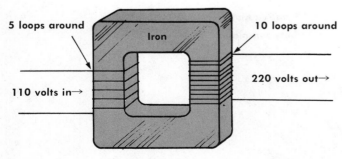

Step-up Transformer   **Figure 2-28.**

GRAPHS—CONSTRUCTION AND ANALYSIS

Graphs are one of the most important devices for the communication of data and of the interrelationships of data in science. Pick up any scientific journal today, and you will see a number of graphs throughout. It is often the technician's job to represent data graphically, and to do this properly there are a few standard procedures.

1. Be sure to mark the divisions between points on each axis *uniformly* so that you will have a valid representation of your data. For example, if you were to plot the mass of an object against its volume, you might get the following chart:

	Data
mass	volume
2 g	5 ml
4 g	10 ml
10 g	25 ml
50 g	125 ml

A direct proportionality produces a *straight line* through the *origin* (see Fig. 2-29).

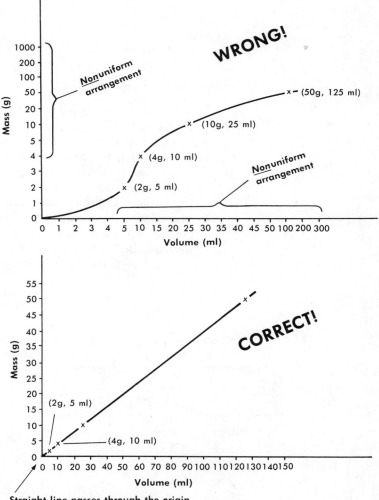

Figure 2-29.   Straight line passes through the origin

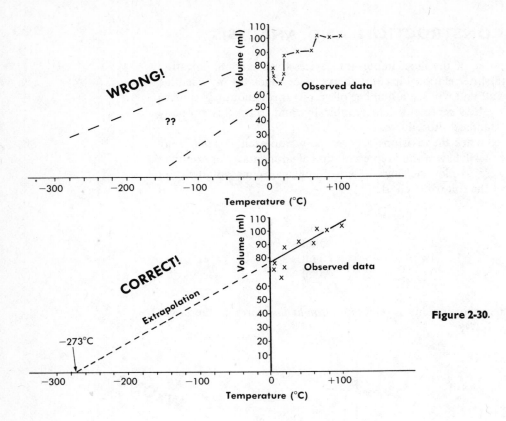

Figure 2-30.

2. If the points do not fall perfectly on the line, draw through the area of greatest point density. Graphically represented data are not supposed to look like a business sales graph (Fig. 2-30).
3. **Extrapolation** is a process of extending a line *beyond* experimentally determined points. An example is found in the **volume-temperature** graph for gas relationship in which the volume of a gas is directly proportional to the absolute temperature (Fig. 2-30).
4. The relationship between the volume occupied by a gas and the pressure on the gas is inversely proportional (Fig. 2-31). In other words, as the pressure increases, the volume decreases. The relationship may be developed mathematically:

$$V \propto \frac{1}{p}$$

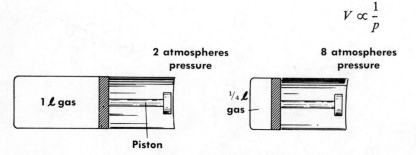

Figure 2-31. The inverse relationship between pressure and volume.

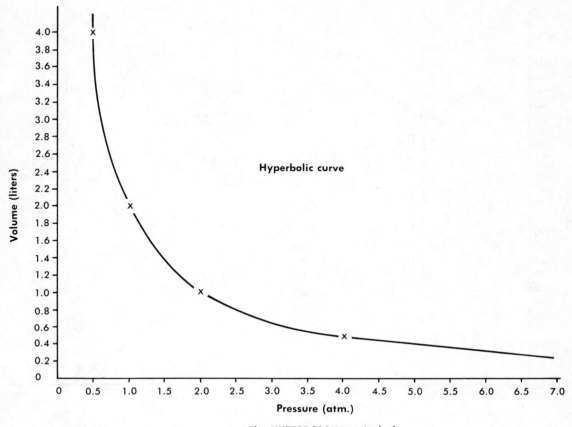

Hyperbolic curve

The **HYPERBOLA** is typical of
an inverse relationship.

Figure 2-32.

Volume is inversely proportional to pressure

$$V = k \frac{1}{p} = \frac{k}{p}$$ Proportionality constant

Experimental Data

V (l)	P (atm)	k
1	2	2
2	1	2
4	$\frac{1}{2}$	2
$\frac{1}{2}$	4	2
$\frac{1}{4}$	8	2

5. The **slope** of the line can help in analyzing a graph. Sometimes a graphical representation of data can provide you with a particularly valuable piece of information. When a value of the vertical line change (ΔY) divided by a value of the horizontal line change (ΔX) yields a constant, you have a fact most useful in solving equations (see Fig. 2-33).

For example, as radioactive material decays, measurable atomic particles can be rated, i.e., you can count the number of sub-atomic

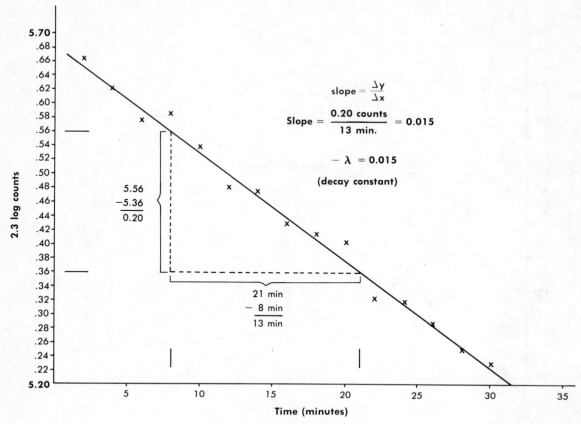

Figure 2-33. The slope of a line.

particles shooting out per unit of time by using the proper instruments. If the logarithm to the base e, which is the same as 2.3 times the ordinary log to the base 10, is plotted against time, the slope of the curve (vertical value change divided by horizontal value change) gives a **decay constant** ($\lambda = $ lambda)

$$-\lambda = \frac{2.3 \log \text{counts of radioactive particles}}{\text{time (minutes)}}$$

The negative sign ($-\lambda$) means the *decrease* in counts with respect to the passing of time.

Time (min)	2.3 log counts
0–2	6.66
3–5	6.62
6–8	6.58
9–11	5.59
12–14	5.54
15–17	5.43
18–20	5.41
21–23	5.32
24–26	5.30
27–29	5.27
30–32	5.25

6. Whenever the observed data include some uncertainty because of limited precision, the probable variations can be graphically shown

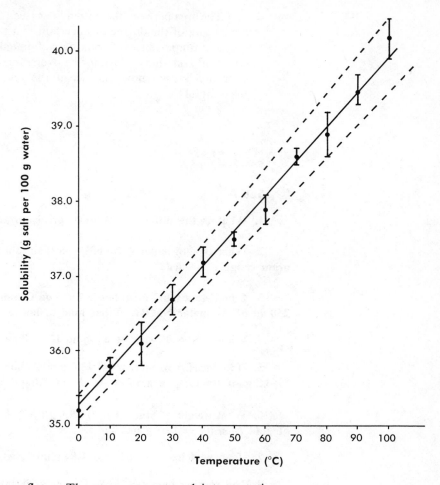

Figure 2-34. Use of error flags in graphing.

by the use of error flags. The measurement and interpretation of this error will be discussed in Chapter Five under the heading of Elementary Statistical Analysis.

For example, suppose the increasing water solubility of sodium chloride (common salt) is measured as the temperature rises (Fig. 2-34). The balance is such that the weight cannot be read with greater precision than 0.2 to 0.3 of a gram above or below the most careful observation. This is represented by writing " $\pm$ " (plus or minus) 0.2 or 0.3.

The Data

Solubility in grams of salt per 100 grams of water	Temperature °C
35.2 ± 0.2	0°
35.8 ± 0.1	10°
36.1 ± 0.3	20°
36.7 ± 0.2	30°
37.2 ± 0.2	40°
37.5 ± 0.1	50°
37.9 ± 0.2	60°
38.6 ± 0.1	70°
38.9 ± 0.3	80°
39.5 + 0.2	90°
40.2 ± 0.3	100°

The area between the dotted lines (see Fig. 2-34) gives an immediate picture of the degree of uncertainty in the measurements.

The importance of gaining a familiarity with graphical construction and analysis cannot be overemphasized. Leaf through any scientific journal now and see if the graphs are a little bit more meaningful.

PROBLEMS

Set A

1. What is the difference between precision and accuracy?

2. What type of balance would be most suitable if precision in the microgram range is needed?

3. Complete a label for a bottle filled on October 3, 1969 which contains 250 ml of six molar NaOH. What kind of bottle would you choose?

4. What is the density of a liquid if 35.0 ml weighs 22.5 g?

5. If the density of mercury is 13.6 g ml^{-1}, how much of it must be added to 3.2 g of tin to give a total weight of 5.8 g?

6. What weight of brass (density 8.0 g cm^{-3}) is equal in volume to 1 cm^3 of gold (density = 19.3 g cm^{-3})?

7. If an acid has a density of 1.84 g ml^{-1}, what is its specific gravity?

8. The specific heat (sp. ht.) of a metal is inversely proportional to its atomic weight (at. wt.). Write an equation for this relationship.

9. How many ohms are needed in a circuit having a 60 volt battery and an instrument requiring 30 milliamps?

10. What voltage is required by a 100 watt light bulb drawing 0.88 amps?

11. Graph the following data and indicate the relationship.

Temperature °C	Density (g ml^{-1})
0	26 ± 1
10	24 ± 2
20	21 ± 1
30	20 ± 3
40	18 ± 1
50	16 ± 4
60	14 ± 1
70	11 ± 3
80	8 ± 1
90	7 ± 2
100	5 ± 2

Set B

12. Define the following terms:

a. autoclave
b. gravimetric
c. voltage
d. galvanometer

e. extrapolate
f. transfer pipet
g. volumetric flask
h. slope of the line (on a graph)

13. Why would a triple beam balance be unsuitable for weighing out 4.22 g of a substance?

14. Write a label for 500 ml of a stock solution of methanol (MeOH is a common abbreviation) you prepared on February 6, 1970. The methanol is 60 per cent and the rest is water.

15. Calculate the density of a metal cylinder that is 10.0 cm long and has a diameter of 1.0 cm. The cylinder weighs 34.2 g.

16. What volume of aluminum chips (density $= 2.7 \, g \, cm^{-3}$) must be added to 20.0 ml of water in order to produce a total weight of about 50 g?

17. A gold ring weighs 15.4 g in air and 14.9 g when suspended in water. If the specific gravity of gold is 19.3, calculate the volume of the ring.

18. An acid solution is 60 per cent acid by weight. If the density of the solution is $1.4 \, g \, ml^{-1}$, find the weight of 100 ml of pure acid.

19. The product of a gas pressure (p) and volume (V) is directly proportional to the number of gas molecules (N) and the temperature (T). Write an equation relating these terms.

20. How much current may be drawn from a circuit having a potential difference of 6.0 volts and a resistance of $5 \, k\Omega$?

21. What is the watt value of a light bulb if it draws 0.52 amps from a 115 volt line?

22. Graph the following data and calculate the value of the slope. What is significant about the slope value if the substance being investigated is brass?

Volume (x-axis)	Mass (y-axis)
$2 \, cm^3$	16 g
$5 \, cm^3$	40 g
$7.5 \, cm^3$	60 g
$9 \, cm^3$	72 g
$11 \, cm^3$	88 g

SUPPLEMENTARY READING

Furth, R.: The Limits of Measurement. *Sci. Amer.*, July, 1950.

An informative article for the scientist.

Purcell, E. M.: *Electricity and Magnetism.* McGraw-Hill Book Co., New York, 1965.

An excellent text for a thorough review of the principles of electricity.

Kisch, B.: *Scales and Weights.* Yale University Press, New Haven, 1966.

A beautifully illustrated book for the science history buff.

Middleton, R. G.: *Electrical and Electronic Signs and Symbols.* Bobbs-Merrill Co., Inc., New York, 1968.

Contains an excellent chapter on how to read schematic diagrams.

Marcus, A.: *Electricity for Technicians.* Prentice-Hall, Inc., Englewood Cliffs, New Jersey, 1968.

A text emphasizing practice more than theory.

Cooper, W. D.: *Electronic Instrumentation and Measurement Techniques.* Prentice-Hall, Inc., Englewood Cliffs, New Jersey, 1970.

A thorough discussion of electronic measurement techniques.

Some Fundamental 3
Laws and Concepts
Concerning Matter,
Energy, and the Atom

The field of chemistry is very broad. Considering some of its aspects, such as inorganic, organic, physical, analytical, and their subdivisions, an author or teacher faces the problem of scope and content. The question of approach—the broad survey, lacking depth, versus the limited topics in depth—is an unavoidable burden set squarely on the shoulders of the one who must choose. The choice must be dictated by clearly formed objectives that are realistically geared to the needs of the student. The needs of a research laboratory technician are of a decidedly practical nature. For this reason, discussions of theory will be simplified as much as possible without grossly distorting the truth. However, enough substance should be provided to enable the more curious to have a solid base for further reading in standard college chemistry texts.

Many fascinating stories about the historical development of theories, laws, discoveries, inventions, and great experiments will have to be omitted. It grieves the author to have to ask students to accept scientific principles "on faith," but this may have to be done occasionally; it is hoped that the reader will be stimulated to further his knowledge with supplementary study.

MATTER

Chemistry is often defined as the investigation of matter and the changes that matter undergoes. What, then, is matter and what is meant by change?

Definition

Matter is usually described as anything having mass and occupying space. In other words, matter has a measurable density. Minerals, air, milk, plants, and animals are a few of the countless things composed of matter. But it isn't always so simple to place a label. For example, electrons and "particles" of light blur the lines that scientists have drawn for their convenience.

Variability and Stability

Considering familiar types of matter first, it is commonly observed that its state is variable. A **solid** can usually be changed to a **liquid** by an increase in temperature and further changed to a **gas** as the temperature goes still higher. This change in state is necessarily linked to energy because it is energy that enables molecules to vibrate more vigorously—which is what a temperature rise indicates. Change in state is a story of forces in conflict. There is a tendency for matter to achieve maximum stability; at the same time, it mysteriously strives for maximum randomness in arrangement.

The forces of attraction between the unit particles of matter (atoms or molecules) may be described as nuclear (very powerful), electrical or magnetic, and gravitational. These very real forces pull unit particles together in an orderly geometric arrangement that characterizes stability (Fig. 3-1). This tendency of matter to sacrifice energy in its drive toward stability is a characteristic that all of us have observed many times. A ball on a hill rolls *down* and finally stops. An automobile wears out and finally comes to an end on the scrap pile. A living organism grows old, breaks down, and finally dies. However, nature seems to provide a counterforce for every one-directional tendency just as each muscle in the body has another muscle operating in opposition to it. In this condition of tension, there is smoothness, control, and balance.

When the unit particles have some source of energy that they can absorb, there is a movement toward disorder, made possible since the absorption

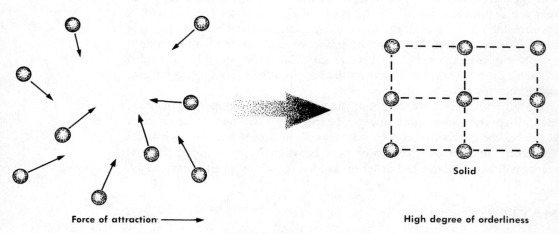

Force of attraction ———➤   High degree of orderliness

Solid

Figure 3-1.

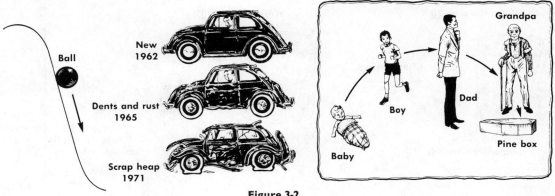

Figure 3-2.

of this energy permits the particles to overcome the force of attraction
(Fig. 3-3). With the absorption of still more energy, the unit particles can
overcome forces of attraction by having the energy to move at high speed
to great distances apart. This is the gaseous state (Fig. 3-4). This tendency
toward randomness is often called an increase in the **entropy** of a system.

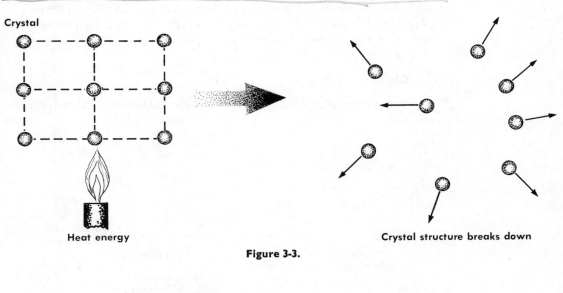

Figure 3-3.

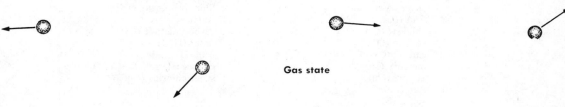

Figure 3-4.

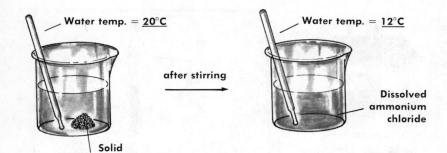

Water temp. = <u>20°C</u>

after stirring

Water temp. = <u>12°C</u>

Dissolved
ammonium
chloride

Solid
ammonium chloride

Figure 3-5. Absorption of heat energy in the drive toward randomness.

This is not entirely accurate, but it should be noted that many simple definitions, although useful, are less than sacred.

Sometimes this drive toward an increase in entropy is so powerful that a highly ordered solid may spontaneously absorb energy in order to achieve greater randomness. An example would be the remarkable drop in water temperature when solid ammonium chloride is dissolved (Fig. 3-5). This illustrates the transfer of heat energy from the water to the solid for the process of dissolving. The dissolved particles are much more random in their arrangement than the original solid they composed.

Changes of Matter

The changes of matter are conveniently divided into **physical** and **chemical** change.

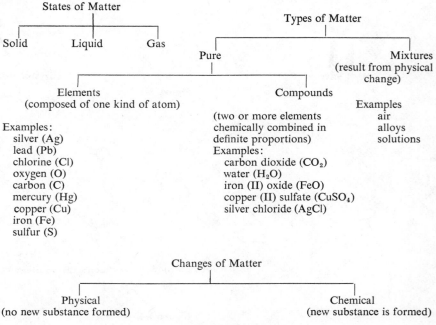

States of Matter

Solid   Liquid   Gas

Types of Matter

Pure

Mixtures
(result from physical change)

Elements
(composed of one kind of atom)

Examples:
  silver (Ag)
  lead (Pb)
  chlorine (Cl)
  oxygen (O)
  carbon (C)
  mercury (Hg)
  copper (Cu)
  iron (Fe)
  sulfur (S)

Compounds
(two or more elements
chemically combined in
definite proportions)
Examples:
  carbon dioxide (CO_2)
  water (H_2O)
  iron (II) oxide (FeO)
  copper (II) sulfate ($CuSO_4$)
  silver chloride (AgCl)

Examples
  air
  alloys
  solutions

Changes of Matter

Physical
(no new substance formed)

Chemical
(new substance is formed)

Figure 3-6.

Physical change. The previous discussion of changes of state would serve as one excellent example of physical change. The principal factor involved is that no new substance was formed, regardless of the fact that the outward appearance was altered. This is the essence of physical change. Water, ice, and steam are the same substance—they are all composed of exactly the same unit particles (the water molecule). While the arrangement of particles of matter may vary from orderly to random, it is only when the original unit particles are broken and new types of particles are formed that a change other than physical has resulted.

If a glass pane is ground into powder it is still glass. If salt is dissolved in water, mere evaporation of the water permits the recovery of that same salt. If blackish iron powder is blended with yellow sulfur to form a homogeneous gray pile, the iron could be removed with a magnet or the sulfur could be dissolved in carbon disulfide liquid, which is easily evaporated. Mashing, melting, dissolving, or blending are all physical changes because no *new* substance is formed.

Chemical change. The chemical change is one which begins with **reactants** that have a set of unique characteristics (properties) quite different from the **products** that finally result. When a **reaction** (a chemical change) goes to completion, i.e., when *all* of the reactants are converted to products at the end, the reaction is said to be **stoichiometric**. Stoichiometry is concerned with very specific amounts of reactants and products.

$$\underset{\text{reactants}}{A + B}\xrightarrow{\text{stoichiometric}}\underset{\text{products}}{C + D}$$

Reactions that have products interacting in such a way so as to re-form the original reactants until a balance is established are called **equilibrium** reactions.

$$A + B\underset{\text{equilibrium}}{\rightleftharpoons}C + D$$

Equilibrium reactions are usually much more difficult to observe visually, and the quantitative (mathematical) considerations are more complicated. For the sake of expediency, the following discussions of evidence of and conditions for chemical change will apply mostly to stoichiometric reactions.

Evidence for the occurrence of a chemical change is most obvious when one or more of the following occurs:

1. A *gas* is produced.
2. A *precipitate* (visible solid particles) is formed.
3. A *large energy change* is observed in the form of light or heat or both.
4. A distinct *color change* occurs.

The conditions necessary for a chemical change to take place may be:

1. Simple *contact* between two reacting solids at room temperature:

yellow phosphorus + iodide crystals

→ red phosphorus triiodide + flames

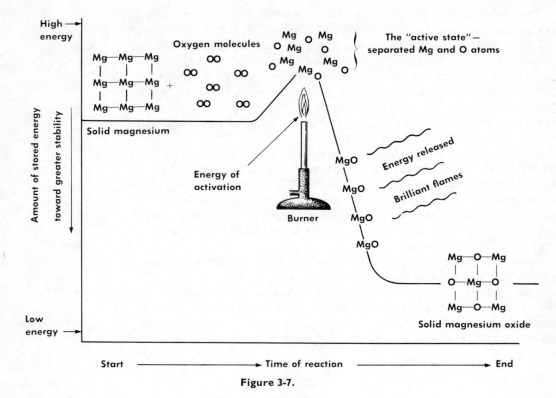

Figure 3-7.

2. *Supplying enough energy* to jostle sluggish reactants into an active condition which allows the breaking of old bonds of attraction and the formation of new ones (Fig. 3-7):

$$\text{flexible, silvery ribbon} \quad \text{Mg} + \tfrac{1}{2}\text{O}_2 \rightarrow \text{MgO is white powder}$$
odorless, colorless gas

3. Getting the reactants into water **solution** is often necessary (Fig. 3-8). If white crystals of silver nitrate are mixed with white crystals of

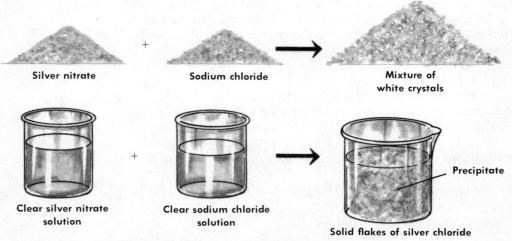

Silver nitrate   +   Sodium chloride   →   Mixture of white crystals

Clear silver nitrate solution   +   Clear sodium chloride solution   →   Precipitate

Solid flakes of silver chloride

Figure 3-8. Water solution as a condition for chemical change.

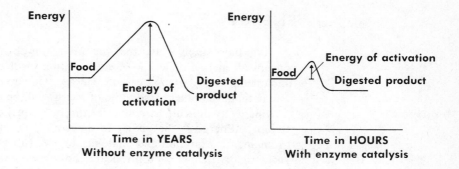

Figure 3-9.

table salt, nothing is observed except a larger amount of white crystals. However, water provides a means for the breakdown of the crystal structure (increase in entropy) which results in the formation of the white precipitate, silver chloride.

4. The use of a **catalyst** is frequently necessary if a reaction is to occur in a reasonable length of time (Fig. 3-9). A catalyst is anything that speeds up a chemical reaction without being permanently altered itself. Large complex substances called **enzymes** often function as catalysts in living organisms. For example, the breakdown of our food into simple substances (the process of digestion) involves complicated chemical changes that must be catalyzed by enzyme action. If it were not for enzymes, you would still be in the process of digesting the first food you ever ate! A catalyst works by acting as a gathering point for normally spread-out, randomly distributed reactants, or it may actually enter into a reaction as a temporary intermediate product which lowers the energy of activation and therefore shortens the time it takes to supply that energy of activation.

5. Just to put this "exact" science of chemistry in its proper perspective, the other conditions that sometimes seem to be necessary for reactions to occur could be a few bits of luck, a prayer, or black magic.

The Law of Conservation

One other consideration of matter, central to the process of stoichiometry, is the fact that under ordinary conditions, matter cannot be created or destroyed. If you start with 10 g of reactants, you must end with 10 g of product—or be able to account for the increase or decrease. The increase may be something like the invisible addition of oxygen from the air. A decrease may be the result of the loss of some invisible gas. Usually, the measurement of matter lost because of its conversion to energy in ordinary chemical change is not measurable by the laboratory balance. Only in *extra*ordinary changes, such as a nuclear bomb explosion, is the "annihilation" of matter a measurable quantity. The term "annihilation" is a little misleading anyway. What happens in this case is the conversion of matter into an equivalent amount of energy. The familiar Einstein equation states that, in reality (although not obviously), matter and energy are the two sides of the same mysterious coin.

$$E = m \quad c^2$$
(energy) (mass) (the proportionality constant, the speed of light squared)

ENERGY

You have surely noted that the discussion of matter often involved the concept of energy. The law of conservation most directly underscores the fundamental unity of matter and energy. However, for purposes of clear systematic development, the topic of energy will be explored.

Energy is difficult to define in any one concise statement. The usual attempt, "Energy is the ability to do work," is, at best, a starting point. It remains for examples and analogies to give substance to a poor skeleton. **Work**, rigidly defined as a physical action, means to *move* an object. When you push or pull an object in the process of moving it, you apply a **force** and forces can only come about if there is energy.

Types and Measurements of Energy

Potential energy. A sharpshooting basketball player is described as a scorer. When actually playing in a game he *is* scoring. Before a game the coach plans his strategy on the basis of his star's **potential** scoring ability. Energy, as the **ability** to do work, is described in much the same way. A book perched on a table edge has the potential to burn and so liberate heat energy, which has the theoretical possibility of doing work. It also has the potential to fall and move a hapless insect closer to the ground.

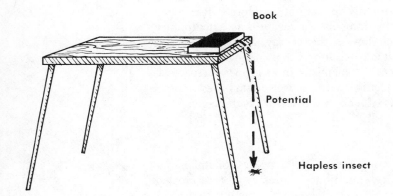

Book

Potential

Hapless insect

Figure 3-10.

Kinetic energy. When the book actually does fall or burn, it is effectively converting **potential** energy into moving energy, which is known as **kinetic energy**. The symbols for these forms of energy are E_p for potential and E_k for kinetic. The interrelationship can be illustrated by a pendulum (Fig. 3-11).

Kinetic energy is a property of matter. In fact, one of the criteria for classifying anything as a particle of matter is its ability to transfer kinetic energy. This can be illustrated by a line of billiard balls that rapidly perform this transfer (Fig. 3-12).

Kinetic energy is described mathematically as being equivalent to one half of the product of the mass of the particle and its velocity squared.

$$E_k = \frac{mv^2}{2}$$

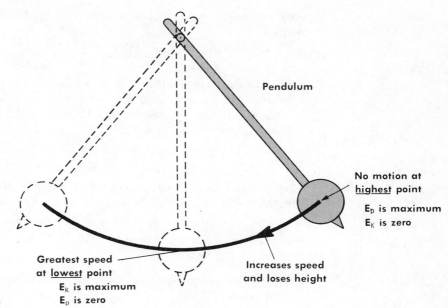

Pendulum

No motion at <u>highest</u> point

E_p is maximum

E_K is zero

Increases speed and loses height

Greatest speed at <u>lowest</u> point

E_K is maximum

E_p is zero

Figure 3-11.

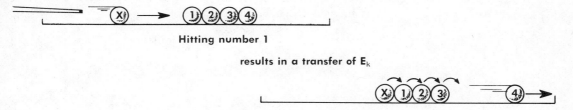

Hitting number 1

results in a transfer of E_k

Figure 3-12.   Transfer of kinetic energy.

Units of energy.   If the units used are mass in grams and velocity in centimeters per second, the product of these units is gram centimeters squared per second squared, or:

$$E_k = \text{g cm}^2 \text{ sec}^{-2}$$

THE ERG.   This cumbersome group of units is more commonly known as **ergs.** One erg is roughly the amount of energy required by a mosquito to do one push-up.

Just as length can be measured by any number of units (centimeters, inches, miles, rods), energy can be treated similarly. There are times when it may not be appropriate to use ergs. The energy required to lift a load of ship's cargo would be so great that ergs would be as awkward as measuring the distance between New York and Paris in centimeters. In other words, energy values can be expressed in a number of equivalent units. If you find or develop a conversion factor, you can express mechanical energy as an equivalent amount of heat or electrical energy with ease.

THE JOULE.   An extremely useful energy unit is the **joule.** As a measure of the kinetic energy of particles, a joule is equal to half the product of a kilogram weight moving at a velocity of one meter squared per second squared.

$$E_k = \frac{\text{kg } (\text{m}^2 \text{ sec}^{-2})}{2} = \text{joules}$$

The great value of the joule is its direct relationship to electrical energy which is the product of one volt and one coulomb.

THE COULOMB.   A **coulomb** is an amount of electric charge carried by 1 ampere in one second. Therefore,

$$1 \text{ joule } = \text{volt-coulomb}$$

$$\text{joules } = (\text{volts} \times \text{coulombs})$$

When the gram and centimeter units of the erg are converted to the kilogram and meter units of the joule, a conversion factor emerges.

$$10^7 \text{ ergs joule}^{-1} \quad \text{(ten million ergs per joule)}$$

Example 1

   If the kinetic energy of a moving ball is 5×10^7 ergs, what is the energy value in joules?
1. Predict a smaller numerical value for an answer because one joule equals so many ergs.
2. Try division by the conversion factor as the operation most likely to yield a smaller number:

$$\frac{5 \times 10^7 \text{ ergs}}{10^7 \text{ ergs joule}^{-1}} = 5 \text{ joules}$$

Example 2

   Electrical measurements on a motor indicate 2.5×10^{-4} joules of energy being used. What is the mechanical equivalent of this in ergs?
1. Predict a larger value for the answer by the same reasoning as Problem 1.
2. Try multiplication:
$$(2.5 \times 10^4 \text{ joules}) (10^7 \text{ ergs joule}^{-1}) = 2.5 \times 10^{11} \text{ ergs}$$

THE CALORIE.   In the research laboratory, the one most useful unit for energy measurement is the **calorie**. This is the unit of heat energy. Chemical reactions, while fundamentally a matter of electrical interactions by virtue of the movement of the atom's electrons, are often easier to measure by the amount of heat absorbed (**endothermic**) or evolved (**exothermic**). On the other hand, there are numerous reactions, especially in the investigations of biological chemistry, where electrical measurements are more practical than direct heat measurements. There is no question about the great practicality of being able to interconvert the heat and electrical units of energy measurement.

Considering the calorie first, it is defined as the amount of heat needed to raise the temperature of one gram of water one degree Celsius. Many calories could be related to a number of grams of water undergoing a significant temperature change.

$$\text{cal} = \overset{\text{(grams of water)}}{g}\ (\Delta t)$$

(temperature change)

Example 3

How many calories are required to raise 5 g of water from 10° C to 25° C?

1. Organize the data:

cal	?
g	5
Δt	$25 - 10 = 15°$

2. Develop the equation:

$$\text{cal} = g\ (\Delta t) = 5\ g\ (15°)$$
$$\text{cal} = 75$$

If calories can be measured directly, the instrument used is called a **calorimeter**. This is a well insulated container that works on the principle of conservation. This time it is the conservation of energy stated as the same kind of law applied to the conservation of matter. The energy absorbed by water surrounding a chemical reaction chamber must be equal to the energy produced by the reaction. In other words,

calories lost in reaction = calories gained by water

A diagrammatic picture of a calorimeter is found in Figure 3-13. A hypothetical example of how you might convert an electrical measurement of energy into a heat equivalent is illustrated in Problem 4.

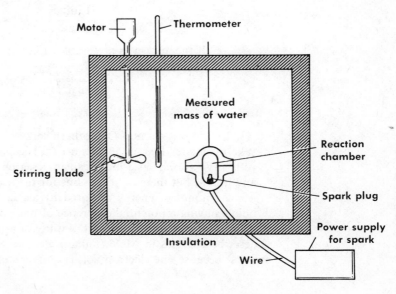

Figure 3-13. Diagrammatic view of a calorimeter.

Example 4

A chemical system produces a potential of 1.2 volts because of the resulting electron pressure. What is the equivalent heat energy yield?

1. The first step of converting the voltage to joules will be discussed later, but assume the result to be 3.0×10^5 joules.
2. Note the conversion factor:

$$4.2 \text{ joules cal}^{-1}$$

3. Since there are several joules per calorie, you would expect a smaller numerical value for the answer. Try division:

$$\frac{3.0 \times 10^5 \text{ joules}}{4.2 \text{ joules cal}^{-1}} = 0.7 \times 10^5 \text{ cal}$$

4. Simplify the answer:

$$7 \times 10^4 \text{ cal}, \quad \text{or} \quad \frac{7 \times 10^4 \text{ cal}}{10^3 \text{ cal kcal}} = 70 \text{ kcal}$$

kilocalorie

Example 5

A chemical reaction in living tissue produces a voltage of 300 mv (millivolts). Convert this to a heat energy equivalent. Assume 300 mv = 6.1 joules.

1. The conversion factor would be used in the same way as in Problem 4.

$$\frac{6.1 \text{ joules}}{4.2 \text{ joules cal}^{-1}} = \sim 1.5 \text{ cal}$$

2. If the answer is preferable in kcal:

$$\frac{1.5 \text{ cal}}{10^3 \text{ cal kcal}^{-1}} = 1.5 \times 10^{-3} \text{ kcal}$$

ELECTRON VOLTS. One other energy unit should be mentioned for the sake of greater literacy in this area. The practical applications of **electron volts** (eV) are not as widespread in research laboratory activity as calories and joules, but there are occasions for the use of this unit. An electron volt is the amount of energy acquired by an electron as it passes between two points having a potential difference of one volt.

Laboratories involved in the study of atoms and their nuclei frequently describe energies in MeV (million electron volts) or BeV (billion electron volts) because one electron volt is quite a small amount of energy. It takes about a hundred trillion trillion electron volts to equal one joule, or close to a million trillion trillion electron volts to equal one kilocalorie.

Table of Energy Conversion Factors

4.2×10^7 ergs cal^{-1}
2.6×10^{19} eV cal^{-1}
4.2 joules cal^{-1}
1.6×10^{-12} erg eV^{-1}
1×10^7 ergs joule^{-1}

Energy as Light

A very important aspect of energy production and measurement, more closely related to heat than to the kinetic energy of material particles, is the topic of **light**. The main reason for this alleged importance is that many instruments used in research laboratories are directly dependent on the understanding and utilization of light. Light is just a part of a huge range of what is called **radiant energy**, or **electromagnetic radiation**.

Electromagnetic radiation encompasses a range, or **spectrum**, from alternating electric current (low energy) to the cosmic radiation of the stars (high energy). Electromagnetic radiation means the production and outward radiation of energy waves because of the vibrations (oscillations) of electrically charged particles. The part of the E-M (electromagnetic) spectrum being radiated depends on the energy involved Since radiant energy is a wave phenomenon, it is necessary to relate the characteristics of waves to the amounts of energy associated with those characteristics.

The wave length. It helps to think of waves in terms of what results when a stone is dropped into a still pond. This old but durable analogy presents a picture of crests and troughs radiating outward from the point of disturbance. The essential difference between the water wave and the E-M wave is that the E-M wave does not require matter in order to move. It can proceed through a vacuum. In fact, precise measurements indicate the speed of light and all other forms of E-M radiation to be 3×10^{10} cm sec^{-1} (more familiarly as 186,000 miles per second). This universal constant is designated by the letter c (remember, $E = mc^2$).

If you measure the distance between the peak of one crest and the peak of the next, this is the wavelength, and it is symbolized by the Greek letter lambda (λ). The other symbol in the diagram (a) represents the height of the wave, called the **amplitude** (Fig. 3-14).

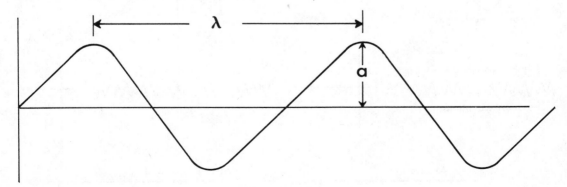

Figure 3-14.

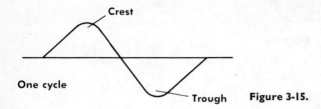

Crest

One cycle

Trough   Figure 3-15.

The frequency. The other essential characteristic of a wave is its **frequency**, symbolized by another Greek letter, nu (ν). Frequency means the number of waves, or crest-trough cycles, that will pass an observation point in one second (Fig. 3-15). The unit of measure for frequency is reciprocal seconds (per second or sec^{-1}), since time is the only unit of measure. "Cycles per second" is being replaced by a modern label called **hertz** (Hz), named after Heinrich Hertz.

Relating wavelength, frequency, and energy. Comparing two "bits" of E-M radiation as they pass an observation point, moving at the same speed (the speed of light), you will see that more cycles will pass the observation point if the wavelength is shorter (Fig. 3-16).

From the idealized diagram, you can see that the wavelength is inversely proportional to the frequency. The equation for this relationship may be developed:

$$\lambda \propto \frac{1}{\nu}$$

$$\lambda = k \frac{1}{\nu}$$

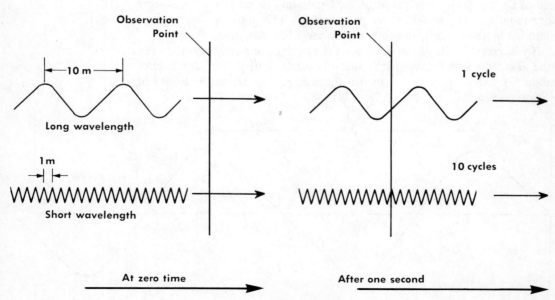

Figure 3-16.

The proportionality constant (k) relating the frequency and wavelength is the speed of light. Therefore, the equation may be rewritten:

$$\lambda = \frac{c}{v}$$

The other relationships derived from this equation are

$$c = \lambda v \quad \left(\begin{array}{c} \text{speed of light} = \text{the product of frequency} \\ \text{and wavelength} \end{array} \right)$$

and

$$v = \frac{c}{\lambda} \quad \left(\text{frequency} = \frac{\text{speed of light}}{\text{wavelength}} \right)$$

Example 6

What is the frequency of radiation having a wavelength of 500 mμ?

1. Organize the data and be sure the units are compatible:

c	3×10^{10} (cm) sec^{-1}	compatibility
λ	500 m$\mu = 5 \times 10^{-5}$ (cm)	
v	?	

$$\frac{500 \text{ m}\mu}{10^7 \text{ m}\mu \text{ cm}^{-1}} = 500 \times 10^{-7} = 5.00 \times 10^{-5} \text{ cm}$$

2. Substitute and solve equation:

$$v = \frac{c}{\lambda} = \frac{3 \times 10^{10} \text{ cm sec}^{-1}}{5 \times 10^{-5} \text{ cm}} = 0.6 \times 10^{15} \text{ sec}^{-1}$$

$$v = 6 \times 10^{14} \text{ sec}^{-1} \quad \text{(or hertz)}$$

Many observations over the years clearly indicate that higher energy radiation has a shorter wavelength and, thus, a higher frequency. For example, radio waves, having wavelengths in the meter or kilometer range, are much less energetic than x-rays. The wavelengths of x-rays are in the angstrom region. This difference is related to the source of the radiation (see Fig. 3-17).

Most of the important optical instruments used in the laboratory will be discussed later in the text. For the most part, they are concerned with the tendency of atoms and molecules to absorb or emit very definite wavelengths of the E-M spectrum. There are such instruments as the infrared (IR) spectrophotometer, visible (vis.) spectrophotometers and colorimeters, ultraviolet (uv) spectrophotometers, fluorescence spectrometers, x-ray spectrophotometers, and atomic absorbance spectrophotometers, to name a few of the most prominently used instruments.

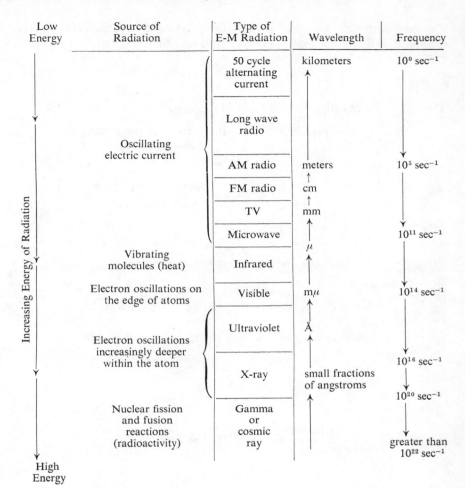

Low Energy	Source of Radiation	Type of E-M Radiation	Wavelength	Frequency
		50 cycle alternating current	kilometers	10^0 sec^{-1}
	Oscillating electric current	Long wave radio		
		AM radio	meters	10^5 sec^{-1}
		FM radio	cm	
		TV	mm	
		Microwave		10^{11} sec^{-1}
	Vibrating molecules (heat)	Infrared	μ	
	Electron oscillations on the edge of atoms	Visible	mμ	10^{14} sec^{-1}
	Electron oscillations increasingly deeper within the atom	Ultraviolet	Å	
		X-ray	small fractions of angstroms	10^{16} sec^{-1}
				10^{20} sec^{-1}
High Energy	Nuclear fission and fusion reactions (radioactivity)	Gamma or cosmic ray		greater than 10^{22} sec^{-1}

Increasing Energy of Radiation

Figure 3-17.

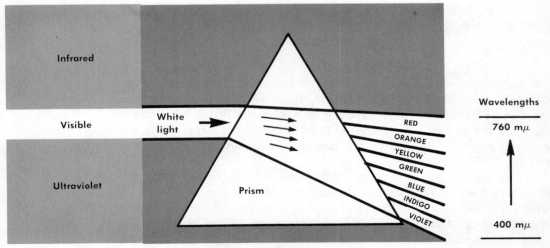

Figure 3-18.

A closer look at the visible light region of the E-M spectrum will be useful before considering the structure of the atom (Fig. 3-18).

The violet part of the spectrum has the shortest wavelength and, therefore, the highest energy. All wavelengths of light traveling through a vacuum have the same speed (c), but through a **quartz prism**, the violet is slowed more than the red and, as a result, it is bent more sharply. The speed increases with the color as the colors move toward red. The optical instruments mentioned before make use of prisms or **diffraction gratings** (a film etched with fine lines producing the same diffusion of the spectrum as the prism) in the process of selecting specific wavelengths with their associated energy values.

The business of relating energy and wavelength was puzzling, and not too useful, for many years. Light was considered to be a continuous wave phenomenon alone. There was no consideration of *thinking* of light as having a particle nature until the German physicist Max Planck suggested it early in this century. Today, light is thought of as being made up of distinct *"pulses"* of energy that are called **photons**. A widely used term that means "a definite quantity of energy" (just described as "pulse" or "photon") is the **quantum**. Planck's proposal is called the **quantum theory**. There is considerable doubt as to the reality of a photon as a "particle," but in some ways these pulses of energy may be thought of as having a few characteristics of material particles. The great difficulty in properly describing the photon and the electron in absolute physical descriptive terms is that they are unlike anything in our common experience. There is no basis for comparison. When photons and electrons are described as waves sometimes and as particles at other times, you are inclined to accept the Principle of Complementarity proposed by the great Danish scientist, Niels Bohr. This principle says, in effect, that both views together—the wave and particle models—probably give a more accurate picture of what is actually true.

Planck expressed the relationship between energy and frequency of radiation by saying that the *energy* of a photon was directly proportional to the *frequency*.

$$E \propto \nu$$

The proportionality constant that emerged is known as Planck's constant of action. It curiously has the dimensions of energy and time. The symbol is h.

$$E = h\nu$$

Energy of photon = Planck's constant × frequency

The value of Planck's constant can be determined from data:

E	3.6×10^{-12} ergs
h	?
ν	5.5×10^{14} sec^{-1} (green light)

$$h = \frac{E}{\nu} = \frac{3.6 \times 10^{-12} \text{ ergs}}{5.5 \times 10^{14} \text{ sec}^{-1}}$$

$$h = 6.6 \times 10^{-27} \text{ erg (sec)}$$

Example 7

What is the energy of a photon of violet light ($\lambda = 420$ mμ)?

1. Organize the data for the relationship $E = h\nu$:

E	?
h	6.6×10^{-27} erg (sec)
ν	$= \dfrac{c}{\lambda} = \dfrac{3 \times 10^{10} \text{ cm sec}^{-1}}{4.2 \times 10^{-5} \text{ cm}}$

$$\frac{420 \text{ m}\mu}{10^7 \text{ m}\mu \text{ cm}^{-1}} = 4.2 \times 10^{-5} \text{ cm}$$

2. Substitute in the equation $E = h\dfrac{c}{\lambda}$:

$$E = \frac{6.6 \times 10^{-27} \text{ erg (sec)} \, (3 \times 10^{10} \text{ cm sec}^{-1})}{4.2 \times 10^{-5} \text{ cm}}$$

$$E = 4.7 \times 10^{-12} \text{ ergs}$$

This relationship between energy and the radiation of light is very helpful in understanding the concept of energy levels in atoms.

ATOMIC STRUCTURE

The part of the atom that is directly involved in the process of chemical change is the electron. The atom as a whole will receive less attention than it deserves, but the amount of discussion needed for an intellectually satisfying study is enormous. Supplementary reading may serve to answer questions that are raised here. The topics to be considered are: (1) a model of the atom, (2) the concept of energy levels, and (3) the idea of valence.

A Model of the Atom

Atoms could not be "seen" until the twentieth century. With the development of special methods called field-ion microscopy and x-ray diffraction, man could see projections of atoms fuzzily enlarged. This is called high magnification and low resolution (focusing). But even that is good. It supports the notion that atoms are remarkably small, imperfect spheres. Not rigid, but dynamic pulsating units of matter so small that 1 gram of coal, for example, is made up of about 1×10^{22}, or 10,000,000,000,000,000,000,000 (ten billion trillion) atoms; so small that if all the atoms composing the head of a straight pin were magnified to basketball size, they could fill the Great Lakes of the United States.

Surely you have seen animated "planetary" models of atoms. Brilliant little electrons with their comet-like tails whizzing around a nucleus that looks like an unripe raspberry. The usefulness of this model is questionable because of the huge distortions it presents in terms of sizes and spatial relationships among the principal parts of the atom. The fuzzy sphere that you can "see" may be more appropriate.

The principal parts of the atom are the **electron** (e⁻), the **proton** (p⁺), and the **neutron** (n⁰). Protons and neutrons compose the **nucleus** Later, when radioactivity is discussed, the nucleus will be taken up in more detail. For the time being, the nucleus can be considered as the very massive, but very tiny, core of the atom. The incredibly small size of the nucleus is such that it would take about 10,000 nuclei in a row to equal the diameter of the whole atom. By analogy, if an atom were magnified in size to Houston's Astrodome ball park, the nucleus of that atom would be about the size of a single baseball suspended ninety feet up in short center field.

Yet, when you consider that the electrons must be composing the rest of the Astrodome, while they have a weight close to nothing, the single baseball must have almost all the weight! In fact, atomic nuclei are so dense that an attempt toward visualization would demand that you try to imagine the ships of the world's navies jammed into a thimble. Something like 100 million tons per cubic centimeter!

Nearly all of the volume of the atom then must be made of electrons, those units of negative charge that have practically no mass or size. If atoms must be described as more than 99 per cent *nothing* (according to cold, disconcerting, and inescapable logic), how do you account for the very effective illusion of solidity presented by people and things—all made of atoms?

The conclusion is that these electrons must be moving at very remarkable speeds—so fast that they must "seem" to be everywhere within the rough dimensions of the atom at once! It is similar to the effect of an airplane propeller (a weak analogy at best) which can move fast enough to create the illusion of a solid circle of metal. The electron, by contrast, moves fast enough to make about seventeen trillion revolutions per second around the nucleus. Their energies (kinetic and potential) are both significant and measureable as *definite* amounts.

Because of the uncertainty of the electron's whereabouts, the planetary model is misleading. Any attempt to simultaneously measure the path and speed of a particular electron is fruitless, because the very application of measuring devices upsets the object of the measurement. As a result, scientists describe the distribution of electrons in atoms in terms of probabilities. The concept of probability suggests that a better model of an atom is a "charge-cloud" picture in which the greatest probability for locating an electron is directly related to its total energy (see Fig. 3-19).

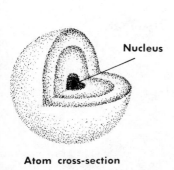

Nucleus

Atom cross-section

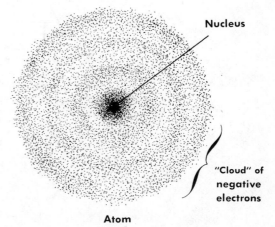

Nucleus

"Cloud" of
negative
electrons

Atom

Figure 3-19.

The Concept of Energy Levels

It was Niels Bohr who brilliantly wedded Planck's quantum theory of energy—that energy is "bundled" into distinct packages having a value proportional to the frequency—to the energies of wavelengths of light observed through spectrum apparatus (spectroscope). Bohr studied the hydrogen spectrum and from these studies came the concept of energy levels of electrons in atoms. The probability concept has long since replaced the Bohr model, but the energy level description persists. Some of the original terminology that related to an electron as a particle on some definite track, or orbit, is now applied to the energy description of negative charge in some region of high probability.

Bohr visualized the hydrogen atom as a single proton nucleus and an electron capable of occupying a variety of energy levels, depending on which *definite* amount of energy it absorbs or emits. As the electron shifts (oscillates) between energy levels, it radiates electromagnetic energy in precise amounts. The energy values are directly related to the energy levels involved in the shift. This means that when an electron shifts from a high energy level (farther from the nucleus) to a lower one, the energy of the radiated photon has a definite value (Fig. 3-20). This energy could be measured in ergs, kcal, eV, or joules, and its value is gained through knowledge of the wavelength of the emitted radiation.

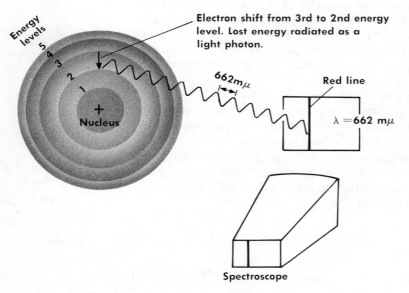

Energy levels
5
4
3
2
1
+
Nucleus

Electron shift from 3rd to 2nd energy level. Lost energy radiated as a light photon.

662mμ

Red line

λ = 662 mμ

Spectroscope

Figure 3-20.

Red line viewed and measured by spectroscope

Example 8

Calculate the energy of a photon of red light having a wavelength of 662 mμ.

1. State the related data as an equation:

$$E = h\nu$$

since $\nu = \dfrac{c}{\lambda}$, substitute

$$E = h\frac{c}{\lambda}$$

2. Organize the data:

E	?
h	6.62×10^{-27} erg (sec)
ν	$\nu = \dfrac{c}{\lambda}$
c	3×10^{10} cm sec^{-1}
λ	662 m$\mu = 6.62 \times 10^{-5}$ cm

$$\lambda = 662 \text{ m}\mu$$

$$\frac{662 \text{ m}\mu}{10^7 \text{ m}\mu \text{ cm}^{-1}} = 662 \times 10^{-7} \text{ cm}$$

$$\lambda = 6.62 \times 10^{-5} \text{ cm}$$

3. Substitute in the equation:

$$E_{\text{photon}} = \frac{6.62 \times 10^{-27} \text{ erg (sec) } (3 \times 10^{10} \text{ cm sec}^{-1})}{6.62 \times 10^{-5} \text{ cm}}$$

$$E_{\text{photon}} = 3 \times 10^{-12} \text{ erg}$$

4. This means that the amount of energy lost by the electron as it "falls" from the third energy level to the second energy level is a quantum having a value of 3×10^{-12} erg.

Relating the observed spectral lines to other energy levels in hydrogen illustrates how these energy levels provide an excellent means of organizing the electrons in an atom (Fig. 3-21).

The maximum number of electrons permitted on an energy level is regulated by the three dimensional space available considering the fact that electrons, all having the same negative charge, tend to repel each other. As the probability regions increase in distance from the nucleus, more space

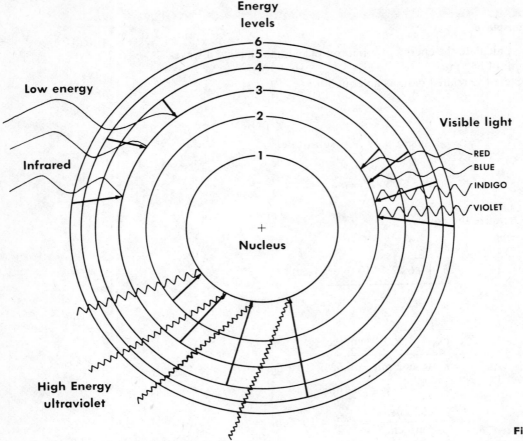

Figure 3-21

is available to accommodate the electrons. The maximum electron populations for each energy level are seen in the following chart:

Principal Energy Level	Maximum Number of Electrons
1	2
2	8
3	18
4	32
5	50

A simplified model is used to visualize this distribution in Figure 3-22.
    As knowledge of the atom's structure advanced beyond Bohr's original
theory, it was discovered that each energy level is in reality subdivided.
These subdivisions have their own maximum electron populations, represented
in the next chart. There is a maximum of four subdivisions (called **orbitals**
from the old terminology) for each principal energy level. It should also
be noted that the lowest energy level which a particular electron can occupy
is commonly called its **ground state**.

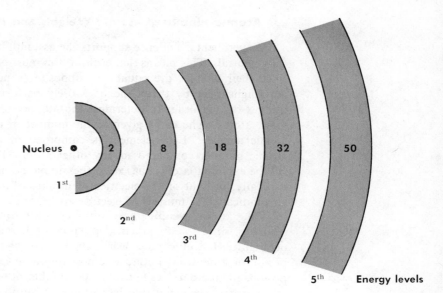

Nucleus ●   2   8   18   32   50

1st

2nd

3rd

4th

5th   Energy levels

Figure 3-22.

Subdivision	Name of Subdivision	Maximum Number of Electrons
1st (lowest energy)	s orbital	2
2nd	p orbital	6
3rd	d orbital	10
4th (highest energy)	f orbital	14

The most significant fact associated with the subdivisions is that the s and p orbitals of *any* energy level are filled to the maximum before electrons begin occupying higher energy levels. The total number of electrons in the s and p orbitals together is *eight*. While this "octet" portends no special magic, it does seem to endow the atom with marked stability. The drive toward stability (and lower energy) once again appears as a fundamental driving force in nature. In view of this tendency toward the achievement of the "octet" arrangement, a great deal of atomic linkage (chemical bond formation) becomes more understandable. The electrons that are involved in the addition or subtraction process leading to the "octet" configuration are naturally the ones farthest from the attractive grasp of the nucleus. These farthest removed electrons are the ones in the outermost energy level of any atom. The outermost energy level of various atoms, having a number of electrons varying from one to eight, must then play the critical role in determining the properties of an atom. In a larger sense, the properties of the atom are the properties of the element which it composes. The outer "shell" (energy level) determines the kind of bond a particular atom will form with other atoms, as well as the numbers of each atom in the larger structure.

An old and somewhat overworked term used to describe the combining ratios of atoms is **valence**. While the concept of valence itself is superficial, it is still commonly applied to the outer energy level insofar as it is often called the **valence shell**.

Atomic Number, Atomic Weight, and Isotopes

Experimental evidence supports the assumption that atoms are electrically neutral. This means that each proton must be balanced by an electron, since their charges are equal and opposite (experimental support again). An Englishman by the name of Mosely performed brilliant experiments using x-rays in order to determine the number of protons in the nuclei of many atoms. The name given to the number of protons in an atom is the **atomic number**. Each element is composed of atoms having a different atomic number compared to any other atom. The periodic chart, which will be explored later, is based on the atomic numbers of the atoms. Automatically, in light of the electrical neutrality of atoms, the atomic number also indicates the number of electrons in an atom.

The **atomic weight** (mass) number is the sum of the protons and the neutrons, since for all practical purposes, the weight of the electrons are negligible. (It would take nearly 2000 electrons to equal the weight of a proton or a neutron—both are almost the same.) Although the number of protons in an atom does *not* vary, the number of neutrons may. The atoms of a particular element that have different numbers of neutrons are called **isotopes**.

Isotopes have a certain notoriety because of their popular association with atomic bombs and deadly radioactive fallout. It is true that some better known isotopes do have the critical number of neutrons that render a nucleus unstable. Unstable nuclei tend to break down at varying rates (from microseconds to thousands of years) with the possible radiation of their lost energy in the deadly gamma ray range of the spectrum.

Today, all the atoms of the more than 100 known elements are assigned atomic weights on the basis of comparison to the isotope of carbon whose atomic weight is set at 12 (6 protons and 6 neutrons). The unit of weight is the **atomic mass unit** (amu). The proton and neutron are each about 1 amu. Another isotope of carbon, carbon 14, is famous for its role in dating ancient civilizations or tribes, while carbon 12 is most abundant and is not radioactive. Isotopes, then, explain why atomic weights are not the simple whole numbers expected when simple whole numbers of protons and neutrons are added. If a volume of chlorine gas, for example, is 75 per cent chlorine 35 and 25 per cent chlorine 37, the *average* atomic weight would be about 35.5. This is the atomic weight observed on an appropriate chart or table.

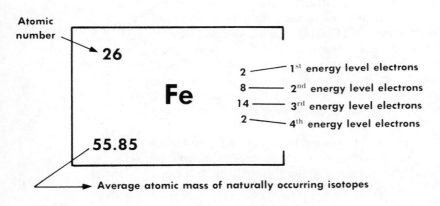

Atomic number

26

Fe

2 —— 1ˢᵗ energy level electrons
8 —— 2ⁿᵈ energy level electrons
14 —— 3ʳᵈ energy level electrons
2 —— 4ᵗʰ energy level electrons

55.85

Average atomic mass of naturally occurring isotopes

Figure 3-23.

The usual method of illustrating these atomic data follows the format of the iron example in Figure 3-23. The number of neutrons is found by rounding off the atomic weight to 56. This number represents the total of the protons and neutrons. Subtract 26 (number of protons) from 56 (total number) to obtain the neutron number.

$$56 \; \text{protons} + \text{neutrons}$$

$$-26 \; \text{protons}$$

$$\overline{\phantom{-26 \; \text{protons}}}$$

$$30 \; \text{neutrons}$$

A few more common examples are found in Figure 3-24.

As you observe these examples, a real correlation is seen between the atomic number and the number of energy levels. This relationship will also be investigated within the framework of the periodic table. It should also be noted that, for practical purposes, the atomic weights of the atoms are usually rounded off: oxygen 16, magnesium 24, silver 108, and lead 207.

The most important observation to make, however, is that the familiar **metals** of silver and magnesium have a low number of electrons in the *valence shell*, while the distinctly *nonmetallic* oxygen atom has a comparatively large number—remember that the range is from zero to eight, from the special "octet" point of view. Curiously, lead is in the middle with four electrons. With this in mind, the label **metalloid** (both characteristics of metals and nonmetals) applied to lead seems reasonable.

The drive toward the stability of the octet is accomplished most easily by the metals "losing" their few electrons which in turn are "gained" by the nonmetals. Incidentally, the stability of the "octet" is a characteristic of the **noble** gases—helium, neon, argon, krypton, xenon, and radon—which are remarkably stable elements. The process of "losing" electrons is called **oxidation**. The actual number of electrons that an atom will usually lose in a chemical reaction is known as its common **oxidation number**. The common oxidation number is the value loosely equated by many people to valence.

The oxidation number of an atom that *gains* electrons is designated by a *negative* sign. It would be a positive number if a system of common reduction numbers were used. **Reduction**, as implied, is the opposite of oxidation. Oxidation and reduction ordinarily occur simultaneously. The oxidation number may also be thought of as the **charge on the ion**. An ion is the term applied to electrically charged (positive or negative) particles. A typical

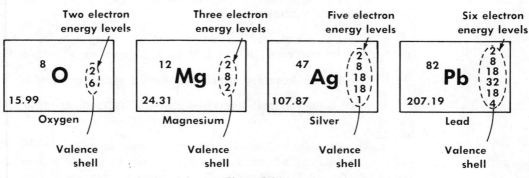

Figure 3-24.

ion is an atom that has lost or gained electrons so that it is no longer an electrically neutral unit. Some examples for comparison follow:

	Atom	Symbol	Electrons lost $(-e^-)$ Electrons gained $(+e^-)$	Common Oxidation Number	Ion
metals	sodium	Na	$-1e^-$	$1+$	Na^+
	magnesium	Mg	$-2e^-$	$2+$	Mg^{2+}
	aluminum	Al	$-3e^-$	$3+$	Al^{3+}
non-metals	oxygen	O	$+2e^-$	$2-$	O^{2-} oxide
	chlorine	Cl	$+1e^-$	$1-$	Cl^{1-} chloride
	phosphorus	P	$+3e^-$	$3-$	P^{3-} phosphide

In the next chapter the connection between the oxidation numbers of atoms and the formation of larger structures, such as molecules, will be developed. The concept of the **radical** (slightly complex ion) will have to be explored before a larger table of common oxidation numbers can be used to the greatest advantage.

PROBLEMS

1. Label the following changes as physical or chemical: rusting iron, souring milk, dissolving sugar, blending iron powder and sulfur, chewing a bite of steak thoroughly.

2. Identify the following as elements, compounds, or mixtures: silver, table salt, air, milk, carbon dioxide, sulfur, tea, hemoglobin.

3. How many kilocalories are required to raise the temperature of 200 g of water from $3°$ C to $17°$ C?

4. How fast must a 2 g object move in order to have a kinetic energy of 8×10^6 ergs?

5. If the energy absorbed in a chemical change is 3×10^{-5} joules, what is the mechanical equivalent in ergs?

6. Scientists use an instrument capable of generating energies measured at 10 million electron volts (10 MeV). What is this energy in calories or kcal?

7. What is the frequency of radiation having a wavelength of 320 mμ?

8. What is the energy (in ergs) of a photon if its frequency is 5.0×10^{15} sec^{-1}?

9. There are three naturally occurring isotopes of element "X". Each one has the following percentage of a typical sample, ^{30}X (20 per cent), ^{32}X (50 per cent), and ^{34}X (30 per cent). Calculate the average atomic weight of element "X".

10. Complet the chart.

Element	Atomic Number	Number of Neutrons	Number of Electrons	Number of Protons	Atomic Weight
$^{27}_{13}$Al					
Chromium	24				52
$_{35}$Br		45			
			79		197

11. Label the following changes as physical or chemical: burning methane gas, solidifying carbon dioxide gas, separating salt from sand by adding water, filtering and evaporating, grinding glass to powder, baking a cake.

12. Identify the following as elements, compounds, or mixtures: mercury, fruit punch, nitric acid, oxygen, pewter, beef broth, brass, ice.

13. If 50 g of water at $18°$ C absorbs 2.5 kcal of heat, what will the final temperature be?

14. What is the kinetic energy of a 2×10^{-4} mg object moving at a rate of 10 m sec^{-1}?

15. A chemical change released 4×10^4 cal. What is this energy in joules?

16. If an electron absorbs 3.5×10^{-12} erg in shifting energy levels, calculate the same value in electron-volts.

17. What wavelength of E-M radiation has a frequency of 1.6×10^{14} sec^{-1}?

18. A certain molecule absorbs light having a wavelength of 1.2×10^3 mμ. What is the frequency of this radiation and what is the energy (calories) of a photon?

19. Calculate the average atomic weight of element "Q" if it naturally occurs as ^{50}Q (34 per cent) and ^{53}Q (66 per cent).

20. Complete the chart.

Element	Atomic Number	Number of Neutrons	Number of Electrons	Number of Protons	Atomic Weight
		61	47		
$^{45}_{21}$Sc					
^{32}S					
				82	207

SUPPLEMENTARY READING

Moore, R.: *Niels Bohr.* Alfred A. Knopf, New York, 1966.

An absorbing biography of the great Danish physicist. Excellent explanations of developing atomic theories during the first third of this century.

Gamow, G. A.: *Thirty Years That Shook Physics.* (paperback) Anchor Books, New York, 1966.

Superb discussion of the theories of atomic structure during the first thirty years of the 20th century.

Guillemin, V.: *The Story of Quantum Mechanics.* Charles Scribners Sons, New York, 1968.

An extremely readable story of the historical development of quantum mechanics.

Wilson, M.: *Energy.* Life Science Library, Time, Inc., New York, 1960.

An introduction to energy that is pleasurable reading. Beautiful illustrations.

Andrade, E. N. da C.: *Rutherford and the Nature of the Atom.* (paperback) Doubleday & Co., Inc., New York, 1966.

A fascinating biography of the life and work of Niels Bohr's mentor.

Fundamentals of 4
Chemistry

One of the most useful tools available to chemists is the **periodic chart** of the elements. A vast amount of basic information is presented in a highly organized format. Anyone attempting to see the orderly relationships among elements as a basis for understanding their behavior needs to become familiar with this chart. The **periodic law** says that the properties of elements, when arranged according to increasing atomic number, differ until a noble gas appears, and then the same properties (with some modification) appear again. This periodic *recurrence* of similar properties happens seven times. An overview of the entire chart points up questions to be discussed in more detail (Fig. 4-1).

The periodic chart, in much the same way as a map, can be divided into convenient regions based on similarities and contrasts. The chart may be used to fullest advantage when broad generalizations in addition to detailed inspection of particular elements are made. Broad generalizations provide the perspective that shows regional relationships; detailed inspection yields instant information about the distribution of electrons in the energy levels. The following diagrams of the chart are designed to emphasize the broad generalizations (Figs. 4-2, 4-3, 4-4, and 4-5).

From this point on, the special section labeled **rare earth elements** will be omitted. This is a unique group of very similar elements that have limited practical value in the laboratory.

The Roman numerals at the top of the chart indicate **groups** while the subheadings A or B indicate the closely related families that subdivide the group. The elements within a group have marked similarities in their physical and chemical properties.

IA	IIA	IIIB	IVB	VB	VIB	VIIB	VIII	VIII	VIII	IB	IIB	IIIA	IVA	VA	VIA	VIIA	VIIIA
1 **H** 1.00797 ±0.00001																	2 **He** 4.0026 ±0.0005
3 **Li** 6.939 ±0.0005	4 **Be** 9.0122 ±0.00005											5 **B** 10.811 ±0.003	6 **C** 12.01115 ±0.00005	7 **N** 14.0067 ±0.00005	8 **O** 15.9994 ±0.0001	9 **F** 18.9984 ±0.0005	10 **Ne** 20.183 ±0.0005
11 **Na** 22.9898 ±0.00005	12 **Mg** 24.312 ±0.0005											13 **Al** 26.9815 ±0.00005	14 **Si** 28.086 ±0.001	15 **P** 30.9738 ±0.0005	16 **S** 32.064 ±0.003	17 **Cl** 35.453 ±0.001	18 **Ar** 39.948 ±0.0005
19 **K** 39.102 ±0.0005	20 **Ca** 40.08 ±0.005	21 **Sc** 44.956 ±0.0005	22 **Ti** 47.90 ±0.005	23 **V** 50.942 ±0.0005	24 **Cr** 51.996 ±0.001	25 **Mn** 54.9380 ±0.00005	26 **Fe** 55.847 ±0.005	27 **Co** 58.9332 ±0.00005	28 **Ni** 58.71 ±0.005	29 **Cu** 63.54 ±0.005	30 **Zn** 65.37 ±0.005	31 **Ga** 69.72 ±0.005	32 **Ge** 72.59 ±0.005	33 **As** 74.9216 ±0.0005	34 **Se** 78.96 ±0.005	35 **Br** 79.909 ±0.002	36 **Kr** 83.80 ±0.005
37 **Rb** 85.47 ±0.005	38 **Sr** 87.62 ±0.005	39 **Y** 88.905 ±0.0005	40 **Zr** 91.22 ±0.005	41 **Nb** 92.906 ±0.0005	42 **Mo** 95.94 ±0.005	43 **Tc** (99)	44 **Ru** 101.07 ±0.005	45 **Rh** 102.905 ±0.0005	46 **Pd** 106.4 ±0.005	47 **Ag** 107.870 ±0.003	48 **Cd** 112.40 ±0.005	49 **In** 114.82 ±0.005	50 **Sn** 118.69 ±0.005	51 **Sb** 121.75 ±0.005	52 **Te** 127.60 ±0.005	53 **I** 126.9044 ±0.0005	54 **Xe** 131.30 ±0.005
55 **Cs** 132.905 ±0.0005	56 **Ba** 137.34 ±0.005	57 **La** 138.91 ±0.005	72 **Hf** 178.49 ±0.005	73 **Ta** 180.948 ±0.0005	74 **W** 183.85 ±0.005	75 **Re** 186.2 ±0.05	76 **Os** 190.2 ±0.05	77 **Ir** 192.2 ±0.05	78 **Pt** 195.09 ±0.005	79 **Au** 196.967 ±0.0005	80 **Hg** 200.59 ±0.005	81 **Tl** 204.37 ±0.005	82 **Pb** 207.19 ±0.005	83 **Bi** 208.980 ±0.0005	84 **Po** (210)	85 **At** (210)	86 **Rn** (222)
87 **Fr** (223)	88 **Ra** (226)	89 †**Ac** (227)	104 (257)														

Lanthanum Series

58 **Ce** 140.12 ±0.005	59 **Pr** 140.907 ±0.0005	60 **Nd** 144.24 ±0.005	61 **Pm** (147)	62 **Sm** 150.35 ±0.005	63 **Eu** 151.96 ±0.005	64 **Gd** 157.25 ±0.005	65 **Tb** 158.924 ±0.0005	66 **Dy** 162.50 ±0.005	67 **Ho** 164.930 ±0.0005	68 **Er** 167.26 ±0.005	69 **Tm** 168.934 ±0.0005	70 **Yb** 173.04 ±0.005	71 **Lu** 174.97 ±0.005

Actinium Series

90 **Th** 232.038 ±0.0005	91 **Pa** (231)	92 **U** 238.03 ±0.005	93 **Np** (237)	94 **Pu** (242)	95 **Am** (243)	96 **Cm** (247)	97 **Bk** (247)	98 **Cf** (249)	99 **Es** (254)	100 **Fm** (253)	101 **Md** (256)	102 **No** (253)	103 **Lw** (257)

Figure 4-1. Periodic chart of the elements.

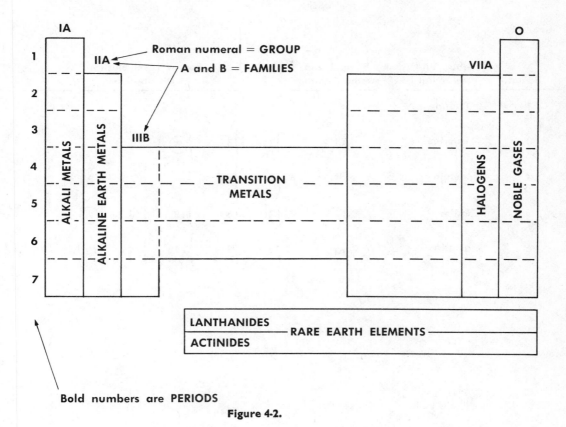

Figure 4-2.

Bold numbers are PERIODS

Period	Group 1 A Metal	Symbol and Atomic Number	Energy Level Electron Distribution						
2	lithium	$_3$Li	)2	)①					
3	sodium	$_{11}$Na	)2	)8	)①				
4	potassium	$_{19}$K	)2	)8	)8	)①			
5	rubidium	$_{37}$Rb	)2	)8	)18	)8	)①		
6	cesium	$_{55}$Cs	)2	)8	)18	)18	)8	)①	
7	francium	$_{87}$Fr	)2	)8	)18	)32	)18	)8	)①

Figure 4-3.   The alkali metals.

Period	Group VIIA Nonmetal	Symbol and Atomic Number	Energy Level Electron Distribution					
2	fluorine	$_9$F	)2	⑦				
3	chlorine	$_{17}$Cl	)2	)8	⑦			
4	bromine	$_{35}$Br	)2	)8	)18	⑦		
5	iodine	$_{53}$I	)2	)8	)18	)18	⑦	
6	astatine	$_{85}$At	)2	)8	)18	)32	)18	⑦

Figure 4-4.   The halogens.

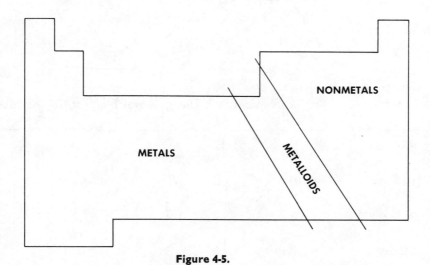

Figure 4-5.

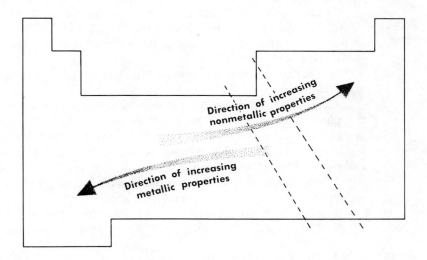

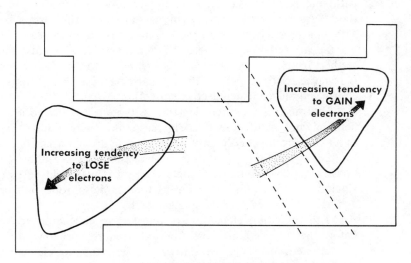

Figure 4-5. *Continued.*

For example, the metals in group I, the **alkali metals**, are very much alike in physical appearance and chemical behavior (Fig. 4-3). The key to this similarity lies in the number of electrons in the outer energy level. Not only do the alkali metals all have one electron in the outer energy level, but there are also several other significant relationships to be noted.

1. Each alkali metal occurs immediately after a noble gas. This is the periodic reoccurrence characteristic of the chart.
2. The number of the **period** is the same as the number of **energy levels** occupied by electrons.
3. The Roman numeral (group I) agrees with the common oxidation number of all the alkali metals, $1+$.

Group VII, the **halogens**, provides an interesting contrast. Although the physical properties of the decidedly nonmetallic halogens are not as similar as alkali metals, their chemical behavior is much alike (Fig. 4-4). Astatine is often ignored because of unavailability. Both astatine and francium (group I) undergo rapid radioactive decay.

In addition to the observation that each halogen has seven electrons in the outer energy level, the following points are noted again.

1. Each halogen occurs on the chart just *before* a noble gas.
2. The number of the period is still the same as the number of energy levels occupied.
3. The Roman numeral (group VII) points to the number of electrons needed for the completion of the "octet." The halogens lack one electron and this is the value of the common oxidation number in each case $(1-)$.

The bold numbers to the left on the periodic chart represent the periods, as mentioned earlier. The main characteristic of a period is *change*. Start with metals on the left and gradually proceed toward nonmetals on the right. As you move downward in the chart, the atoms become larger. The outermost electrons on the larger atoms are more easily "lost" (usually via the attractive force exerted by nonmetallic elements), since they are farthest from their own nuclei. This tendency to "lose" electrons (oxidation) is a characteristic of metals. By summarizing these observations, the following charts will aid your perspective (Fig. 4-5).

Hydrogen is a unique element. Although it is placed in group IA, it is definitely not an alkali metal. The reason it is there is that it has one electron in its one normal energy level. However, hydrogen needs only two electrons for the completion of its valence shell so that there is no concern for an "octet" arrangement. The single electron is also held very strongly by its single proton nucleus. This means that oxidation does not occur as easily as it does in the case of the alkali metals.

The transition elements are another special case. It was mentioned before that procession across a period from group I toward the noble gases is characterized by change. However, the transition elements have definite similarity. This apparent contradiction is explained by the fact that while the number of electrons in the transition elements are increasing as the atomic number increases, the valence shell is generally fixed at two electrons while the next to outer energy level is adding electrons (Fig. 4-6).

The point to be remembered is that the properties of an element are mainly dependent on the number of electrons in the *outermost* energy level. The examples of manganese, iron, and cobalt indicate that the reason for their similarity is the presence of two electrons in the valence shell of each one.

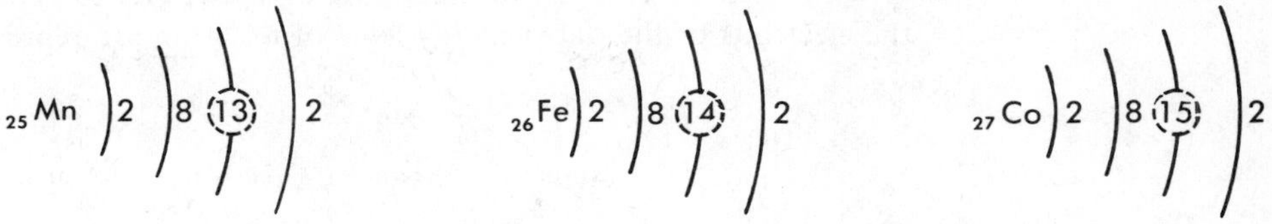

$_{25}$Mn $\Big)$2 $\Big)$8 (13) $\Big)$2 $_{26}$Fe $\Big)$2 $\Big)$8 (14) $\Big)$2 $_{27}$Co $\Big)$2 $\Big)$8 (15) $\Big)$2

Figure 4-6.   Manganese   Iron   Cobalt

Another notable characteristic of the transition elements is the *variability* of the common oxidation number. Because of the slight energy difference between the electrons of two outer energy levels, a transition element such as iron might "lose" two or three electrons. This experimental evidence indicates that iron has more than one oxidation state. It might be Fe^{2+} (iron II) or Fe^{3+} (iron III). A number of elements exhibiting more than one common oxidation number will appear on the chart in this chapter.

THE NATURE OF THE CHEMICAL BOND

A **chemical bond** is a union between atoms resulting in a structure having greater stability and lower energy than the isolated atoms. A bond is also described as a pair of electrons jointly shared by two atoms. Two or three shared pairs are called double bonds and triple bonds respectively. Still another description of a chemical bond is an attraction between ions bearing opposite charges. Chemical bonding results in the formation of molecules, or large orderly arrangements of ions. The molecule is a distinct particle. It has a specific size, shape, and weight, and it may be polar or nonpolar with regard to its electrical charge distribution. The large orderly arrangements of ions, on the other hand, have no specific dimensions. They form what is called a **crystal lattice** structure in which oppositely charged ions hold each other quite strongly.

The atomic orbital method is the simplest one for describing chemical bonding. This method focuses on the valence shell of the atom and the tendency toward achieving the "octet" of electrons there. Some examples may be helpful at this time. Consider the formation of the hydrogen molecule, H_2 (Fig. 4-7).

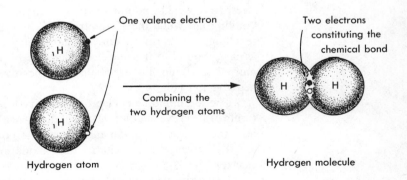

One valence electron

Two electrons constituting the chemical bond

Combining the two hydrogen atoms

Figure 4-7. The formation of the hydrogen molecule.   Hydrogen atom   Hydrogen molecule

A convenient alternative to diagramming is the **electron-dot** formula. Here the electrons of the outer energy level of an atom are represented by a dot,

$$H^\bullet \qquad Na^\bullet \qquad {}^\bullet\!\overset{\bullet}{\underset{\bullet}{C}}{}^\bullet \qquad {}^\bullet\!\overset{\bullet\bullet}{\underset{\bullet\bullet}{Cl}}{:}$$

hydrogen   sodium   carbon   chlorine

The hydrogen molecule can then be represented by an electron-dot formula.

$$H^\bullet + H^\bullet \longrightarrow H{:}H$$

The placement of the dots is purely a matter of convenience and has nothing whatsoever to do with the actual positions of the electrons. Take hydrogen chloride as another example.

hydrogen   chlorine

the shared pair provides hydrogen with TWO electrons and chlorine with the "octet"

In the electron-dot format

$$H^\bullet + {}_\circ\overset{\circ\,\circ}{\underset{\circ\,\circ}{Cl}}{}_\circ \qquad\qquad H^\bullet_\circ\overset{\circ\,\circ}{\underset{\circ\,\circ}{Cl}}{}_\circ$$

Water is a slightly more complicated example, since oxygen, having a common oxidation of 2⁻ in compounds, requires two hydrogen atoms.

$$\begin{array}{c} H^\bullet \\[2pt] + \;\overset{\circ\circ}{\underset{\circ\circ}{O}}{}^{\circ}_{\circ} \\[2pt] H^\bullet \end{array} \longrightarrow \overset{\circ\,\circ}{\underset{\displaystyle H \quad H}{{}^\bullet\overset{}{O}{}^\circ_\circ}}$$

HOH = H_2O
water

As an illustration of bonding that results in nonmolecular structures, consider table salt, NaCl (sodium chloride). The single outer energy level electron is weakly bound to the nucleus because there are only 11 protons "pulling in" while the 10 electrons of the first and second energy levels are repelling, or "pushing out." The seven outer energy level electrons are *strongly* bound to the nucleus which has 17 protons, while there are still no more than 10 repelling electrons. The positive ion is strongly attracted to the negative ion (Fig. 4-8).

The sodium ion has 11 protons (units of positive charge) and 10 electrons (units of negative charge). The net difference is one positive charge, which is both the charge on the ion and the common oxidation number. The chloride ion, by contrast, has a charge (and oxidation number) reflecting the extra electron.

Using the electron-dot method the formula appears as follows:

transfer

$$Na^\bullet + {}_\circ\overset{\circ\,\circ}{\underset{\circ\circ}{Cl}}{}^\circ \longrightarrow Na^+, \quad \left[{}_\circ^\circ\overset{\circ\,\circ}{\underset{\circ\,\circ}{Cl}}{}^\circ_\circ\right]^-$$

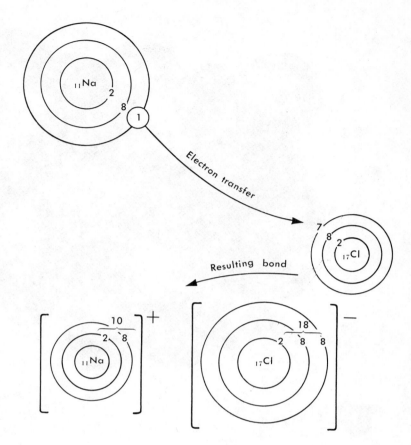

Figure 4-8. A nonmolecular chemical bond.

The result is only a fragment of the gigantic arrangement of many millions of ions. It is convenient to separate the types of bonds into two main categories—**ionic** and **covalent**.

Ionic Bonds

Ionic bonds are formed when metal atoms **transfer** electrons. The recipients of the transferred electrons are usually atoms or groups of atoms that have large positive nuclear charges and a small requirement for the completion of the octet. There is no rigid boundary that separates ionic bonds from others. It is a matter of convenience to organize compounds into various categories. The rules involved would be considered "rules of thumb" and exceptions will be expected. In fact, chemists often describe some compounds as covalent but with a great deal of ionic character. For purposes of clarity, the examples selected will be very typical members of their type. Rule: Ionic bonds are principally formed between metals exhibiting common oxidation numbers of 1+ or 2+ and nonmetals.

Sodium chloride, which was mentioned earlier, is a typical ionic compound. The crystal lattice structure results from a stable alignment of the ions where positive sodium ions and negative chloride ions surround

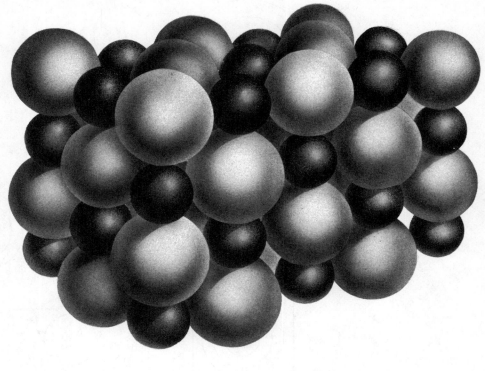

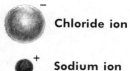

Chloride ion

Sodium ion

Figure 4-9. The crystal lattice structure of salt.

each other (Fig. 4-9). The three dimensional aspect is also represented by the diagram in Figure 4-10.

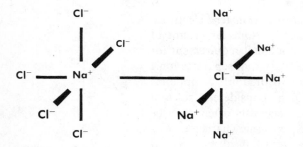

Figure 4-10.

The crystal lattice structure for sodium chloride serves as a model for many other ionic compounds even though the ratios of ions involved and the basic geometry may vary.

Finally, a few more examples of ionic bonding include:

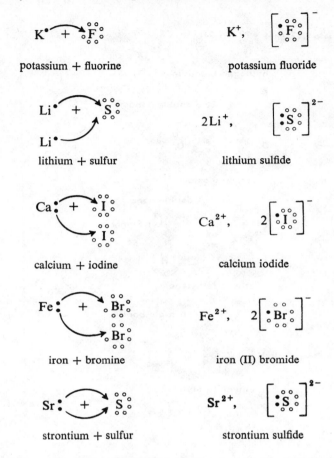

Notice the net effect of electrical neutrality in every example. Rule: The sum of the positive and negative charges must equal zero in a compound. In other words, the sum of the common oxidation numbers in a compound must equal zero. This rule applies to *all* compounds, whether they are ionic or not.

K^+, F^-	plus one and minus one = zero
$2Li^+$, S^{2-}	two ions at plus one and one ion at minus two = zero
Ca^{2+}, $2I^-$	one ion at plus two and two ions at minus one = zero
Fe^{2+}, $2Br^-$	one ion at plus two and two ions at minus one = zero
Sr^{2+}, S^{2-}	plus two and minus two = zero

Covalent Bonds

Covalent bonds are formed when pairs of electrons are *shared* between atoms. The extent of the sharing may be unequal, since one atom can have

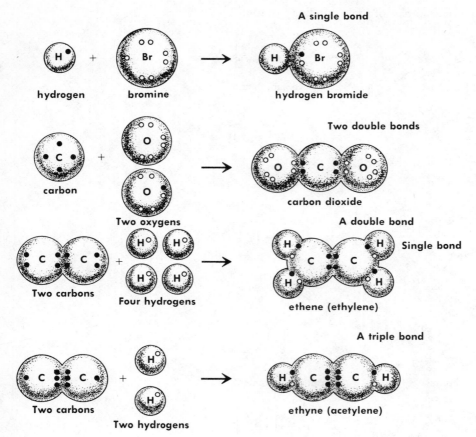

Figure 4-11.

a much stronger attraction for the electron pair, but this is still not the same as the complete transfer characteristic of ionic bonds.

Most compounds that exhibit covalent bonding involve the atoms and groups of atoms that have oxidation numbers of metals higher than 2+ combined with nonmetals. When the transition elements combine in the 2+ oxidation state they are usually classified as ionic, but in the 3+ or higher state, they are more likely to be described as covalent with some degree of ionic character. Another rule of thumb: Covalent compounds are formed between nonmetallic atoms. The other types of bonds will be described briefly after the topic of radicals.

A distinguishing characteristic of covalent compounds is that they do exist in the form of distinct molecules.

Figure 4-11 shows some typical examples of covalent compounds. Note that the sharing process moves in the direction of the "octet" formation for all the atoms involved with the exception of hydrogen.

A very special and important aspect of covalent bonding involves the **radical**. A radical is a group of covalently bound atoms that bears an electrical charge because of an imbalance between the total number of nuclear protons and the number of electrons. Although a radical and a

Name of Radical	A Possible Electron-dot Formula	Radical Formula	Oxidation Number (net charge) of Radical
nitrate ion	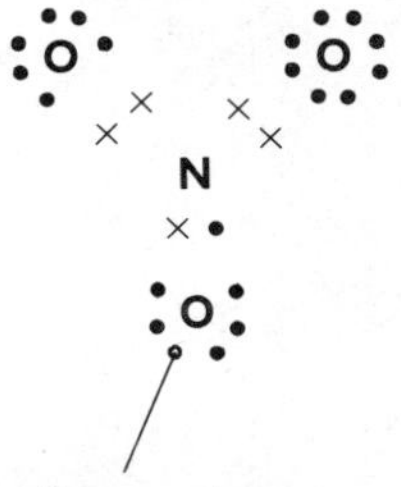 Extra electron supplied by the bonded metal atom	$NO_3{}^-$	1−
sulfate ion	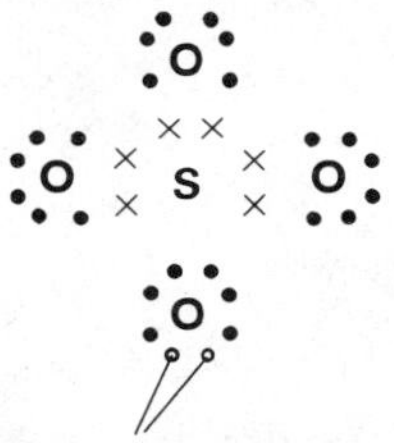 Two extra electrons from transfer	$SO_4{}^{2-}$	2−
phosphate ion	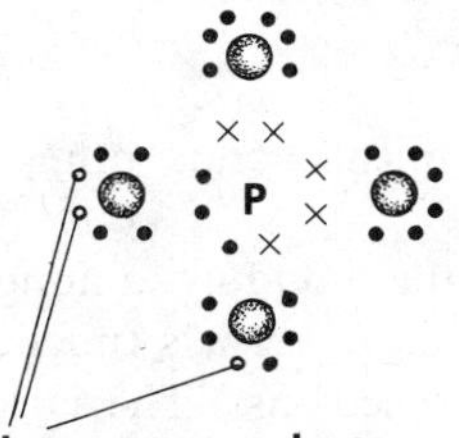 Three extra electrons from transfer	$PO_4{}^{3-}$	3−

Figure 4-12.

compound may have the same apparent formula, the compound is electrically neutral and the radical is not.

$$SO_3 \qquad\qquad SO_3^{2-}$$

formula for the compound sulfur trioxide

formula for the sulfite ion (radical)

Radicals do not normally exist alone. They are usually found in combination with metal ions that have transferred their electrons. This is what causes the imbalance between the total positive and negative charges. The compounds involving radicals are often ionic because of the *transfer* of electrons, but the radical itself exhibits covalent bonding. (See Figure 4-12 for some examples of radicals.)

Some examples of compounds involving radicals are:

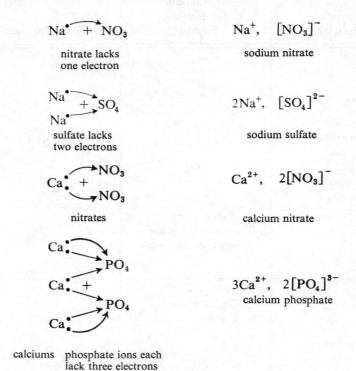

$$Na^\bullet + NO_3 \qquad Na^+, \quad [NO_3]^-$$

nitrate lacks
one electron

sodium nitrate

$$Na^\bullet \atop Na^\bullet \searrow + SO_4 \qquad 2Na^+, \quad [SO_4]^{2-}$$

sulfate lacks
two electrons

sodium sulfate

$$Ca^{\bullet}_{\bullet} + {\nearrow NO_3 \atop \searrow NO_3} \qquad Ca^{2+}, \quad 2[NO_3]^-$$

nitrates

calcium nitrate

$$Ca^{\bullet}_{\bullet} \atop Ca^{\bullet}_{\bullet} \atop Ca^{\bullet}_{\bullet} + {PO_4 \atop PO_4} \qquad 3Ca^{2+}, \quad 2[PO_4]^{3-}$$

calcium phosphate

calciums   phosphate ions each
lack three electrons

Note that the sum of the positive and negative charges equals zero in every compound.

Other Types of Bonds

A brief summary of chemical bonds other than ionic and covalent follows.

Metallic bond. This is the bonding among atoms in metals. It is usually described as positive atomic kernels (nucleus plus lower energy level electrons) in a sea of valence electrons. The lack of a rigid crystal lattice type of geometry endows the metal with its characteristic flexibility (see Fig. 4-13).

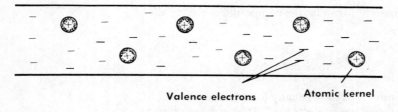

Valence electrons   Atomic kernel

Metal wire   Figure 4-13. Metallic bonding.

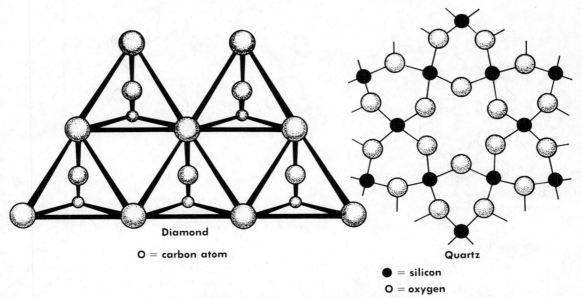

Diamond

O = carbon atom

Quartz

● = silicon
O = oxygen

Figure 4-14. Network bonding.

Network bond. This is a special case of complex sharing of electron pairs. The covalent nature is so extensive that huge crystals result rather than molecules. The combination of carbon and silicon with each other or with nitrogen and oxygen often results in network bonding. (Compounds formed by network bonding are usually very hard.) Typical examples are diamond (a network of bonds between carbon atoms) and quartz (silicon dioxide) (Fig. 4-14).

Hydrogen bond. This type of bond describes the tendency of small nonmetallic atoms—notably fluorine, oxygen, and nitrogen—to share a hydrogen atom. The small non-metallic atoms exert a powerful attraction on the hydrogen because its electron is so far removed that it is almost a hydrogen ion. Remember that the hydrogen ion is a proton—a unit of positive charge.

The empirical formula for hydrogen fluoride is HF. In reality, HF forms a chain because of hydrogen bonding. Its actual structure might be H_6F_6 or any other number of atoms in a 1:1 ratio (Fig. 4-15).

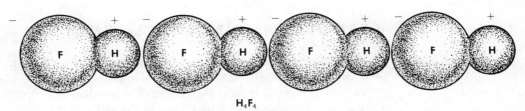

H_4F_4

Figure 4-15. Hydrogen bonding.

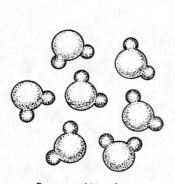

Dense packing of water
molecules at low temperature

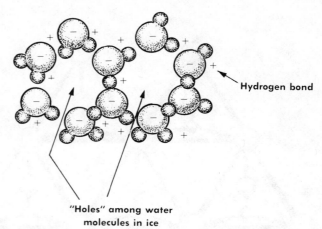

Hydrogen bond

"Holes" among water
molecules in ice

Figure 4-16.

Ice is another example of hydrogen bonding. The fact that ice floats in water is due to a reduction in density resulting from "hole" formation as hydrogen bonding spreads the molecules (Fig. 4-16).

A final example of hydrogen bonding is found in proteins (Fig. 4-17). Amino acids join to form peptides, and polypeptides link together through hydrogen bonding to form proteins.

Polypeptide
fragment

Hydrogen bond

Two small atoms having a strong
attraction for the proton.

Figure 4-17.

FORMULAS AND NOMENCLATURE

The development of skill and confidence in the writing and naming of chemical formulas for compounds is absolutely necessary for the laboratory worker. The formula is more than a shorthand method of writing: it implies very specific quantities of a compound. A "trivial" error in formula writing may be more disastrous than total ignorance. Use the chart of common oxidation numbers in Figure 4-18 to systematically develop this essential skill.

Common Oxidation Numbers ("Valence")

1+	2+	3+	4+	3−	2−	1−
H hydrogen	Mg magnesium	Al aluminum	C carbon	N nitride	O oxide	OH hydroxide
Li lithium	Ca calcium	Cr chromium (III)	Si silicon	P phosphide	S sulfide	F fluoride
Na sodium	Ba barium	Bi bismuth (III)	Pb lead (IV)	PO_3 phosphite	CO_3 carbonate	Cl chloride
K potassium	Cd cadmium	Sb antimony (III)	Mn manganese (IV)	PO_4 phosphate	SO_3 sulfite	Br bromide
NH_4 ammonium	Zn zinc	As arsenic (III)	Sn tin (IV)	AsO_3 arsenite	SO_4 sulfate	I iodide
Hg_2 mercury (I)	Co cobalt (II)	Co cobalt (III)		AsO_4 arsenate	CrO_4 chromate	CN cyanide
Cu copper (I)	Hg mercury (II)	Ni nickel (II)			Cr_2O_7 dichromate	OCN cyanate
Ag silver	Ni nickel (II)	Fe iron (III)			HPO_4 monohydrogen phosphate	ClO_3 chlorate
	Fe iron (II)				C_2O_4 oxalate	NO_2 nitrite
	Cu copper (II)					NO_3 nitrate
	Pb lead (II)					HCO_3 hydrogen carbonate
	Mn manganese (II)					MnO_4 permanganate
	Sn tin (II)					H_2PO_4 dihydrogen phosphate
	Cr chromium (II)					HSO_4 hydrogen sulfate
						CH_3COO acetate

Figure 4-18.

You will note the Roman numerals written next to the atoms having variable oxidation numbers. The older "classic" nomenclature used different suffixes to distinguish these atoms. For example,

Fe^{2+}, or iron II, was called ferr*ous*

and

Fe^{3+}, or iron III, was ferr*ic*

Because of increasing ambiguity, this older method is gradually being replaced by the more descriptive Roman numeral method of indicating the oxidation number of an atom. Learning the modern system thoroughly should enable anyone to quickly master the classic suffix method if it seems advisable.

Some examples of formula writing follow:

Example 1

Write the formula for copper (I) chloride.

1. Set down the components with their respective oxidation numbers.

$$Cu^+ \qquad Cl^-$$

2. See if the sum of the electrical charges equals zero

$$\underset{\text{copper}}{1+} \quad \text{and} \quad \underset{\text{chloride}}{1-} = 0$$

3. The original number of ions is accepted. The formula is

$$CuCl$$

Example 2

Write the formula for copper (II) chloride.

1. As in Example 1:

$$Cu^{2+} \qquad Cl^-$$

2. In order to balance the electrical charges, two chloride ions are required:

$$Cu^{2+} \qquad Cl_2^-$$

3. Balancing formulas *always* involves *subscripts*. The formula is

$$CuCl_2$$

Example 3

Write the formula for potassium sulfate.

1. $K^+ \qquad SO_4^{2-}$

2. Two potassiums are needed for balance:

$$K_2^+ \qquad SO_4^{2-}$$

3. The formula:

$$K_2SO_4$$

4. Note: There is no parenthesis used in the case of a single radical in the formula.

Example 4

Write the formula for iron (III) oxalate.

1. Fe^{3+} $C_2O_4^{2-}$

2. The balancing of electrical charges requires a total of six positive and negative charges (the lowest common denominator):

$$Fe_2^{3+} \quad (C_2O_4)_3^{2-}$$

3. A parenthesis *must* be placed around the oxalate radical. The subscript (3) multiplies *both* the number of carbons and oxygens composing the radical.

4. The finished formula is

$$Fe_2(C_2O_4)_3$$

Summarizing the rules for writing formulas:

1. Careful attention must be paid to capital and small case letters in the symbols used.

   Co means the element cobalt

   CO means the compound carbon monoxide

2. Roman numerals are not used in formulas.

   $FeCl_2$ is iron (II) chloride

   ~~$FeIICl_2$~~ is confusing

3. Parentheses are used when *more* than one radical is indicated in the formula. A parenthesis around a single radical is not usually necessary.

   $Ca(OH)_2$ 1 calcium   is calcium hydroxide
                   ② oxygens
                   2 hydrogen

   ~~$CaOH_2$~~   1 calcium
                   ① oxygen   is wrong
                   2 hydrogen

A short and snappy method is possible for writing formulas. It is really a "gimmick" that can only be justified after a fundamental understanding of formula writing has been developed. The method is described in the following examples:

sodium chromate Na^+ CrO_4^{2-}

switch the oxidation
number values

Na_2 CrO_4 $_1$

the final form is Na_2CrO_4

cobalt (III) sulfide

Co^{3+} S^{2-}

the final formula is Co_2S_3

lead (IV) sulfide

Pb^{4+} S^{2-}

this ratio should be reduced from

Pb_2S_4 to PbS_2

ammonium monohydrogen phosphate

NH_4^+ HPO_4^{2-}

the final formula is $(NH_4)_2HPO_4$

nickel (III) hydroxide

Ni^{3+} OH^-

the final formula is $Ni(OH)_3$

Naming formulas correctly is usually accomplished by starting with the subscripts and moving them back along the same imaginary diagonal arrows used in writing formulas from oxidation numbers. Some examples:

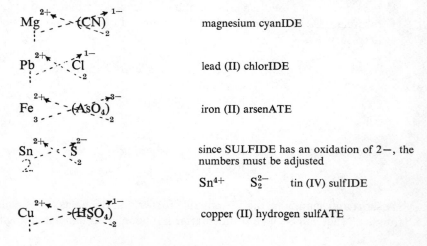

Mg^{2+} $(CN)^{1-}_2$ magnesium cyanIDE

Pb^{2+} Cl^{1-}_2 lead (II) chlorIDE

Fe^{2+}_3 $(AsO_4)^{3-}_2$ iron (II) arsenATE

Sn^{2+}_2 S^{2-}_2 since SULFIDE has an oxidation of 2−, the
numbers must be adjusted

Sn^{4+} S_2^{2-} tin (IV) sulfIDE

Cu^{2+} $(HSO_4)^{1-}_2$ copper (II) hydrogen sulfATE

The Special Nomenclature of Acids

Formula	Standard Nomenclature	Acid Nomenclature
HCl	hydrogen chlorIDE	HYDROchlorIC acid
HCN	hydrogen cyanIDE	HYDROcyanIC acid
HBr	hydrogen bromIDE	HYDRObromIC acid
HF	hydrogen fluorIDE	HYDROfluorIC acid

When the suffix is IDE, the acid is HYDRO-root-IC acid.

HNO_2	hydrogen nitrITE	nitrOUS acid
H_2SO_3	hydrogen sulfITE	sulfurOUS acid
H_3PO_3	hydrogen phosphITE	phosphorOUS acid
H_3AsO_3	hydrogen arsenITE	arsenOUS acid

When the suffix is ITE, the acid is root-OUS acid.

HNO_3	hydrogen nitrATE	nitrIC acid
H_2CrO_4	hydrogen chromATE	chromIC acid
H_2CO_3	hydrogen carbonATE	carbonIC acid
H_2SO_4	hydrogen sulfATE	sulfurIC acid

When the suffix is ATE, the acid is root-IC acid.

HClO	hydrogen HYPOchlorITE	HYPOchlorOUS acid
HBrO	hydrogen HYPObromITE	HYPObromOUS acid
$HMnO_4$	hydrogen PERmanganATE	PERmanganIC acid
$HClO_4$	hydrogen PERchlorATE	PERchlorIC acid

The prefixes HYPO and PER remain in the acid nomenclature.

Example 5

Write the formula for oxalic acid.
1. Note that the ending is oxalIC. This corresponds to hydrogen oxalATE.

2. H+

3. The formula is $H_2C_2O_4$

Example 6

What is the name of $HClO_2$?
1. Identify the ClO_2^- as the chlorITE radical.
2. The corresponding name for hydrogen chlorITE is chlorOUS acid.

The Special Nomenclature Used for Nonmetallic Compounds

In the case of nonmetallic compounds (no metals at all), a simple descriptive method is often used. A Greek prefix describes the number of atoms in the right-hand member of a formula. Examples are listed:

Prefix	Example	Name
1-mono	CO (1 oxygen)	carbon MONoxide
2-di	CO_2 (2 oxygen)	carbon DIoxide
3-tri	SO_3 (3 oxygen)	sulfur TRIoxide
4-tetra	CCl_4 (4 chlorine)	carbon TETRAchloride
5-penta	N_2O_5 (5 oxygen)	nitrogen PENToxide
6-hexa	XeF_6 (6 fluorine)	xenon HEXAfluoride
7-hepta	Cl_2O_7 (7 oxygen)	chlorine HEPToxide

The nomenclature of organic compounds is very special. Because of the thousands of compounds known and those yet to be made, a very definite system was developed by the International Union of Pure and Applied Chemistry (I.U.P.A.C.). This system will be discussed in detail in the chapter on Organic Chemistry.

THE MOLE CONCEPT

Formulas, and the equations that describe their changes, provide definite quantitative information. The formula tells you *what* is involved and, more importantly, it tells you *how much*. Obviously, the atomic mass unit (amu) is a totally impractical unit for laboratory application. There are no instruments available to measure objects having weights for which the order of magnitude is 10^{-24} grams. Since the common unit of weight is the gram, the formula weight of compounds must be expressed accordingly.

The solution is simple enough. Chemists *substitute* the gram for the amu for all practical work. The justification for this convenient substitution lies in the fact that the basic ratios of weights are unchanged. Just as you can compare the weights of one dime, one nickel, and one quarter, the ratios of their weights are the same if you compare a thousand of each type of coin (Fig. 4-19).

In chemistry today, the standard of comparison is carbon 12. The number of carbon-12 atoms needed to add up to 12 g is 6.02×10^{23}. This fundamental relationship is the core of the **mole concept**. The number is known as **Avogadro's number** in honor of the man who did pioneer work that indirectly led to its determination. In order to make valid mass comparisons among the elements, the Avogadro number of atoms of each element has to be used. In each case, as expected, the Avogadro number of atoms of an element approximates its average atomic weight (amu) very closely in grams. This weight is known as a **mole** of the element.

$^{12}_{6}C$ 12 g carbon = 1 mole of carbon atoms

$^{27}_{13}Al$ 27 g aluminum = 1 mole of aluminum atoms

$^{108}_{47}Ag$ 108 g silver = 1 mole of silver atoms

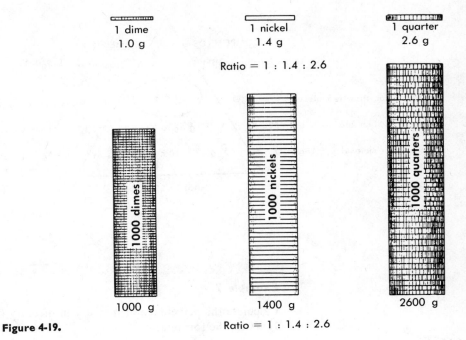

1 dime
1.0 g

1 nickel
1.4 g

1 quarter
2.6 g

Ratio = 1 : 1.4 : 2.6

1000 dimes

1000 nickels

1000 quarters

1000 g

1400 g

2600 g

Figure 4-19.

Ratio = 1 : 1.4 : 2.6

A direct laboratory method of determining the mole is possible. A common experiment involves finding the weight of silver deposited on an electrode when 6.02×10^{23} electrons are gained by silver ions. The experiment is described in the chapter on Electrochemistry.

$$Ag^+ \text{ (ions)} + 1 e^- \rightarrow Ag^0 \text{ (atoms)}$$

1 mole   1 mole   ·1 mole
the weight is $\cong 108$ g

Since compounds are made of elements, the mole concept still applies:

$$Na \quad + \quad Cl \longrightarrow NaCl$$

1 mole   1 mole   1 mole
6.02×10^{23} atoms 6.02×10^{23} atoms 6.02×10^{23} units

$$C \quad + \quad O_2 \longrightarrow CO_2$$

1 mole   1 mole   1 mole
6.02×10^{23} atoms 6.02×10^{23} molecules 6.02×10^{23} molecules

One other important aspect of the mole concept, hypothesized by Avogadro and proved experimentally, is the fact that one mole of any gas at 0° C and 1 atmosphere pressure (called STP for Standard Temperature and Pressure) occupies 22.4 liters of volume. This 22.4 liters is called a **molar volume** because it contains a mole of gas molecules at STP.

The mole concept is central to most quantitative chemical work. The diagram in Figure 4-20 summarizes the fundamental relationships.

Introduction to Stoichiometry

Stoichiometry is the process of measuring quantitative chemical relationships based on the mole. One of the simpler calculations is the determination of the percentage composition of compounds.

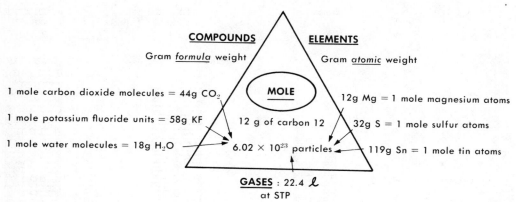

Figure 4-20

Example 7

Calculate the percentage of mercury in mercury (II) oxide.
1. Write the formula:

$$HgO$$

2. Obtain the total formula weight from the sum of the atomic weights:

$$Hg + O = 217 \text{ g}$$
$$201 \quad 16$$

one mole of HgO weighs 217 g

3. Mercury is ? per cent of the total:

$$201 = \text{ ? per cent of } 217$$

4. $\dfrac{201}{217} = 0.926 = 92.6$ per cent mercury

Example 8

Find the percentage of water in $Na_2SO_4 \cdot 10H_2O$ (sodium sulfate—ten hydrate).
1. Calculate the formula weight:

$$Na_2 \qquad S \qquad O_4 \qquad 10H_2O \ = \ 322 \text{ g}$$
$$(2 \times 23) + 32 + (4 \times 16) + (10 \times 18) \qquad \text{per mole}$$

2. Water composes 180 g out of a total of 322 g. The percentage is

$$\frac{180}{322} = 0.56 = 56 \text{ per cent}$$

A very important calculation related to many problems is the determination of the *number of moles*, or the *fraction of a mole*.

Example 9

How many moles are there in 80 g of carbon dioxide?

1. Calculate the weight of one mole of CO_2 from the formula:

$$C \qquad O_2 \ = 44 \text{ g mole}^{-1}$$
$$12 + (2 \times 16)$$

2.
$$\underset{\text{(number of moles)}}{n} = \frac{g}{\text{g mole}^{-1}}$$

$$n = \frac{80 \text{ g}}{44 \text{ g mole}^{-1}} = 1.8 \text{ moles}$$

Example 10

How many moles (or fraction of a mole) of carbon dioxide are there in 5.0 liters at STP?

1. By definition: one mole of gas occupies 22.4 liters at STP.

2. $n = \dfrac{\text{liters}}{22.4 \text{ liters mole}^{-1}}$

3. $n = \dfrac{5.0 \text{ liters}}{22.4 \text{ liters mole}^{-1}}$

$n = 0.22$ mole

Empirical Formulas

The term **empirical** refers to knowledge gained directly through observation and experimentation. An empirical formula is one in which the simplest ratios of the elements in a compound are determined from the weight percentages provided by analysis. It is a matter of finding out the number of moles of each element in one mole of the compound.

Example 11

A compound is analyzed and found to be 21.6 per cent sodium, 33.3 per cent chlorine, and 45.1 per cent oxygen. Find the empirical formula.

1. Quickly convert the percentages to grams by assuming that 100 grams were analyzed:

	grams per 100 g
Na	21.6 g
Cl	33.3 g
O	45.1 g

2. Find the number of moles of each element by using the gram atomic weight:

$$\text{Na} \quad \frac{21.6\ g}{23\ g\ mole^{-1}} = 0.94\ mole$$

$$\text{O} \quad \frac{45.1\ g}{16\ g\ mole^{-1}} = 2.82\ moles$$

$$\text{Cl} \quad \frac{33.3\ g}{35.5\ g\ mole^{-1}} = 0.94\ mole$$

3. Since a formula must be in whole number ratios, the formula $Na_{0.94}Cl_{0.94}O_{2.82}$ will not do. Convert this ratio to whole numbers by dividing all the values by the *smallest* one. This assures you of having no number less than 1.0.

$$\left.\begin{array}{l} \text{Na} \quad \dfrac{0.94}{0.94} = 1 \\[2mm] \text{Cl} \quad \dfrac{0.94}{0.94} = 1 \\[2mm] \text{O} \quad \dfrac{2.82}{0.94} = 3 \end{array}\right\} Na_1Cl_1O_3$$

4. The empirical formula then is

$$NaClO_3 \quad \text{sodium chlorate}$$

Example 12

A compound was analyzed and found to contain 0.965 g of sodium, 1.345 g of sulfur, and 2.690 g of oxygen. Find the empirical formula.

1. Combine the separate steps in Example 11 into one chart:

Element	Grams	Multiply all values by 100 to simplify arithmetic	Number of Moles	Mole Ratios	Whole Number Ratios
Na	0.965 g	96.5 g	$\dfrac{96.5\ g}{23\ g\ mole^{-1}} = 4.2$		$\dfrac{4.2}{4.2} = 1$
S	1.345 g	134.5 g	$\dfrac{134.5\ g}{32\ g\ mole^{-1}} = 4.2$		$\dfrac{4.2}{4.2} = 1$
O	2.690 g	269.0 g	$\dfrac{269.0\ g}{16\ g\ mole^{-1}} = 16.8$		$\dfrac{16.8}{4.2} = 4$

2. The empirical formula is

$$NaSO_4$$

3. If the *true* formula weight is known to be 238, what is the actual formula?

$$\frac{Na + SO_4}{23 + 32 + (4 \times 16)} = 119 \text{ (formula weight of the empirical formula)}$$

$$\frac{238 \text{ (actual weight)}}{119 \text{ (empirical weight)}} = 2$$

The true formula must be double the empirical formula, since the actual formula weight is double the weight of the empirical formula.

$$Na_2S_2O_8$$

Example 13

A hydrocarbon is 82.76 per cent carbon and 17.24 per cent hydrogen. Its density is 2.59 g liter^{-1} at STP. Calculate the true molecular formula.

1. Find the empirical formula:

	Convert per cent to Grams	Number of Moles	Mole Ratio	Whole Number Ratio
C	82.76 g	$\dfrac{82.76 \text{ g}}{12 \text{ g mole}^{-1}} = \dfrac{6.90}{6.90} = 1$		2
H	17.24 g	$\dfrac{17.24 \text{ g}}{1 \text{ g mole}^{-1}} = \dfrac{17.24}{6.90} = 2.5$		5

Empirical formula is C_2H_5 and the formula weight is 29:

$$[(2 \times 12) + 5 = 29].$$

2. Find the true molecular weight from the density (2.59 g liter^{-1}):

$$2.59 \text{ g liter}^{-1} (22.4 \text{ liter mole}^{-1}) = 58 \text{ g mole}^{-1}$$

58 g mole^{-1} means the molecular weight is 58

3. $\dfrac{58 \text{ (actual molecular weight)}}{29 \text{ (empirical molecular weight)}} = 2$

The true formula must be double the empirical

$$C_4H_{10}$$

Equations

A chemical equation uses formulas to state *what substances react* and the *products that finally result*. The equation also tells *how much* of each substance is involved. Equations are derived from experimental observations. When a chemist writes the formulas for reacting substances (**reactants**) and then proceeds to predict the resulting substances (**products**), it is not a matter of some mysterious manipulation of the symbols. The prediction of products is a skill developed through considerable experience and knowledge of the chemical properties of substances.

A **balanced** equation represents the amounts of reactants and products that truly reflect the invariable laws of conservation. If there is one mole of element X in a reactant, the equation must indicate one mole of element X in some product—regardless of how dramatic the chemical change may be. The rules governing the writing of balanced equations will be developed through examples.

Example 14

A mixture of hydrogen gas and oxygen gas can be activated by a spark to produce water (Fig. 4-21).

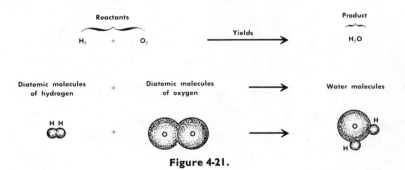

Figure 4-21.

1. Observe *two* units of oxygen as a reactant and *one* unit of oxygen in water. This nonconformity to the conservation law must be corrected by multiplying the *whole* formula by a numerical coefficient. The subscripts, which indicate the ratios of elements in a compound, *cannot* be altered because each formula for a particular compound has a definite composition. Water is always H_2O (or HOH) and it may not be written as HO, HO_2, or anything else. The simplest coefficient is often a fraction:

$$H_2 + \tfrac{1}{2}O_2 \rightarrow H_2O$$

2. The coefficient of $\tfrac{1}{2}$ multiplies the two units of oxygen, resulting in one unit. Now the oxygens are balanced. The equation now states:

$$H_2 + \tfrac{1}{2}O_2 \rightarrow H_2O$$

1 mole of + $\tfrac{1}{2}$ mole of → 1 mole
hydrogen   oxygen   of water
molecules   molecules

1 mole of H_2 (molecular weight of 2) = 2 g

$\frac{1}{2}$ mole of O_2 (molecular weight of 32) = 16 g

1 mole of H_2O (molecular weight of 18) = 18 g

$$H_2 + \tfrac{1}{2}O_2 \rightarrow H_2O$$
$$2\,g + 16\,g \rightarrow 18\,g$$
$$18\,g = 18\,g$$

The conservation law is obeyed.

3. Under standard conditions of temperature and pressure, more information can be obtained

$$H_2 \quad + \quad \tfrac{1}{2}O_2 \quad \rightarrow H_2O$$
$$1\ \text{mole gas} + \tfrac{1}{2}\ \text{mole gas} \rightarrow (1\ \text{mole})$$
$$22.4\ \text{liters} + 11.2\ \text{liters} \rightarrow \quad 18\,g$$

4. These mole values will be discussed further in this chapter under the heading Stoichiometric Equations.

Example 15

Calcium reacts directly with water to produce calcium hydroxide and hydrogen gas.

1. Write all the unalterable formulas:

$$Ca + H_2O \rightarrow Ca(OH)_2 + H_2$$

2. The condition, or state, of each substance *may* be indicated parenthetically after the formula:

solid (s)

liquid (l)

water solution (aq) = aqueous

gas (g)

$$Ca(s) + H_2O(l) \rightarrow Ca(OH)_2(aq) + H_2(g)$$

3. Inspect the equation for conservation of metals, nonmetals (including radicals), hydrogen, and oxygen—in that order.

Reactants	Products
one calcium unit	*one* calcium unit
two hydrogen units	*four* hydrogen units

Balance the hydrogens

now

$$Ca + \boxed{2H_2O} \rightarrow Ca(OH)_2 + H_2$$

four hydrogen units   *four* hydrogen units

$$Ca + 2H_2O \qquad\qquad Ca(OH)_2 + H_2$$

two oxygen units   *two* oxygen units

The equation is balanced.

Example 16

When ethane gas (C_2H_6) is burned (combined with oxygen), carbon dioxide and water are produced.

$$C_2H_6 + O_2 \rightarrow CO_2 + H_2O$$

Reactants   Products

two carbon units   *one* carbon unit

carbon
adjusted $C_2H_6 + O_2 \rightarrow 2CO_2 + H_2O$

six hydrogen units   *two* hydrogen units

hydrogen
adjusted $C_2H_6 + O_2 \rightarrow 2CO_2 + 3H_2O$

two oxygen units   4 + 3
                        seven oxygen units

oxygen
adjusted $C_2H_6 + \frac{7}{2}O_2 \rightarrow 2CO_2 + 3H_2O$

seven oxygen units   *seven* oxygen units

The equation is balanced.

Example 17

Iron (III) chloride reacts with sodium carbonate to form the precipitate iron (III) carbonate and sodium chloride.

$$FeCl_3(aq) + Na_2CO_3(aq) \rightarrow Fe_2(CO_3)_3(s) + NaCl(aq)$$

Reactants   Products

one iron unit   *two* iron units

iron
adjusted $2FeCl_3 + Na_2CO_3$ $Fe_2(CO_3)_3 + NaCl$

six chloride units   *one* chloride unit

chloride
adjusted $2FeCl_3 + Na_2CO_3 \rightarrow Fe_2(CO_3)_3 + 6NaCl$

two sodium units   *six* sodium units

sodium
adjusted $2FeCl_3 + 3Na_2CO_3 \rightarrow Fe_2(CO_3)_3 + 6NaCl$

three carbonate units   *three* carbonate units

The equation is balanced.

Stoichiometric Equations

A stoichiometric reaction is one in which all the reactants are converted to final products. The evidence for this type of reaction is usually seen as the evolution of gas, the formation of a precipitate, or a large energy change such as burning or an explosion. The stoichiometry of a reaction, i.e., the mole to mole ratios of reactants and products, is clearly indicated by the balanced equation:

$$Al(s) + 3HCl(aq) \rightarrow AlCl_3(aq) + \tfrac{3}{2}H_2(g)$$

The stoichiometry of this reaction is 1 mole of aluminum + 3 moles of hydrochloric acid yields 1 mole of aluminum chloride and $1\tfrac{1}{2}$ moles of hydrogen gas. Since the gas escapes (is evolved as bubbles), it cannot react with $AlCl_3$ to produce a reverse reaction. The equation provides quantitative information.

$$
\begin{array}{ccccccc}
Al & + & 3HCl & \rightarrow & AlCl_3 & + & \tfrac{3}{2}H_2 \\
1 \text{ mole} & + & 3 \text{ moles} & \rightarrow & 1 \text{ mole} & + & 1\tfrac{1}{2} \text{ moles} \\
27 \text{ g} & + & (3 \times 36.5 \text{ g}) & \rightarrow & 133.5 \text{ g} & + & 3 \text{ g}
\end{array}
$$

or

$$(1\tfrac{1}{2} \times 22.4 \text{ liters} = 33.6 \text{ liters at STP})$$

Example 18

How many grams of hydrogen would result if 9 g of aluminum were used in the reaction with hydrochloric acid?

1. What fraction of a mole is 9 g?

$$n = \frac{9 \text{ g}}{27 \text{ g mole}^{-1}} = 0.33 \text{ mole or } \tfrac{1}{3} \text{ mole}$$

2. If one mole of aluminum yields $1\tfrac{1}{2}$ moles of hydrogen, then $\tfrac{1}{3}$ mole of aluminum would yield $\tfrac{1}{3}$ of $1\tfrac{1}{2}$ moles of hydrogen:

$$\tfrac{1}{3} \times \tfrac{3}{2} = \tfrac{1}{2} \text{ mole.}$$

A half mole of H_2 is 1 g.

3. The arithmetic can most easily be arranged and handled by using the equation to take advantage of the proportionality:

$$
\begin{array}{cc}
9 \text{ g} & X \text{ g} \\
Al \rightsquigarrow & \tfrac{3}{2}H_2 \\
27 \text{ g} & 3 \text{ g}
\end{array}
$$

$$\frac{9 \text{ g}}{27 \text{ g}} = \frac{X}{3 \text{ g}}$$

$(\tfrac{1}{3} \text{ mole})$

4. Solve for X by cross multiplication:

$$\frac{9}{27} \times \frac{X}{3} \quad \text{or} \quad 27X = 27$$

$$\frac{27}{27} = X$$

$$X = 1 \text{ g of hydrogen}$$

Example 19

What volume of hydrogen gas can be produced from 4 g of aluminum and hydrochloric acid under standard conditions? How many moles of HCl are required?

$$\begin{array}{cc} 4 \text{ g} & X \text{ liters} \\ \text{Al} + 3\text{HCl} \rightarrow \text{AlCl}_3 + \tfrac{3}{2}\text{H}_2 \\ 1 \text{ mole} & 1\tfrac{1}{2} \text{ moles} \\ 27 \text{ g} & 33.6 \text{ liters} \end{array}$$

1. Since volume of gas is the objective, it is more appropriate to use a volume to volume ratio for hydrogen. This expresses exactly the same fraction of a mole as a gram to gram ratio.

2. $$\frac{4}{27} \times \frac{X}{33.6} \quad \text{(cross multiply)}$$

$$27X = 4(33.6)$$

$$X = \frac{4(33.6)}{27}$$

$$X = 4.98 \text{ liters of hydrogen at STP}$$

3. Just as mole ratios can be expressed as grams to grams, or volume to volume, they can also be set up as mole to mole ratios. In every case, the units in a ratio are compatible. Incompatibility would be represented by a grams to volume ratio, for example. This would be as meaningless as a ratio comparing pounds to quarts. To obtain the number of moles of HCl, the equation works again:

$$\begin{array}{cc} 4 \text{ g} & X \text{ moles} \\ \text{Al} + 3\text{HCl} \rightarrow \text{AlCl}_3 + \tfrac{3}{2}\text{H}_2 \\ 1 \text{ mole} + 3 \text{ moles} \\ 27 \text{ g} \end{array}$$

$$\frac{4 \text{ g}}{27 \text{ g}} \times \frac{X \text{ moles}}{3 \text{ moles}} \quad \text{(cross multiply)}$$

$$27X = 4(3) \quad \text{(solve for } X\text{)}$$

$$X = \frac{12}{27} = 0.44 \text{ mole of HCl}$$

4. In order to set the previous examples in proper perspective, it should be remembered that the *balanced equation is the stoichiometry* and all the calculating based on the equation is just arithmetic.

PROBLEMS

Set A

1. What are the outstanding characteristics of the GROUP and the PERIOD in the periodic table?

2. Explain the change in the size and weight of atoms as you move from left to right across a period and top to bottom within a group.

3. Name the following compounds:

a. KF

b. $NaHCO_3$

c. $FeCl_2$

d. $H_2C_2O_4$

e. $MgSO_3$

f. $Ca(OH)_2$

g. HNO_3

h. Li_3AsO_4

i. PCl_3

j. NH_4Cl

k. K_2HPO_4

l. CBr_4

m. HgO

n. PbS_2

o. H_2SO_3

p. $K_2Cr_2O_7$

q. SiO_2

r. AlN

s. $NaCN$

t. SO_3

4. Write the formulas for the following compounds:

a. copper (II) sulfate

b. carbon disulfide

c. magnesium hydroxide

d. aluminum arsenite

e. phosphoric acid

f. cobalt (II) bromide

g. barium chlorate

h. zinc iodide

i. acetic acid

j. mercury (I) chloride

k. phosphorus pentoxide

l. potassium permanganate

m. calcium oxalate

n. nickel (III) sulfate

o. hydrochloric acid

p. tin (IV) nitrate

q. cadmium nitrate

r. carbon monoxide

s. chromium (II) hydroxide

t. ammonium carbonate

5. Describe the type of chemical bonding in the following substances.

a. KBr

b. CO_2

c. ice

d. $NaNO_3$

e. HCl

f. CH_4 (methane)

g. copper

h. $C_{12}H_{22}O_{11}$ (sucrose)

i. $CaCl_2$

j. SiC (carborundum)

6. What is the weight of 1.5 moles of sodium hydroxide?

7. How many atoms are there in 0.5 mole of calcium? What is the weight in grams of one calcium atom?

8. What volume is occupied by 0.02 mole of hydrogen gas (H_2) at STP?

9. Calculate the percentage composition of

a. CO_2 b. Na_2CO_3 c. C_2H_5OH

10. Find the empirical formula of a compound that by analysis is 55.8 per cent carbon, 7.0 per cent hydrogen, and 37.2 per cent oxygen.

11. Complete and balance the following equations:

a. silver oxide → silver + oxygen
b. lithium + water → lithium hydroxide + hydrogen
c. lead (II) nitrate + ammonium sulfide → lead (II) sulfide +
   ammonium nitrate
d. iron (III) chloride + sodium carbonate → iron (III) carbonate +
   sodium chloride
e. acetylene (C_2H_2) + oxygen → carbon dioxide + water

12. Aluminum reacts with hydrochloric acid to produce aluminum chloride and hydrogen gas. How many grams of aluminum are needed to produce 400 ml of hydrogen at STP?

Set B

13. What is the relationship between the periodic chart group number and the oxidation number of an element?

14. What arrangement of the electrons characterizes the transition metals?

15. Name the following compounds:
a. NH_4OH
b. HCN
c. As_2S_3
d. SiC
e. $KMnO_4$
f. CdI_2
g. $Ag_2Cr_2O_7$
h. $Ba(ClO_3)_2$
i. SiF_4
j. $Na_2SO_4 \cdot 10H_2O$
k. H_3PO_3
l. $Sn(CO_3)_2$
m. $NaHCO_3$
n. $XeCl_6$
o. HI
p. $Ni_2(C_2O_4)_3$
q. $BiCl_3$
r. $Cr(OH)_3$
s. Hg_2Cl_2
t. $Ni(OCN)_2$

16. Write the formulas for the following compounds:
a. carbon tetrachloride
b. silver acetate
c. iron (III) monohydrogen
   phosphate
d. arsenic acid
e. copper (II) sulfate · five
   hydrate
f. cobalt (III) hydroxide
g. ammonium sulfide
h. zinc hydrogen carbonate
i. lead (IV) chromate
j. oxalic acid
k. nickel (III) sulfite
l. mercury (II) hydroxide
m. nitrous acid
n. barium chlorate
o. phosphorus triiodide
p. chromium (II) hydrogen sulfate
q. ammonia
r. permanganic acid
s. fluorine
t. chlorine heptoxide

17. Describe the type of chemical bonding in the following substances:

a. CS_2

b. $CHCl_3$ (chloroform)

c. NaBr

d. silver

e. SiO_2 (sand)

f. H_2SO_4

g. AlN (alundum)

h. K_2SO_4

i. Li_3PO_4

j. Li_2O

18. How many grams are there in 0.75 mole of CaF_2?

19. How many atoms are there in 2.2 moles of sulfur? What is the weight of a million sulfur atoms?

20. What volume is occupied by 0.4 mole of carbon dioxide gas at STP?

21. Calculate the percentage composition of

a. HF

b. $Cd(NO_3)_2$

c. $Na_2S_2O_3$

22. Find the true molecular formula of a compound that has a molecular weight of 284 when a 9.2 g sample is analyzed and found to contain 4.02 g of phosphorus and 5.18 g of oxygen.

23. Complete and balance the following equations:

a. sodium chlorate → sodium chloride + oxygen

b. calcium + water → calcium hydroxide + hydrogen

c. cobalt (III) chloride + sodium oxalate → cobalt (III) oxalate + sodium chloride

d. sulfuric acid + zinc → zinc sulfate + hydrogen

e. ammonium phosphate + mercury (II) nitrate → ammonium nitrate + mercury (II) phosphate

24. Mercury (II) oxide is easily decomposed by heat into mercury and oxygen. What volume of oxygen can be obtained at STP from the decomposition of 0.03 mole of mercury (II) oxide?

SUPPLEMENTARY READING

Sienko, M. J.: *Stoichiometry and Structure, Part I.* (paperback) W. A. Benjamin, Inc., New York, 1964.

Contains many excellent solved problems as examples.

Companion, A. L.: *Chemical Bonding.* (paperback) McGraw-Hill Book Co., New York, 1964.

One of the most readable books available on the topic of chemical bonding.

Brown, G. I., *A New Guide to Modern Valency Theory.* J. Wiley & Sons, Inc., New York, 1967.

Clear and up to date.

Pauling. L.: *The Chemical Bond.* Cornell University Press, Ithaca, New York, 1967.

Clearly written with fine illustrations.

5 The Laboratory Notebook

The notebook of the laboratory technician is one of the most valuable documents in the research laboratory. The research team is very dependent upon the methods, data, calculations, and general observations of the technician as he has recorded them in the notebook. The success or failure of experiments may possibly depend on the thoroughness and accuracy of the records.

The notebook should be maintained in two hardbound sections. Pages are numbered, and space should be provided for a table of contents. The first section includes the outlining of methodology articles from appropriate scientific journals and up-to-date texts on chemical methods. This section should also contain carefully outlined procedures, called **protocols**, developed in the laboratory. The conscientious technician may experience pleasure from seeing a superb textbook on chemical methods develop from his own hand.

The second section of the notebook should contain records of all experiments. Each experiment should contain the following information:

1. The title, date, and purpose.
2. A list of required materials.
3. An outlined procedure.
4. Tables for the organization of observed data.
5. Graphical representation of data when it is appropriate.
6. Equations and calculations used to process the data.
7. Experimental results, including curves obtained from instrumental methods of measurement.
8. Observations of any unusual events that may have affected the results.

Sample pages from a technician's notebook and an annotated reprint, may serve to illustrate the essentials of good record keeping more effectively than rules and generalizations.

INTRACELLULAR DISTRIBUTION OF ENZYMES

III. THE OXIDATION OF OCTANOIC ACID BY RAT LIVER FRACTIONS*

By WALTER C. SCHNEIDER

(From the McArdle Memorial Laboratory, Medical School, University of Wisconsin, Madison)

(Received for publication, June 19, 1948)

Insoluble residues prepared by low speed centrifugation of tissue homogenates made in isotonic saline have recently been used to study the enzyme systems for the oxidation of fatty acids (6–9, 11), the oxidation of the acids of the Krebs tricarboxylic acid cycle (3), oxidative phosphorylation (10), p-aminohippuric acid synthesis (2), and citrulline synthesis (1). Our own studies on the differential centrifugation of rat tissue homogenates have shown that isotonic saline produces extensive agglutination of the mitochondria with the result that the fraction obtained at low speed was a mixture of all of the nuclei and 40 to 80 per cent of the mitochondria of the homogenate in addition to any cells that had not been broken during the homogenization (5, 14). It is, therefore, of interest to determine whether the enzymatic reactions mentioned above are associated with the nuclei, the mitochondria, or the unbroken cells, or whether a combination of two or more fractions is necessary for enzymatic activity. Such a study was made possible by the finding that agglutination was greatly decreased when the homogenates were made in isotonic or hypertonic sucrose solutions (5). Thus a more efficient separation of nuclei and mitochondria was possible in the sucrose homogenates than had hitherto been possible in the saline homogenates.

The present report describes the study of the oxidation of octanoic acid by fractions obtained from rat liver and shows that the oxidation of this fatty acid is associated mainly with the mitochondria.[1]

Materials and Methods

Preparation of Livers—Stock rats originally of the Sprague-Dawley strain obtained from the Holtzman Company, Madison, were fasted overnight to eliminate glycogen from the liver. The rats were killed by decapitation and the livers were removed and chilled in ice-cold isotonic KCl. After

* This work was aided by a grant from the American Cancer Society on the recommendation of the Committee on Growth of the National Research Council.

[1] This finding confirms the preliminary report of Kennedy and Lehninger (6). No comparison is possible between their results and those obtained in the present work because the tissue concentration used by these authors was not stated.

cooling, the livers were blotted, weighed, and homogenized (12) in ice-cold isotonic sucrose (8.5 gm. of sucrose per 100 ml. of solution). Isotonic sucrose was used rather than the hypertonic sucrose necessary to maintain the morphological integrity of the mitochondria (5), because it was found in preliminary experiments that preparations made in the hypertonic solutions had a much lower rate of fatty acid oxidation than did preparations made in isotonic sucrose. The latter gave activities in the same range as those reported by Potter (11).

Calculate molar concentration of sucrose. Prepare in liter volumes.

Differential Centrifugation of Rat Liver Homogenate[2]—10 ml. of homogenate (1 gm. of rat liver per 10 ml. of homogenate) were centrifuged for 10 minutes at $600g$ to sediment the nuclei. The supernatant was removed and the sediment was washed twice by resuspension in 2.5 ml. portions of isotonic sucrose and recentrifugation at $600g$ for 10 minutes. The washed sediment was made up to 2.5 ml. with isotonic sucrose and labeled the nuclear fraction, N_W. The supernatant and washings from the nuclear fraction were combined and centrifuged 10 minutes at $8500g$ to sediment the mitochondria. The mitochondria were washed twice by resuspension in 2.5 ml. of isotonic sucrose and recentrifugation at $8500g$ for 10 minutes. The washed mitochondria were made up to 2.5 ml. with isotonic sucrose and labeled the mitochondrial fraction, M_W. The supernatant and washings from the mitochondrial fraction were combined and centrifuged for 60 minutes at $18,000g$ to sediment the submicroscopic particles. The submicroscopic particles were washed once by resuspension in 2.5 ml. of isotonic sucrose and recentrifugation at $18,000g$ for 60 minutes. The washed particles were made up to 2.5 ml. with isotonic sucrose and labeled the submicroscopic particles, P_W. The supernatant and washing from the submicroscopic particles were combined, made up to 20 ml., and labeled the supernatant, S_2.

Determine average weight of whole liver.

Use nomograph to calculate rpm on Servall centrifuge for 600 g and 8500 g. SS-34 or GSA?

Omit!

Measurement of "Octanoxidase"[3] Activity—Optimum concentrations of the components of the octanoxidase system were determined by Potter (11) and were those employed in this paper. The measurements of oxygen uptake were made in the Warburg apparatus at 38°. The flasks contained 0.01 M potassium phosphate buffer, pH 7.4, 0.01 M potassium malonate,[4]

[2] All of the centrifugations were made at 0° in the International refrigerated centrifuge PR-1. The centrifugations at $600g$ were made with the horizontal yoke No. 269 and those at $8500g$ and $18000g$ were made with the multispeed attachment and the No. 295 conical head. The centrifugal forces designated refer to the centers of the tubes employed.

Note!

[3] The term "octanoxidase" will be used to refer to that system of enzymes involved in the oxidation of octanoic acid.

[4] It has been claimed that malonate produces varying degrees of inhibition of octanoxidase activity depending upon the strain of rat employed (9). Since we have

a reaction mixture containing 0.5–1 mg of mitochondrial protein, 5 mM ATP, 50 mM Tris-HCl buffer, pH 7.5, and 100 mM KCl, in a final volume of 1 ml. MgCl$_2$ (5 mM), dinitrophenol (0.1 mM), and oligomycin (2 μg/mg protein) may be added, according to the aim of the experiment. Incubation is made at 30° for 5 and/or 10 minutes. The samples are fixed with 1 ml of cold 1 M perchloric acid, and P$_i$ liberated is determined colorimetrically in 1 ml of the perchloric acid extract.

Energy-linked reduction of endogenous NAD$^+$ in skeletal muscle mitochondria is assayed essentially as described in this volume [112]. Details regarding assay systems used particularly for skeletal muscle mitochondria are described by Klingenberg and Schollmeyer.[7]

[15] Isolation of Liver or Kidney Mitochondria

By Diane Johnson *and* Henry Lardy

Principle. The selected tissue is disrupted by homogenization in cold isotonic sucrose. Differential centrifugation is then employed to separate the mitochondria from cell debris, red blood cells, nuclei, microsomes, and soluble components.[1]

Reagents and Equipment

Sucrose, 0.25 M, CO$_2$-free, or 0.2 M mannitol — Try sucrose and mannitol combination.

Substitute. — International refrigerated centrifuge, Model PR-2, with multispeed — Use Servall.
attachment and head No. 296 — SS-34 head.

Potter-Elvehjem homogenizers with either glass or Teflon pestles. — Collect homogenizers.
    The pestles should be 0.006–0.008 inch smaller in diameter than the inside bore of the homogenizer tube.

Stirring motor with shaft rotation of 600 rpm

All glassware used for preparation of mitochondria should be routinely cleaned by immersion in a hot sulfuric–nitric acid bath. It should be thoroughly rinsed with tap water, distilled water and finally with water that has been deionized and distilled from an all-glass or quartz apparatus.

All solutions that come in contact with the mitochondria must be made with highest purity reagents and water. Suitable water can be prepared by distilling from a glass or stainless steel vessel, passing

— Absolutely *clean* glassware!

[1] W. C. Schneider, *J. Biol. Chem.* **176**, 259 (1948).

[15]   MITOCHONDRIA FROM LIVER OR KIDNEY   95

through a mixed-bed ion exchange resin, and then distilling from an all-quartz still. On storage, such water will absorb CO_2 from the atmosphere and should therefore be boiled before use or may be neutralized with a small amount of tris(hydroxymethyl)aminomethane, triethylamine, or alkali metal hydroxide. Alternatively, distilled water may be purified sufficiently by passing through a suitable mixed-bed ion exchange resin on the all-plastic container such as that supplied by Continental H_2O, Chicago, Illinois.

Omit tris buffer (try some later if respiratory rate seems too low).

Preparation of Liver Homogenate. The animal is stunned and decapitated. Profuse bleeding is encouraged with flowing cold water. The liver is removed, blotted, and immediately placed in sucrose solution at 0°. All subsequent operations are carried out at 0° (vessels are kept immersed in chipped ice). Ten grams of liver is homogenized in 25–50 ml of cold 0.25 M sucrose by passing the homogenizer tube up and down past the rotating pestle. The tube is immersed in ice during this procedure. As little as 3 g or as much as 10 g at a time may be disrupted depending on the size of the homogenizer. Caution must be exercised to avoid pulling a vacuum in the tube. Homogenization is to be discontinued when liver tissue is no longer discernible. Excessive homogenization damages mitochondria and should be avoided. The homogenate is adjusted to a volume of 80 ml with cold sucrose.

Wash in extra sucrose rather than blotting.

Get plastic tray for ice.

Note!

Isolation of Liver Mitochondria (M_w). The homogenate is distributed into Lusteroid centrifuge cups and centrifuged at 600 g for 10 minutes. The supernatant fraction is decanted and saved. The pellets, containing cells, tissue fragments, and some mitochondria, are washed once with a total volume of 20 ml of sucrose. The pellets may be dispersed by using the side of a stirring rod against the wall of the cup or by hand-operating the homogenizer. The resuspended material is centrifuged at 600 g for 10 minutes. The supernatant fractions are combined. The pellets are discarded.

Collect enough Teflon centrifuge tubes.

Stirring rod or policeman.

This washing contributes not only to the yield of the final mitochondrial preparation, but also to its integrity, apparently by permitting the recovery of the larger mitochondria. The combined supernatants are centrifuged at 15,000 g for 5 minutes. The resultant supernatant is discarded along with any lightly packed pink microsomes. Lightly packed tan mitochondria are retained. The pellets are resuspended as the nuclear pellet was, quantitatively collected in one cup, centrifuged at 15,000 g for 5 minutes, and washed twice with 15–20 ml of sucrose. During each decanting operation any question is decided in favor of the purity of the mitochondria-containing fraction at the expense of complete mitochondrial recovery.

Use nomograph to find rpm equal to 600 g and 15,000 g.

It is imperative that the isolation be completed without any delay.

Work rapidly!

THE PROTOCOL

Consider, for example, the development of a protocol for the preparation of a suspension of biologically active, or "live," mitochondria from rat liver tissues. Mitochondria, incidentally, are found in all living cells, and they are the particles responsible for the energy-producing process of respiration. Studies made on living tissue in a container separate from any animal are described as **in vitro**. The live animal presents an **in vivo** type of investigation. The original article describing the methods for the preparation of rat liver mitochondria (RLM_w) is found in the October 1948 issue of The Journal of Biological Chemistry. The marginal notes are a preliminary step in the preparation of the protocol.

After some experience with the original protocol, the technician's task is to exhibit some initiative by modifying and streamlining. Sometimes changes are necessitated because of differences in available apparatus. A final protocol emerges.

RAT LIVER MITOCHONDRIA (RLM_w) PREPARATION

Methods of W. C. Schneider, J. B. C., 1948, and Johnson and Lardy, Vol. 10, Methods in Enzymology modified by S. L. Wolfgang, January, 1963.

Standard Procedure

1. *Avoid* water or ice contamination of preparation.
2. *Sign up* for use of Servall refrigerated centrifuge several hours in advance if the instrument is used jointly by several laboratories.
3. Maintain temperature of centrifuge slightly *above* 0° C.
4. *Rinse* all glassware in preparation medium before using.
5. Keep all materials on *ice* or in the *cold room* at all times.

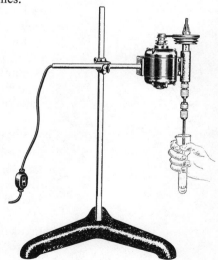

Figure 5-1.

Stock Solutions

0.5 M sucrose (171 g liter^{-1})
0.5 M mannitol (91 g liter^{-1})

Preparation Medium

0.225 M mannitol (450 ml of stock)
0.075 M sucrose (150 ml stock)
dilute to one liter

Materials

1 guillotine for rats
1 liter sucrose—mannitol preparation medium
1 tray for crushed ice
1 scissors (dissecting)
1 scissors (mincing)
1 50 ml beaker
1 50 ml tissue grinder (homogenizer)
1 loose-fitting teflon pestle
4 plastic centrifuge tubes for SS-34 head of Servall refrigerated centrifuge
1 rubber policeman
1 small glass homogenizer
1 10 ml labeled glass vial with screw cap

Procedure

Preliminary steps

1. *Decapitate* rat, *remove* liver, and *quickly* place liver in about 10 ml of preparation medium. Trim excess tissue from the liver.
2. *Mince* liver with scissors, *rinse twice* (decanting bloody supernatant each time).
3. *Homogenize* minced suspension until no large particles remain. Use motor driven teflon pestle at 75 to 100 RPM.
4. Fill grinding vessel to neck with preparation medium and continue grinding for about *one minute*.
5. Suspend homogenate in enough preparation medium to fill *two* plastic centrifuge tubes. *Balance* the tubes by weight.

Centrifugation (SS-34 Servall head)

1. First spin: Low speed (480 g = 2000 rpm) for 8 *minutes*. Carefully transfer supernatant to new tubes. *Discard* the remaining pellets that are composed of blood, cell walls, and unbroken cells.
2. Second spin: High speed (7710 g = 8000 rpm) for 10 *minutes*. *Quickly discard* the supernatant that contains fluffy microstructures. Break mitochondria pellets with the policeman and carefully resuspend in preparation medium in order to wash the mitochondria.
3. Third spin: 8000 rpm for 8 *minutes*. Discard the cloudy supernatant and resuspend the pellet in *one tube* with preparation medium for final washing. Use a second tube of water for balancing.
4. Fourth spin: 8000 rpm for 8 *minutes*. Discard the clear supernatant

Final steps

1. ... pellet and suspend in a small (about 10 ml)
   ... homogenizer. Use two to three ml of preparation medium.
2. ... about ... te and transfer to the small labeled vial to
   ... stored on ice in ...e.

Notes on the Protocol

1. The list of materials prevents lost time in preparation that could
   result when the technician has to stop work to find, wash, and cool
   missing items.
2. Advance preparations and directions to reserve frequently used
   apparatus are essential for successful work.
3. The technician has time to identify new apparatus and define strange
   terminology in the protocol. The following list includes some
   examples (Fig. 5-2).

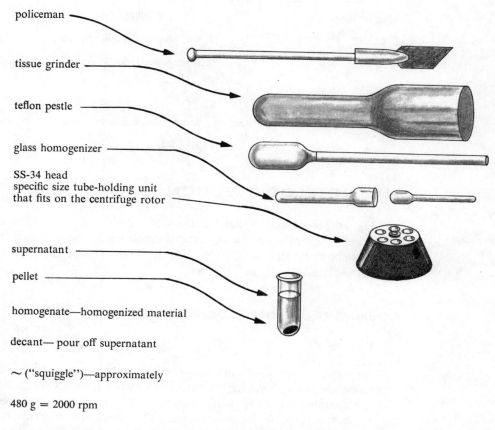

policeman

tissue grinder

teflon pestle

glass homogenizer

SS-34 head
specific size tube-holding unit
that fits on the centrifuge rotor

supernatant

pellet

homogenate—homogenized material

decant— pour off supernatant

~ ("squiggle")—approximately

480 g = 2000 rpm

7710 g = 8000 rpm The value with the g units describes the multiplication of the
normal force of gravity. The speed in rpm is variable. It
depends on the type of centrifuge used. 480 g = 2000 rpm
in the Servall model.

Figure 5-2.

4. Reasons for various steps are included in the protocol so that the technician can operate knowledgeably. He knows what to look for; thus, he can record unusual events.

5. A final item, as a summary, in the protocol section of the notebook is a schematic diagram of the methods (Fig. 5-3).

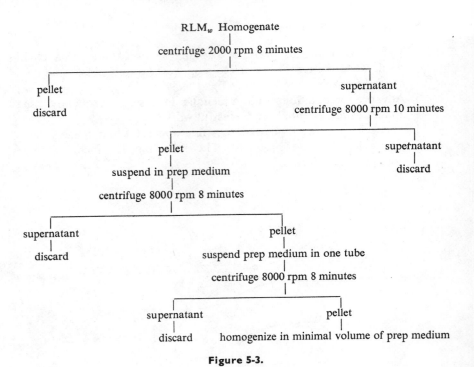

Figure 5-3.

THE NOTEBOOK SECTION

The notebook section that functions as a daily log of experiments can vary greatly. There are so many kinds of experiments that no single format is likely to be completely satisfactory. For illustrative purposes, two different types of experiments will be described.

EXPERIMENT I: PROTEIN DETERMINATION

October 3, 1969

Theory

Proteins are composed of polypeptides. Any compounds containing peptide bonds give a characteristic purple color when treated with an alkaline solution containing $CuSO_4$ (biuret reagent). The purple color deepens as the number of peptide bonds increase. Therefore, the concentration of protein, in milligrams per milliliter, can be measured colorimetrically. The amount of light having a wavelength of 545 mμ which is absorbed by the protein sample is compared to the light absorption of a standardized protein solution. The degree of light absorption is measured in **optical density** (O.D.) units on the instrument (colorimeter or spectrophotometer).

Preparation of Biuret Reagent

1.5 g $CuSO_4 \cdot 5H_2O$
6.0 g KNa tartrate $\cdot 4H_2O$
42.8 ml saturated NaOH diluted to 300 ml
  (the NaOH solution should be aged for a week to permit impurities
  to settle out)
Add the 300 ml of NaOH slowly to the $CuSO_4$-tartrate mixture and
  dilute to one liter. Store the biuret reagent in a polyethylene bottle.

Preparation of Standardized Protein

1. Dissolve 50 mg of crystalline bovine plasma albumin in 1 ml of water.
2. Find the ratio of the O.D. to the concentration of protein at 555 mμ.

$$\underset{(ratio)}{\mathcal{E}} = \frac{O.D.}{mg\ ml^{-1}\ of\ protein}$$

Determination of the value for $\mathcal{E}$ permits the finding of the concentration of protein:

$$mg\ ml^{-1}\ protein = \frac{O.D.}{\mathcal{E}}$$

3. If 1 ml is not used, and if the original sample is diluted, these factors modify the equation:

$$\mathcal{E} = \frac{O.D.\ (dilution\ factor)}{mg\ ml^{-1}\ protein\ (ml\ of\ protein)}$$

4. Assume 0.3 ml standardized protein is diluted to 1.5 ml and 1.5 ml of biuret reagent is added. The O.D. at 555 mμ is 1.350.

$$\mathcal{E} = \frac{1.350}{(50\ mg\ ml^{-1})(0.3\ ml)} = 0.09$$

Procedure

1. Dilute 1 ml of the thick RLM_w homogenate to 10 ml. Prepare samples as outlined in the table below and measure the optical densities.

Tube Number	Volume of Diluted RLM_w	Volume of Water	Volume of Biuret	Observed O.D.
1	blank	1.5 ml	1.5 ml	0.000
2	0.2 ml	1.3 ml	1.5 ml	0.038
3	0.4 ml	1.1 ml	1.5 ml	0.077
4	0.6 ml	0.9 ml	1.5 ml	0.114

2. Use ordinary graph paper to plot the O.D. against the volume of protein in order to verify the proportionality of light absorbance to protein concentration. The slope of the line, multiplied by 10 (the

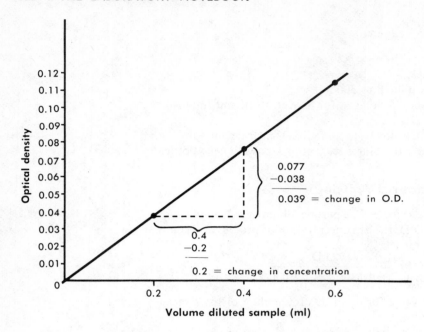

Figure 5-4. Determination of the slope of the line.

dilution of the original sample), should agree with the calculated concentration (Fig. 5-4).

$$\text{slope} = \frac{0.039}{0.2\ (0.09)} \times \underset{\text{(dilution factor)}}{10} = \sim 22 \text{ mg ml}^{-1} \text{ protein}$$

Calculation

$$\text{mg ml}^{-1} \text{ protein} = \frac{\text{O.D. (dilution)}}{\varepsilon \text{ (ml sample)}}$$

$$\text{mg ml}^{-1} \text{ protein} = \frac{(0.114)(10)}{(0.09)(0.6 \text{ ml})} = \sim 21 \text{ mg ml}^{-1} \text{ protein}$$

Conclusion

The slope of the curve is in reasonable agreement with the calculated value for the protein concentration. The calculated value is generally accepted as the recorded value.

The label on the vial should resemble the one in Figure 5-5.

SMC   October 3, 1969
RLM$_w$
21 mg ml^{-1}
protein

Figure 5-5.

Notes on the Protein Determination Experiment

1. Water used in experimentation is always understood to be *distilled* water.

2. A more detailed description of what optical instruments do (such as colorimeters and spectrophotometers) and how they work will be discussed in the chapter on instrumentation.

3. Once a particular experiment has been performed, any repetition of the same experiment could certainly omit the theory and explanatory sections. The title, date, observed data, calculations, and unusual observations are always the heart of the recorded experiment.

4. A discrepancy between the protein concentration obtained from the graph and the calculated value can be expected. The possible error in preparing a small standardized protein is great when a weight as small as 50 mg is used. The *per cent of error* is small.

$$\text{per cent of error} = \frac{22 \text{ mg} - 21 \text{ mg}}{21 \text{ mg}} = \frac{1}{21}$$

$$0.0476 \times 100 = 4.76 \text{ per cent} = \text{less than 5 per cent error}$$

EXPERIMENT 2: THE EFFECT OF A–27 ON RAT OVARIES (PART I)

Background

During the 1950's and early 1960's, extensive research for the development of an oral contraceptive was carried on by many pharmaceutical companies. The vast number of drugs synthesized had to be tested on rats in order to determine the degree of unfavorable side effects. One **parameter** of drug behavior was the physical effect on rat's ovaries. A parameter is a variable factor which may affect the results of an experiment. Separate experiments are often necessary to understand the significance of parameters. The extent to which the ovaries swelled or atrophied was considered significant. A typical experiment included *title, drug code number* and *formula, procedure, results, statistical analysis*, and *conclusions*.

A–27 (code)   August 12, 1960

Purpose

Evaluation of the effect of drug A-27 on rat ovary weight.

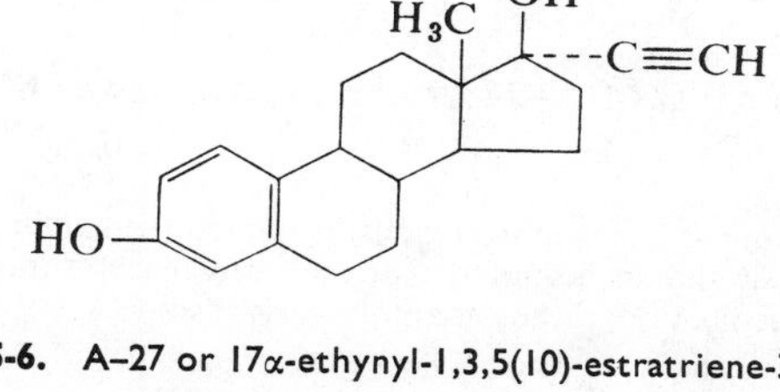

Figure 5-6. A–27 or 17α-ethynyl-1,3,5(10)-estratriene-3,17-diol.

Materials

1 2.0 ml syringe, no. 22 needle
50 ml of A–27, concentration of 5 μg per ml of mineral oil
50 ml of pure mineral oil
1 animal scale (0–500 g)
40 Sprague-Dawley female rats weighing between 200 g and 250 g

Procedure

1. Inject rats, numbers 1 through 20, each with 0.02 μg A–27 per gram of body weight. The equation is

$$\frac{(0.02 \ \mu g \ g^{-1})(\text{weight of rat in grams})}{5 \ \mu g \ ml^{-1}} = \text{ml of A–27}$$

Example 1

What volume of A–27 (5 μg ml^{-1}) should be injected into a 220 g rat in order to adjust the dosage to 0.02 μg g^{-1} of body weight?

$$\text{ml of A–27} = \frac{(0.02 \ \mu g \ g^{-1})(220 \ g)}{5 \ \mu g \ ml^{-1}}$$

$$\text{ml of A–27} = 0.88 \ \text{ml injected}$$

2. Inject rats, numbers 21 through 40, each with approximately 1 ml of mineral oil.
3. Replace all rats in cages, 5 rats per cage. Notch the right ear of the twenty **placebos*** for identification.
4. Continue feeding with rat chow and sugar water for one week.

THE EFFECT OF A–27 ON RAT OVARIES (PART II)
A–27   August 19, 1960

Purpose

To observe the effect of A–27 on the rat ovary weight.

---

* Placebo is the name often given to organisms treated with an innocuous, or harmless, medication. However, it should be noted that the term placebo more accurately applies to the non-effective medication rather than the organism. A sugar pill is a common human placebo.

Materials

dissecting scissors
top-loading analytical balance
small forceps

Procedure

1. Sacrifice rats, one at a time, by hitting them at the base of the skull on the table edge. This severs the brain from the spinal cord.
2. Remove the right ovary from the lower abdominal cavity of each rat. Use forceps to quickly transfer each ovary to the analytical balance.
3. Record the weight of the ovaries as indicated by the analytical balance.
4. Calculate the average difference between the treated rats and the placebos.
5. Include the standard deviation in the results.

Results

Rat Number	Ovary Weight (milligrams)
1	125
2	122
3	122
4	123
5	126
6	129
7	124
8	130
9	129
10	124
11	125
12	124
13	126
14	128
15	128
16	128
17	127
18	122
19	124
20	125
21	107
22	105
23	106
24	109
25	111
26	112
27	106

Rat Number	Ovary Weight (milligrams)
28	105
29	107
30	110
31	109
32	106
33	108
34	111
35	104
36	112
37	106
38	106
39	108
40	107

The average weight of the ovary of the treated rats is 125.55 mg compared to the average placebo weight, which is 107.75 mg. There seems to be a significant difference. However, before any conclusions should be made, a measure of confidence in the data is necessary. If most of the observed weights are very close to the average, it is described as a slight variation. This is a very encouraging result. It indicates a good reason for accepting the measurements as reliable, and it promotes confidence in the direct effect of the drug.

The average by itself is not sufficient evidence for a cause and effect relationship. For example, $2 + 3 + 7 + 8 =$ an average of 5. However, each individual number varies considerably from the average. The difference between 8 and the average is 3. This represents a variation of 60 per cent, since the difference, 3, is 60 per cent of the average, 5. If the set of four values is expected to be very similar, the 60 per cent variation for two values and the 40 per cent variation for the other values destroy confidence. The tentative conclusion is that the data are not reliable. By contrast, four values of $4 + 5 + 5 + 6$, yielding an average of 5, appear to be much more significant. Two values do not vary at all from the average, and the other values only vary by 20 per cent.

Standard Deviation

The method of standard deviation is specifically designed to measure the **reliability** of a set of observed data. It is a deviation from the average that is represented by a plus or minus sign ($\pm$) attached to the average. When the overall variation from the average is small, the data are considered to be very reliable. When confidence in the data is established, reasonable conclusions may be drawn. There is no sacred percentage of variation that applies to all sets of measurements for evaluation purposes. The experimenter must finally make a carefully reasoned value judgment in light of the variables and purposes of the experiment.

The equation for calculating the standard deviation is:

$$s = \sqrt{\frac{\Sigma(X_i - \bar{X})^2}{n - 1}}$$

$s =$ the standard deviation

Σ = **sum**
(sigma)

X_i = the individual measurement

$\overline{X}$ = the average of all the measurements

n = the total number of measurements

$(X_i - \overline{X})$ = the difference between each individual measurement and the average. This value is always a positive number, obtained by subtracting the smaller number from the larger

$(X_i - \overline{X})^2$ = each difference is squared

$\Sigma(X_i - \overline{X})^2$ = the sum of all the squared differences

$n - 1$ = one less than the total number of measurements

The experimental data from the rat ovary measurements will be processed as an example of obtaining the standard deviation.

Rat Number	X_i Ovary Weight (milligrams)	$X_i - \overline{X}$	$(X_i - \overline{X})^2$
1	125	1	1
2	122	4	16
3	122	4	16
4	123	3	9
5	126	0	0
6	129	3	9
7	124	2	4
8	130	4	16
9	129	3	9
10	124	2	4
11	125	1	1
12	124	2	4
13	126	0	0
14	128	2	4
15	128	2	4
16	128	2	4
17	127	1	1
18	122	4	16
19	124	2	4
20	125	1	1

$\overline{X} = 126$ $\qquad\qquad$ $\Sigma(X_i - \overline{X})^2 = 123$

$$s = \sqrt{\frac{\Sigma(X_i - \overline{X})^2}{n - 1}}$$ $\qquad\qquad$ $n = 20$

$$s = \sqrt{\frac{123}{19}} = \sqrt{7}$$

$$s = 2.7 = {\sim}3$$

The final answer for the average ovary weight of the treated rats is

$$126 \pm 3 \text{ mg}$$

The three milligram standard deviation from the average is an indication of very good reliability. It is 3/126 or, approximately, a 2 per cent variation. When normal differences in animal organ weights is considered, the reliability is enhanced to the point of excellence. When the same calculation is made for the untreated rats, the result for the average ovary weight is 108 ± 2 mg. Since the reliability of these data is excellent, some tentative conclusions may be made after the parameters are established.

Defining the Parameters

The data seem to indicate a direct effect by the drug A–27 on the increase in weight of the rat ovary from an average of 108 ± 2 mg for the untreated rats compared to 126 ± 3 mg for the treated rats. A difference of about 16 mg is significant. However, before a conclusion is made that the drug has the undesirable side-effect of causing a 15 per cent weight increase in ovaries (absolutely horrendous when scaled to human proportions), other possible contributing factors must be considered. These are the parameters.

1. What effect might pure mineral oil have on rat ovary weight?
2. Was the increase in ovary weight a temporary or permanent change?
3. Was the increase due to fluid absorption or cancerous growth?
4. Did the rats respond in a manner peculiar to their own species? What effect would A–27 have on mice, guinea pigs, or monkeys?

Drawing Conclusions

The function of the conclusion is to evaluate the data and calculations as well as to consider steps that might control the parameters. Conclusions to the rat ovary experiment should include the following points:

a. Longer time periods are needed to determine the permanence of the increase in the mass of the ovary.
b. Microscopic study of ovary sections is required to learn if the growth is cancerous, or if it is the result of body fluid absorption or some other factor.
c. A similar experiment should be performed using mice. The dosage in μg A–27 per g of animal body weight should be reduced proportionately.

Notes on the Experiment

1. Ear notching is a relatively painless method of nipping off a tiny bit of ear tissue for identification purposes.
2. The position of the ovaries should be determined before the actual dissection.
3. *All* observed data should be recorded. Use a symbol, such as a star or asterisk, to note any data that are the result of what appears to be an obvious error in procedure.

There is no universal format that is ideal for all kinds of experiments in the variety of laboratories that exist. The examples in this chapter are meant to point out the kinds of items most likely to be essential to laboratory technicians. The individual laboratory will indicate the various modifications, additions, and deletions that fit its particular program.

PROBLEMS

Set A

1. Define the following terms:

a. protocol
b. placebo
c. isotonic solution
d. mannitol

e. decant
f. O.D. units
g. parameter
h. per cent of error

2. Why should a hardbound notebook be used for protocols and experiments rather than a loose-leaf notebook?

3. A thick protein suspension is diluted from 1 ml to 5 ml. The O.D. values at 555 mμ are as follows:

Tube Number	Volume of Diluted Protein	Observed O.D.
1	0.2 ml	0.102
2	0.4 ml	0.205
3	0.6 ml	0.310

a. Graph the data, O.D. vs. diluted protein.
b. Calculate the protein concentration in mg ml^{-1} from the slope of the curve and multiply by the dilution factor.

4. For the same data in Problem 3, calculate the milligrams per milliliter of protein from the equation. Use 0.09 as the value of ε.

$$\text{mg ml}^{-1} \text{ protein} = \frac{\text{O.D. (dilution factor)}}{\varepsilon \, (0.4 \text{ ml of sample})}$$

5. Compare the results of Problem 4 and Problem 3b and calculate the per cent of error. Assume the answer in Problem 4 to be the actual value.

6. A particular drug is prepared in a concentration of 10 mg ml^{-1} of water. If the dosage required is 10 mg of drug per 500 g of animal body weight, how many milliliters should be injected into (a) a 300 g animal? (b) A 680 g animal?

7. What volume of drug, whose concentration is 80 mg ml^{-1}, should be injected into a 400 g animal if the dosage is supposed to be 0.04 mg g^{-1} of body weight?

8. Calculate the average ovary weight and standard deviation for the untreated rats numbered from 21 to 40 on page 141.

9. Define the following terms:

a. hypertonic solution
b. atrophy
c. mitochondria
d. homogenate

e. supernatant
f. pellet (residue)
g. RLM$_w$
h. standard deviation

10. Why should distilled water be used for solution preparation?

11. A mitochondria suspension is diluted tenfold. The O.D. values at 555 mμ are as follows:

Tube Number	Volume of Diluted Suspension	Observed O.D.
1	0.3	0.216
2	0.6	0.440
3	0.9	0.678

a. Graph the data, O.D. vs. mitochondria volume
b. Calculate the protein concentration in mg ml^{-1} from the slope of the curve, and multiply by the dilution factor.

12. Given the O.D. of 0.305 for 0.4 ml of sample, calculate the protein concentration from the equation where ε = 0.092.

13. Calculate the percent of error between the actual value for the protein concentration in Problem 12 and the slope value in Problem 11b.

14. A drug in solution has a concentration of 25 mg ml^{-1}. A 200 gram animal requires 5 mg of the drug. How many milliliters should be administered to (a) a 150 g animal? (b) A 420 g animal?

15. What volume of drug (concentration of 20 mg ml^{-1}) should be given to a 250 g animal if the required dosage is 0.05 mg g^{-1} of body weight?

16. Given the following list of densities for a particular mineral, calculate the average density and the standard deviation. Are the data sufficiently reliable to identify the mineral from its density? Why?

Experimental densities for
the unknown mineral (g ml^{-1})

4.2	4.1
4.7	5.0
5.1	4.8
4.0	4.5
3.8	4.5

SUPPLEMENTARY READING

Huff, D.: *How to Lie with Statistics.* W. W. Norton & Co., New York, 1954.

Discusses the uses and abuses of statistical data. This is very relevant to the scientific interpretation of data.

Kline, M., *Mathematics—A Cultural Approach*, Addison-Wesley Publishing Co., Inc., Reading, Massachusetts, 1962.

Very fine chapters on statistics and probability.

Woodford, F. P.: Sounder Thinking Through Clearer Writing. *Science*, May 12, 1967, p. 743.

An excellent article that all laboratory workers should read.

Trelease, S. F.: *How to Write Scientific and Technical Papers.* (paperback) The M.I.T. Press, Cambridge, Massachusetts, 1969.

Schultz, H., and Webster, R. G.: *Technical Report Writing.* McKay Co., Inc., New York, 1962.

The technician should find both of the above-mentioned texts very helpful.

Information Retrieval 6

The twentieth century has witnessed an incredible gain in scientific knowledge. The reality of the common phrase "knowledge explosion" has promoted a drastic change in a scientific investigator's approach to factual information. Scholarship is no longer measured in terms of ability to memorize facts. There are just too many facts for the ordinary mind to cope with. With each passing year comes an increase in the number of scientific journals, although there are about 7000 already. Many thousands of articles and abstracts of published papers engulf the scientific community. The presently existing reference texts, specialized encyclopedias, special dictionaries, and handbooks all serve to boldly underline the need to understand the mechanics of information retrieval. Faced with the common task of *finding out*, it is critically important for the laboratory team to have some clear ideas about where to start looking. The science library has moved from its old role as a convenient addition to a laboratory building to a prominent and integral part of any research complex.

It is the responsibility of a laboratory technician to constantly try to understand the scientific principles related to the objectives of his experimental activities. It is the author's contention, based on years of laboratory research, that an effective technician is one who functions intelligently, rather than mechanically. A knowledgeable technician can discuss his observations and contribute the fruit of his own initiative to the research team. In order for the technician to make this kind of contribution, he must be able to find out what the library has for his continuing education. The knowledge gained can form a permanent and valuable part of the technician's notebook.

This chapter is written only as an introduction to the subject of information retrieval; however, there are books available that treat this subject in much greater detail. One prominent example would be the second edition of *How to Find Out in Chemistry* by C. R. Burman, published by Pergamon Press Ltd. (London) in 1966.

Quite often, a technician is assigned the task of doing a literature search. In order to avoid wasteful duplication in experimentation and to accumulate a foundation of related data, the technician may have to survey the literature of recent years. If there are no specific references available, the logical procedure is to move from the general to the specific. Most likely, the card catalog is the starting point.

THE CARD CATALOG

The heart of any library is the alphabetically arranged file of cards for available materials—books, records, microfilm, pictures, journals, and magazines. This card catalog lists materials which are in the library or are obtainable on loan from other libraries. The organization of cards is based on *title* and *author*. This enables the researcher to find the publications of a prominent author or to find out what scientists have done work in a particular field. For example, a textbook on biochemistry by Harrow and Mazur might be found under biology, chemistry, biochemistry, biological chemistry, Harrow, and Mazur. This multiplication of cards for a single text is known as **cross-reference**. The card contains a variety of information as seen in the example in Figure 6-1.

The sample card contains two sets of code numbers which are the **Dewey decimal system** and the **Library of Congress** system (LC) (Fig. 6-1). The LC system is generally accepted as the preferable one because of its

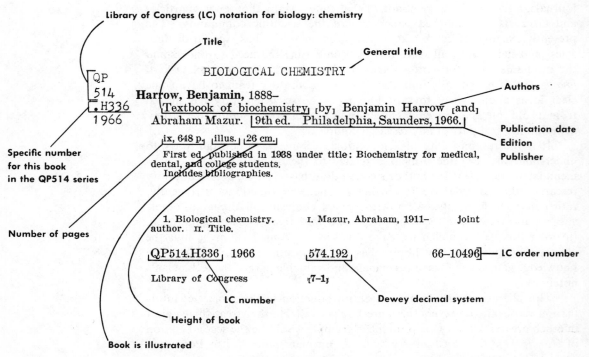

Figure 6-1.

greater flexibility and scope. It has become increasingly difficult to place all knowledge in the restricting confines of the ten divisions of the Dewey decimal system; science occupies the 500 to 599 division in this system that ranges from 0 to 1000.

The LC system utilizes twenty-one alphabet letters to categorize human knowledge. A partial listing is seen below.

LIBRARY OF CONGRESS CLASSIFICATION NUMBERS FOR SCIENCE AND TECHNOLOGY

Subject	LC	Subject	LC
(Chemical) industries economic	HD 9650–9660	Useful arts; therapeutics and pharmacy	RS 402–431
Natural Science	Q	Engineering: chemistry	QC 603–605
Chemistry	QD	Agriculture: chemistry	S 583–588
Chemistry: practical and experimental	QD 43–64	Chemical technology	TP
Chemistry: analytical, general	QD 71–80	Chemical technology manufacturing	TP 155
Chemistry: analytical, qualitative	OD 81–100	Chemical technology foods	TP 370–465
Chemistry: analytical, quantitative	QD 101–150	Chemical technology oils, gas	TP 343–360
Chemistry: inorganic	QD 151–199	Chemical technology metallurgy	TN 600–799
Chemistry: organic	QD 241–499	Manufactures articles of leather	TS 940–1030
Chemistry: mineralogy	QE 351–399	Manufactures articles of paper	TS 1080–1200
Geology	QE	Manufactures articles of textiles	TS 1300–1781
Biology: general	QH 301–705	Manufactures articles of plastics	TP 986
Biology: chemistry	QP 514		

The LC number, title, and author should be recorded from all the cards that may be related to the topic under investigation. The library will probably be arranged in sections that follow the order of the alphabet. The next step is the actual investigation of the literature and the recording of pertinent data.

GENERAL SOURCES

The starting point in a literature search will often concern books of a very general nature that present broad surveys of subjects. This kind of book usually lists a selective **bibliography** at the end of a section or chapter. A good bibliography lists many primary, or original, sources of information used by the author as a source for his survey. A bibliography lists the author, title, edition, publisher, place of publication, and year of publication for each book. For example:

Mazur, A., and Harrow, B., Biochemistry: A Brief Course, 9th ed. Philadelphia, W. B. Saunders Company, 1968.

Reference Books

Some common types of references for general information are:
1. *General textbooks* on chemistry, biochemistry, biology, physics.
2. *Encyclopedias*
   a. General—*Encyclopaedia Brittanica*, *World Book*, and others.
   b. Scientific—*Encyclopedia of Science* (McGraw-Hill), *Encyclopedia of Spectroscopy*, *Encyclopedia of Science and Technology* (Van Nostrand), *Recent Methods in Biochemistry*, and the *Encyclopedia of Biochemistry* (see Fig. 6-2).

```
QP              BIOLOGICAL CHEMISTRY --DICTIONARIES
512
.W5   Williams, Roger John, 1893-
         The encyclopedia of biochemistry,edited by
      Roger J. Williams and Edwin M. Lansford, Jr.
      New York, Reinhold [1967]

         xvii, 876 p. illus. 27 cm.
      Includes bibliographies


      1.  Biological chemistry--Dictionaries

      I. Lansford, Edwin M., joint author II. Title.

      QP512.W5            574.1'92'03        67-15466
```

Figure 6-2.

Popular Review Periodicals

Monthly or weekly scientific reviews are readily available and can be easily surveyed. These magazines are an additional source of bibliography. Some outstanding examples are:

1. Scientific American
2. Nature (Great Britain)
3. Science (American Association for the Advancement of Science)
4. Chemical and Engineering News
5. Analytical Chemistry (publishes a volume of reviews each spring)
6. American Scientist
7. Chemistry (American Chemical Society)

Although the articles in these reviews are often two or three years behind the original publication times of the articles, they serve effectively as starting points for more detailed investigation. It should be mentioned also that *Nature* and *Science* function as primary sources as well as review journals. (See page 156).

Abstracts and Indexes

Some of the most valuable sources of current information are **abstracts** and **indexes**.

An **abstract** is a one paragraph summary of a much longer article published in a scientific journal. The abstract attempts to describe the essence of the article so that any researcher will know if the topic is related to his own work. The abstract contains the author, title, journal, volume number, page number, and date.

For purposes of illustration, a few of the following prominent examples will be reviewed:

1. Chemical Titles (Published biweekly by the American Chemical Society.)
2. Chemical Abstracts (Also by the American Chemical Society; published twice a month.)
3. Bioresearch Index (Published monthly by Biosciences Information Service.)
4. Biological Abstracts (Also published monthly by Biosciences Information Service.)
5. Current Contents (An Institute for Scientific Services (ISI) weekly publication. Lists the tables of contents from many selected journals. Current Contents contains subject and author indexes, structural diagrams, and address directory.)
6. Index Chemicus (A weekly publication of the ISI reporting on new organic and inorganic compounds found in current journals. It contains structural diagrams, journal citation, and abstracts. There are nearly a million compounds in the *Index Chemicus* file.)
7. Science Citation Index (Published annually by ISI. This index includes and relates chemical, physical, and biological sciences. It contains a Citation Index Source Index, Corporate Index, Patent Index, and a Key Word Index.)
8. Chemical Titles (Lists thousands of titles from 700 journals. It is arranged in a manner that permits the rapid location of a specific title from a key word or from the author's name.)

Example I *

Find a recent article on the effect of tri ethyl tin on mitochondrial ion accumulation.
1. List the key words:
   a. tri ethyl tin
   b. mitochondrial
   c. tin

---

* From *Chemical Abstracts*

```
HE HARDENING OF COPPER-  TIN ALLOYS.=    +PHYSICAL FACTORS ON   ANCRAI-0059-1086
ROPERTIES OF PALLADIUM-  TIN ALLOYS.=  +OF THE THERMODYNAMIC   AMETAR-0018-0101
OPERTIES OF THE COPPER-  TIN AND COPPER-GOLD SYSTEMS BY MASS   MTGTBF-0001-0415
F ADDUCTS OF TRI PHENYL  TIN CHLORIDE WITH OXYGEN DONOR SUBST  CJCHAG-0048-0071
CAL STABILITY OF ORGANO  TIN COMPOUNDS BY NUCLEAR MAGNETIC     JACSAT-0092-0365

OF PYRIDINE UPON ORGANO  TIN COMPOUNDS OF THE DI ORGANO TIN    BCSJA8-0043-0266
=    VIBRATION MIXING OF  TIN CONCENTRATES IN A LIQUID MEDIUM. TVMTAX-42-12-015
ING THE REPROCESSING OF  TIN CONTAINING ORES.=  +GRINDING DUR  TVMTAX-42-12-058
LUORITES FROM A KHINGAN  TIN DEPOSITE.=     +INCLUSIONS IN F   ZVMOAG-0098-0748
O TIN OXIDE - DI ORGANO  TIN DI HALIDE MIXED TYPE.=+ DI ORGAN  BCSJA8-0043-0266
LDEHYDES THROUGH ORGANO  TIN ENOLATES OF THE ALDEHYDES.=+OF A  CHDCAQ-0270-0100
TERMINATION OF DI ALKYL  TIN GROUPS IN DI BUTYL AND DI OCTYL   ZACFAU-0249-0035
N ALLENES B+UNSATURATED  TIN HYDROCARBONS.  PREPARATION OF TI  ZOKHA4-0039-2673
APPROXIMATION FOR EVEN   TIN ISOTOPES.=     +BCS-TAMM-DANCOFF  PHRVAG-0187-1306
N.  EFFECT OF TRI ETHYL  TIN ON MITOCHONDRIAL ION ACCUMULATIO  FEBLAL-0005-0331
POUNDS OF THE DI ORGANO  TIN OXIDE - DI ORGANO TIN DI HALIDE   BCSJA8-0043-0266
SOLUBILITY OF HYDRATED   TIN OXIDE COMPLEXES AND CASSITERITE   GEOKAQ-1970-0035
D ADSORPTION IN REDUCED  TIN OXIDE CRYSTALS.=   +AND INFRARE   BJAPAJ-0002-1503
ICAL PROPERTIES OF LEAD  TIN SELENIDE IN THE LOW CARRIER CONC  SSCOA4-0007-1777
LECTRICAL PROPERTIES OF  TIN SELENIDE.=      OPTICAL AND E     BJAPAJ-0002-1507
RYSTALLIZATION OF WHITE  TIN SINGLE CRYSTALS.=     +RE C       MRMTAU-0066-0731
IN SINTERED SAMPLES OF   TIN SULFIDE AND LEAD SULFIDE.=+CIENT  CHDCAQ-0270-0131
WTH MECHANISM OF PROPER  TIN WHISKER.=          GRO           JJAPA5-0008-1404
TION OF VINYL TRI ETHYL  TIN WITH STYRENE.=   +CO POLYMERIZA   VYSAAF-0012-0182
ALPHEN EFFECT IN WHITE   TIN.=              DE HAAS - VAN      JPCSAW-0031-0117
IC COMPOUND NIOBIUM(3)-  TIN.=  +DEPOSITION OF THE INTERMETAL  ZENAAU-0025-0064
BY COMPLEX JUNCTIONS IN  TIN.=+ APERIODIC OPEN ORBITS CLOSED   PYLAAG-0030-0510
ILIBRIUMS ON GERMANIUM,  TIN, AND LEAD.=     +EXCHANGE EQU     ANYAA9-0159-0056
I, STRONTIUM+ANALOGY OF  TIN(II) ION WITH BARIUM(II), LEAD(II CHDCAQ-0270-0052
DINE CHLORIDE, BROMINE,  TIN(IV) CHLORIDE, AND PHOSPHORUS(III  JCPSA6-0051-5467

ND HEXA COORDINATION IN  TIN(IV) ORGANO METALLIC COMPLEXES.=   ANCRAI-0059-1152
XY ACIDS.  COMPLEXES OF  TIN(IV) WITH GLUCONIC AND SACCHARIC   RRCHAX-0014-1231
DE CORDYCEPIN) IN ROOT-  TIP CELLS OF ALLIUM CEPA.=  +RIBOS    ECREAL-0059-0085
UALLY ORIENTED TUNGSTEN  TIPS AND THEIR GROWTH MECHANISM.=+US  JJAPA5-0009-0015
RESENCE OF+OXIDATION OF  TIRON BY HYDROGEN PER OXIDE IN THE P  ZACFAU-0249-0022
RESONANCE ABSORPTION IN  MIRRORS BY OVERCOATING WITH DIELECTR  APOPAI-0008-2556
NENTS ON THE ISOMORPHIC  MISCIBILITY RANGES.=  +OF PURE COMPO  GEOKAQ-1970-0059
ION OF SODIUM +STATE OF  MITOCHONDRIA IN PERFUSED LIVER.  ACT  FEBLAL-0005-0319
TRANSPORT OF NADH INTO   MITOCHONDRIA OF RAT WHITE ADIPOSE TI  BIJOAK-0116-0229
OF ION TRANSLOCATION IN  MITOCHONDRIA.  ACTIVE TRANSPORT AND   EJBCAI-0012-0310
OF ION TRANSLOCATION IN  MITOCHONDRIA.  COUPLING OF POTASSIUM  EJBCAI-0012-0301
OF ION TRANSLOCATION IN  MITOCHONDRIA.  COUPLING OF POTASSIUM  EJBCAI-0012-0319
OF ION TRANSLOCATION IN  MITOCHONDRIA.  COUPLING OF POTASSIUM  EJBCAI-0012-0328
O ACIDS AND+INTRANEURAL  MITOCHONDRIA.  INCORPORATION OF AMIN  JBCHA3-0245-0363
NS BY ISOLATED OX HEART  MITOCHONDRIA.=         +PROTEIN FRACTIO BIJOAK-0116-0269

OSOMES AND SYNAPTOSOMAL  MITOCHONDRIA.=     +BY ISOLATED SYNAPT  JBCHA3-0245-0363
NSLOCATION IN RAT LIVER  MITOCHONDRIA.=  +BIVALENT CATION TRA   EJBCAI-0012-0387
BOVINE ADRENO CORTICAL   MITOCHONDRIA.=+CYTOCHROME P-450 FROM   BBRCA9-0038-0184
MERIC CIRCULAR FORMS OF  MITOCHONDRIAL DNA FROM HUMAN LEUKEMI   JMOBAK-0047-0137
NTROL OF THE LATENCY OF  MITOCHONDRIAL ENZYMES.=+DEPENDENT CO   EJBCAI-0012-0227
ECT OF TRI ETHYL TIN ON  MITOCHONDRIAL ION ACCUMULATION.= EFF   FEBLAL-0005-0331
ULATION OF HEART MUSCLE  MITOCHONDRIAL MALATE DE HYDROGENASE    BICHAW-0009-0274
ULATION OF HEART MUSCLE  MITOCHONDRIAL MALATE DE HYDROGENASE    BICHAW-0009-0285
S OF+REDOX STATE OF THE  MITOCHONDRIAL NAD SYSTEM IN CIRRHOSI   BBACAQ-0201-0009
ITHIN THE LIV+ORIGIN OF  MITOCHONDRIAL PHOSPHATIDYL CHOLINE W   EJBCAI-0012-0399
ESIS OF ADRENO CORTICAL  MITOCHONDRIAL PROTEINS.=  +THE SYNTH   ENDOAO-0086-0444
SELECTIVE INHIBITION OF  MITOCHONDRIAL SYNTHESIS IN SACCHAROM   JMOBAK-0047-0107
INEA PIG LYMPHOCYTES BY  MITOGENS.=  +REACTIVE FACTOR FROM GU   NATUAS-0225-0236
TIAL SYNCHRONIZATION OF  MITOSES IN TOBACCO CELL SUSPENSIONS    CHDDAT-0270-0320
IC X-IRRA+DEPRESSION OF  MITOSIS BY GAMMA GLOBULIN FROM SPLEN   IJRBA3-0016-0501
ON OF DNA SYNTHESIS AND  MITOSIS.=     +CELL FUSION.  REGULATI  NATUAS-0225-0159
EMBRYOS EXPOSED TO DAC+  MITOTIC ABNORMALITIES IN SEA URCHIN    PNASA6-0064-0857
SINUSES.  OCCURRENCE OF  MITOTIC FIGURES.=       +THE SPLENIC   BSIBAC-0045-0610
ON OF DNA SYNTHESIS AND  MITOTIC INDEX BY COLCHICINE IN CULTU   ECREAL-0059-0027
DUCED CYTO PATHOLOGY IN  MITOTIC L-CELLS.=  +OF MENGOVIRUS IN   JOVIAM-0005-0262
L FORCE OF AXIAL ROTARY  MIXERS.=             MIXING.  AXIA     CCCCAK-0034-3673
BINED PROCESS OF DIRECT  MIXING AND HEAT TRANSFER THROUGH       KOKKAI-70-01-025
THE SIGN OF THE HEAT OF  MIXING ENTHALPY FUNCTION.=+BASED ON    ZSTKAI-0010-0977
IMUM NON LINEAR OPTICAL  MIXING IN N-INDIUM ANTIMONIDE.=  OPT   SSCOA4-0008-0117
SENIORITY              MIXING IN ODD-ODD NUCLEI.=            NCLTAX-0003-0098
```

Figure 6-3.

2. Look through the alphabetically arranged titles until the key
   words are found (Fig. 6-3).
3. Note the code terms:

FEBLAL–0005–0331

4. Find the explanations of FEBLAL in the alphabetically
   arranged final section of *Chemical Titles*:

FEBLAL = Federation European Biochemical Society
Letters

0005 = Volume 5

0331 = page 331

```
0747    BODYAGIN DA        SYRKIN AB        LUPINOSOV UV        ZAITSEVA LA
        EFFECT OF HERBICIDES, CHLORO PHEN OXY ACETIC ACID DERIVATIVES, ON
        MAN AND ANIMALS.  REVIEW OF LITERATURE.        747-51
```

FEBLAL   FEBS (Fed. Eur. Biochem. Soc. Lett.), 5, No. 5 (1969)

```
0305    ROSE SPR
        NEUROCHEMICAL CORRELATES OF LEARNING AND ENVIRONMENTAL CHANGE.
                                                     305-12

0313    ALBRECHT JA        ROZENBOOM W        VERMEER C        BOSCH L
        FACTOR REQUIREMENT OF FORMYL METHIONYL TRANSFER RNA BINDING TO
        ESCHERICHIA COLI RIBOSOMES PROGRAMMED WITH A PLANT VIRAL RNA OR A
        PHAGE RNA.                                    313-15

0316    BACHER A           BAUR R             OLTMANNS O        LINGENS F
        BIOSYNTHESIS OF RIBO FLAVINE MUTANTS ACCUMULATING
        6-HYDROXY-2,4,5-TRI AMINO PYRIMIDINE.        316-18

0319    SIES H             BRAUSER B          BUECHER T
        STATE OF MITOCHONDRIA IN PERFUSED LIVER.  ACTION OF SODIUM AZIDE ON
        RESPIRATORY CARRIERS AND RESPIRATION.        319-23

0324    STROSBERG AD       KANAREK L
        IMMUNOCHEMICAL STUDIES ON HEN/S EGG WHITE LYSOZYME. EFFECT OF
        FORMYLATION OF THE TRYPTOPHAN RESIDUES.      324-6

0327    AHRENS ML          MAASS G            SCHUSTER P        WINKLER H
        KINETIC BEHAVIOR OF VITAMIN B(6) COMPOUNDS.  HYDRATION AND PROTON
        TRANSFER.                                    327-30

0331    MANGER JR
        EFFECT OF TRI ETHYL TIN ON MITOCHONDRIAL ION ACCUMULATION.
                                                     331-4
```
_____
```
0335    GROB EC            EICHENBERGER W
        GALACTO LIPIDS AND FATTY ACIDS IN NORMAL AND GLUCOSE STIMULATED
        SPIRODELA OLIGORRHIZA CULTURES.              335-7

0338    ALKONYI I          SANDOR A
        B-METHYL-DELTA-KETO-DELTA(A,B)-HEXENOIC ACID AND MESITYL OXIDE AS
        ACETYL DONORS IN THE ENZYMIC SYNTHESIS OF ACETYL CHOLINE.
                                                     338-40
```

Figure 6-4.

5. Record the bibliographical information in the following way:
author, title, journal name, volume number, page number, year.
This is called *listing* the reference.

> Manger, J. R., "Effect of Tri Ethyl Tin on Mitochondrial Ion Accumulation," Fed. Eur. Biochem. Soc. Lett., *5*, p. 331 (1969).

Example 2

Find a recent article by J. R. Manger on mitochondria.
1. Find Manger in the alphabetical author listing.

```
              RRCHAX-0014-1231    MALYSZKO E  BAPCAQ-0017-0605
MACBRIDE JAH                      MALYSZKO J  BAPCAQ-0017-0605
              TELEAY-1970-0057    MALYUKOV BA ZVDLAU-0035-1448
MACCHI G      BSIBAC-0045-0634    MAMDAPUR VR NATWAY-0057-0040B
MACCHIA B     ANCRAI-0059-1176    MAMEDOV SV  IAFMAF-69-03-011
              CINMAB-0051-1391    MAMMEL WL   MTGTBF-0001-0357
MACCHIA F     ANCRAI-0059-1176    MAMMI M     NATUAS-0225-0380
              CINMAB-0051-1391    MAMONTOV AP IVUFAC-12-12-143
MACDONALD DD                      MAMYKIN PS  ZPKHAB-0042-2829
              JCPSA6-0052-1017    MANATAUOV D VANKAM-25-11-058
MACDONALD J   APOPAI-0009-0035    MANCHERON D APFRAD-0027-0469
MACDONALD SF                      MANCHESTER FD
              CJCHAG-0048-0139                JCPSA6-0051-5437
MACDONALD WM                      MANCINO D   JOIMA3-0104-0224
              PHRVAO-0187-1733    MANCZINGER J
MACH R        PSSABA-0001-005K                POPTAE-0013-0189
MACHI S       JAPNAB-0013-2657    MANDEL LR   CNREA8-029A-2229
MACK E        JCINAO-0049-0232    MANDEL M    CHPLBC-0004-0482
MACKAY D      CJCHAG-0048-0081    MANG HJ     ZEPYAA-0231-0010
MACKELLAR FA                                  ZEPYAA-0231-0026
              JACSAT-0092-0417    MANGALAM MJ BJAPAJ-0002-1643
MACKENZIE AF                      MANGE M     CHALA4-0052-0057
              SOSCAK-0109-0049    MANGER JR   FEBLAL-0005-0331
MACKENZIE JR                      MANGO FD    ADCAAX-0020-0291
              JINCAO-0032-0043    MANGONI L   CINMAB-0051-1382
MACKINNEY G   PHPLAI-0022-1282                CINMAB-0051-1383
MACKINTOSH AR                                 TELEAY-1969-5235
              SSCOA4-0008-0121    MANIV S     PHRVAO-0187-0403
MACKNIGHT WJ                      MANN DL     PNASA6-0064-1380
              ANYAA9-0159-0267    MANN DR     PHPLAI-0022-1139
MACMILLAN BG                      MANN LG     NUPABL-0140-0598
              ANYAA9-0150-0966    MANN SO     BJNUAV-0024-0157
MACMILLAN J   JCSPAC-1970-0337    MANNAFOV TG ZSTKAI-0010-1124
MACNEIL KAG INUCAF-0005-1009      MANNERING GJ
MACOME JC    STEDAM-0015-0181                BCPCA6-0018-2759
```

Figure 6-5.

2. The author's name leads to the FEBLAL code which leads to the title of the journal article.

Biological Abstracts and the Bioresearch Index provide leads to more than 200,000 articles annually from over 7000 journals from nearly 100 countries. Biological Abstracts also provides photocopies, back issues, and microfilm.

The information available in these publications are:
1. Subject index based on key words
2. Author index
3. Cross index to coordinate many related topics
4. An abstract of full length articles appearing in current journals

The index is arranged alphabetically by key words and modifying terms. The actual end of the the the title is noted by a slash (/) (Fig. 6-6).

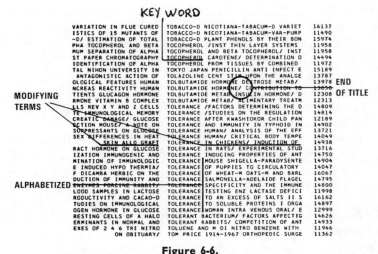

Figure 6-6.

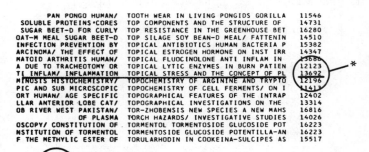

13692. SELYE HANS, ARPAD SOMOGYI, and PAL VEGH. (Inst. Med. and Chir. Exp., Univ., Montreal, Can.) Inflammation, topical stress and the concept of pluricausal disease. In: Upjohn Company: Proceedings of a conference on chemical biology of inflammation, Augusta, Mich., May 31-June 2, 1967. BIOCHEM PHARMACOL Suppl. March. 107-122. Illus. 1968.--Eleven experimental models of acute connective-tissue reactions were designed for comparative investigations in the rat: anaphylactoid edema, ischemic necrosis, formalin-induced pedal inflammation, various forms of calciphylaxis and calcergy, thrombohemorrhagic phenomena (THP), acute conditioned necrosis (ACN), and delayed tissue-clearance of dyes. Several of these reactions can only be produced by the conjoint application of 2 drugs (a conditioner and a challenger), both of which are inactive in

Figure 6-7.

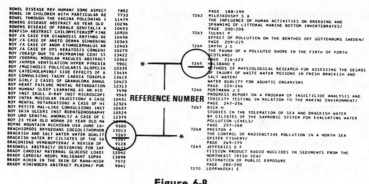

Figure 6-8.

The code number relates to an article abstract in Biological Abstracts (Fig. 6-7).

The reference number in the index leads to the author (Fig. 6-8).

The cross index has the code number of many articles related to a single major subject. One article may appear under several subject headings.

Example 3

An article by Samoilenko discusses methods for determining the nucleotide composition of DNA. Find the abstract of the article and cross references.

1. List the key words:

DNA

Nucleotide

2. Find title in index:

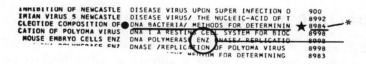

Figure 6-9.

3. Locate abstract of Samoilenko's article by use of code number (Fig. 6-10).

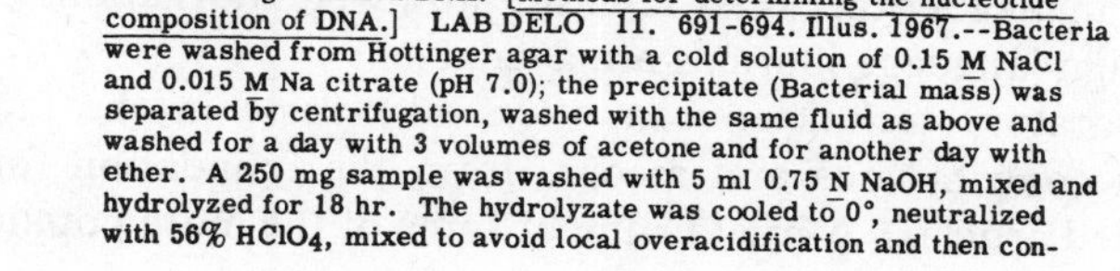

8984. SAMOILENKO, I. I. (N. F. Gamaleya Inst. Epidemiol. Microbiol., Acad. Med. Sci. USSR, Moscow, USSR.) O metodakh opredeleniya nukleotidnogo sostava DNK. [Methods for determining the nucleotide composition of DNA.] LAB DELO 11. 691-694. Illus. 1967.--Bacteria were washed from Hottinger agar with a cold solution of 0.15 M NaCl and 0.015 M Na citrate (pH 7.0); the precipitate (Bacterial mass) was separated by centrifugation, washed with the same fluid as above and washed for a day with 3 volumes of acetone and for another day with ether. A 250 mg sample was washed with 5 ml 0.75 N NaOH, mixed and hydrolyzed for 18 hr. The hydrolyzate was cooled to 0°, neutralized with 56% HClO4, mixed to avoid local overacidification and then con-

Figure 6-10.

4. Find cross references in the cross index section of the journal.

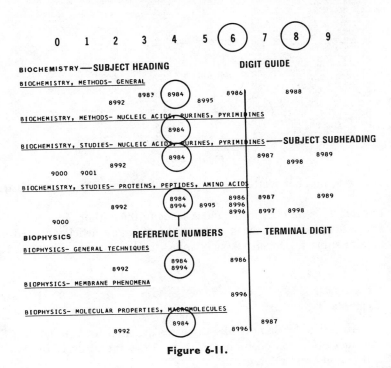

Figure 6-11.

5. The cross index will lead to the author in those cases where he (or they) may not be known.

SPECIALIZED SOURCES

After encyclopedias, general textbooks, popular reviews, and title indexes have provided the researcher with an adequate bibliography, he is ready to review more specialized sources. This often includes original articles which are called **primary sources**. Specialized sources include:

Scientific journals
Monographs
Symposia proceedings
Specialized books
Technical dictionaries and indexes
Data handbooks
Scientific equipment catalogues

Scientific Journals

It is practically impossible to include a complete listing of all the world's reputable scientific journals. There are hundreds of journals, and many of them are in languages other than English; however, translations of the most popular foreign language journals are available.

The laboratory technician can often take pride in the fact that his experimental contributions and records form the foundation for many of these published articles. A brief listing of some of the more popular journals are:

Journal of Biological Chemistry
Analytical Biochemistry
Biochimica et Biophysica Acta
Archives of Biochemistry and Biophysics
Biochemical Journal
Journal of Chemical Physics
Journal of Physical Chemistry
Journal of the American Chemical Society

Monographs

A monograph is usually a lengthy article by a single author which relates the results of many experiments dealing with a particular area of scientific investigation. For example, a monograph describing many experiments with respiratory inhibitors might be organized in the following way:

Title: Inhibitors of Biological Oxidation
By: Author's name and laboratory affiliation
Contents: I. Historical review
        II. Methods
       III. Experimental results
          A. Metal ions
          B. Rotenone
          C. Sodium Amytal
          D. Dinitrophenol
          E. Cyanide and sulfide ion
       IV. Results
        V. Bibliography

The monograph, if it is a significant contribution, often becomes a chapter in a textbook of recent advances in a particular field. Ultimately, the monograph may be expanded into a specialized text.

Symposia

A symposium is a meeting of scientists in which there is a sharing of ideas and readings of recent papers in the process of publication. These

```
   QH            BIOLOGICAL CHEMISTRY
   325     International Symposium on the Origin of Life on
   .15        the Earth, Moscow, 1957.
   1957        Proceedings. Edited for the Academy of Sciences
            of the U.S.S.R. by A.I. Oparin [and others]
         English-French-German ed.,edited for the Interna-
         tional Union of Biochemistry by F.Clark and R.L.M.
         Synge. London, New York, Pergamon Press, 1959.
          xv, 691 p. illus., maps,diagrs.,tables.  26 cm.
          (I.U.B. symposium series, v.1)
          Half title:The origin of life on the earth.
         Includes bibliographies.
         1. Life-Origin.  2.Biological chemistry.
         I.Oparin, Aleksandr Ivanovich,1894- ed.
         (Series:International Union of Biochemistry.
         Symposium  series, v.1)
   QH325.I 5   1957              577            59-12060
```

Figure 6-12.

meetings subject an investigator's work to the critical evaluation of the assembled group. The recorded minutes of these meetings include the presentations of the scientists and the questions, answers, and suggestions that follow. These minutes are usually published and so become an excellent up-to-date reference (see Fig. 6-12).

A few more examples of such annual publications are:

Ciba Foundation Symposium on . . .
Proceedings of the National Academy of Science
Federation Proceedings

Technical Dictionaries and Indexes

There are very important dictionaries in areas where scientists are faced with the chaotic situation of thousands of names, synonyms, formulas, and abbreviations. Indexes provide much the same kind of aid in helping a researcher sift, organize, and coordinate data. Some examples of this kind of reference are:

Dictionary of Organic Compounds
Chemical Elements and their Compounds
Concise Chemical and Technical Dictionary
Stedman's Medical Dictionary
Dictionary of Mathematics

One of the most notable index type reference texts is the *Merck Index*, published by Merck and Company. This index has an extremely comprehensive listing of organic compounds related to the pharmaceutical industry. It contains formulas, synonyms ranging from formal to trivial (common) names, uses, dosage, effects, and side effects.

Data Handbooks

Handbooks are similar to indexes and dictionaries in that they contain a huge amount of organized factual data. The physical properties of thousands of substances are readily available in some of the well-indexed handbooks listed below:

SOCMA Handbook of Commercial Organic Chemical Names
Lange's Handbook of Chemistry
Handbook of Chemistry and Physics
Mathematics Handbook
Electronics and Nucleonics Handbook

SUMMARY

Knowing how to find information is a necessary and important skill for a laboratory technician. He should be able to walk into a library, armed with a pen and a stack of 5 × 8 cards, and proceed systematically from the broad topic, the specific topic, and any authors' names. The 5 × 8 card may have a variety of arrangements. One possibility is illustrated in Figure 6-13.

AUTHOR _____ Ref. No. _____

TITLE _____

_____

SOURCE _____

_____

KEY:

Card No. _____

Figure 6-13.

Keep a list of reference sources as a guide through the literature jungle. Proceed from the general to the specific:

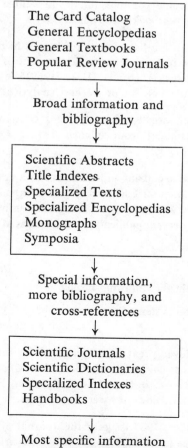

The Card Catalog
General Encyclopedias
General Textbooks
Popular Review Journals

↓

Broad information and
bibliography

↓

Scientific Abstracts
Title Indexes
Specialized Texts
Specialized Encyclopedias
Monographs
Symposia

↓

Special information,
more bibliography, and
cross-references

↓

Scientific Journals
Scientific Dictionaries
Specialized Indexes
Handbooks

↓

Most specific information
and primary sources

PROBLEMS

Set A

1. Define or explain the following terms:

a. abstract

b. symposium

c. monograph

d. bibliography

2. How would you record the following text if it were used as a reference? A second edition book called *Preparative Methods of Polymer Chemistry* was published in 1968 by the New York company of John Wiley and Sons, Inc. It was written by Wayne R. Sorenson and Tod W. Campbell.

3. Use the approved method for citing the following article from the Journal of Biological Chemistry. In issue number one of volume number 238, published in January 1963, Britton Chance and Gunnar Hollunger wrote an article called "Inhibition of Electron and Energy Transfer in Mitochondria." The article appeared on page 418.

4. What LC code would you look for if you wanted to locate the variety of analytical chemistry textbooks in your library?

5. Arrange the following list of four references in the probable order of most recent information to least recent: (a) Journal of Chemical Education (Dec., 1969), (b) Chemical Abstracts (Dec., 1969), (c) Scientific American (Apr., 1970), 8th ed., (d) Vol. VI of Recent Analytical Methods (1970).

6. Assume that you are asked to find a recent method for extracting Coenzyme Q (Plastoquinone) from spinach leaves. Outline your progress through a library search.

7. Assume that you are about to participate in a research effort to determine the effect of alkaloids on Escherichia coli metabolism. How would you go about using the library to gain some background information for your own benefit and to review any recent publications as a means of avoiding duplication?

Set B

8. Define or explain the following terms:

a. LC classification system

b. cross-reference

c. primary source

d. cite a reference source

9. Record the following text as it should be for inclusion into your own bibliography. A second edition textbook called *Chemical Principles* was published in 1969 by the W. B. Saunders Company of Philadelphia. William L. Masterton and Emil J. Slowinski were co-authors.

10. In the December, 1967 issue of the Journal of Chemical Education, volume number 44, issue number 12, W. J. leNoble's article called "Reactions in Solution under Pressure" appeared on page 729. Cite this article correctly.

11. What section of the library would you explore to find some information about EDTA as a chelating agent?

12. Arrange the following references in the probable order of the most recent to the oldest information source: (a) Science (Mar., 1969), (b) Nature (Jan., 1970), (c) Federation Proceedings (Jan., 1969), (d) Biochemical Journal (Jan., 1969).

13. If you were asked to find information about the etiology of hepatitis, where would this take you in the library?

14. Given the key words "brain" and "rat," use a current issue of Chemical Titles to list the related titles and authors.

SUPPLEMENTARY READING

Burman, C. R.: *How to Find Out in Chemistry*, 2nd ed. Pergamon Press Ltd., London, 1966.

An extensive discussion of information retrieval that should be on every laboratory technician's shelf.

Bottle, R. T.: *The Use of Chemical Literature*, 2nd ed. Archon Books, London, 1969.

Another excellent text for sources and methods of information retrieval.

Bourne, C. P.: *Methods of Information Handling*. J. Wiley & Sons, Inc., New York, 1966.

A fine general guide to methods of information retrieval.

7 Solutions (I)

It is very useful to define a solution as a homogeneous mixture of a dissolved substance, called a **solute**, and a medium in which the solute is evenly dispersed, called the **solvent**. The particles of solute that move about randomly in the solvent are submicroscopic in size. If the particle sizes are large enough to reflect light, we call the system a **colloid** rather than a solution. The solute particles will not settle to the bottom of a container, nor can they be filtered out by ordinary filter paper or animal membranes. However, since solutions are mixtures, the solute and solvent components

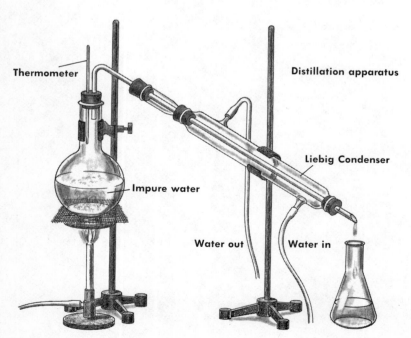

Thermometer

Distillation apparatus

Liebig Condenser

Impure water

Water out   Water in

Figure 7-1.

can be separated by the simple physical means of evaporating the part that has the lower boiling point. Distillation is one very common method of accomplishing this kind of separation (Fig. 7-1).

SOLUBILITY AND MISCIBILITY

If the solute is a solid and the solvent is a liquid, we are concerned about the **solubility** data. Established tables of water solubilities of hundreds of solid substances, both inorganic and organic, may be found in chemistry handbooks and reference texts. These solubilities are usually expressed as the maximum number of grams of a particular solute that can be dissolved in 100 g of water at a stated temperature. Very often, the solubility of a compound may be increased by heating. It seems that the particles of solute (ions or molecules) can more easily overcome the forces binding them in the crystal if they can absorb the additional energy supplied by the external heat source (Fig. 7-2). Grinding the crystals or stirring does not increase the solubility; it merely speeds up the dissolving process by putting more surface area in contact with the solvent as it acts on the crystal to pull it apart. This has been observed as the natural tendency of orderly geometric structures to move toward a greater randomness, or disorder (Fig. 7-3).

When the components of a proposed solution are liquids, there is concern for the "mixability." The extent to which one liquid will dissolve in another is described as the degree of **miscibility**. When two liquids, such as oil and water, are obviously not miscible, a **phase separation** can be observed (Fig. 7-4). A phase separation between liquids can usually be observed

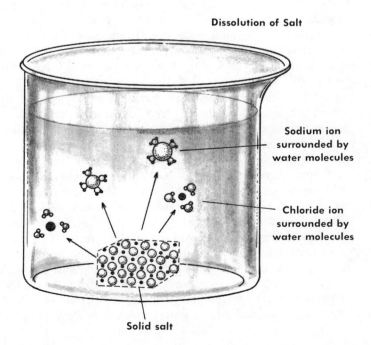

Dissolution of Salt

Sodium ion surrounded by water molecules

Chloride ion surrounded by water molecules

Figure 7-2.

Solid salt

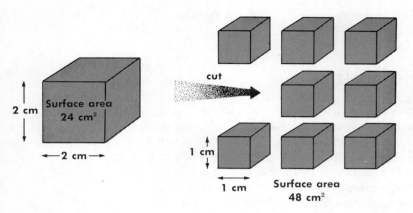

Figure 7-3. The surface area increases as the particle size diminishes.

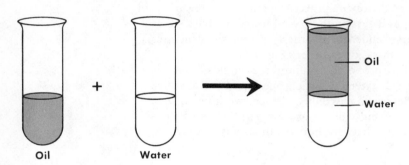

Figure 7-4. The phase separation of immiscible liquids.

as two distinct layers because of the different light refractive properties of the liquids. Immiscibility can often be explained on the basis of structural differences of the liquid molecules. In the case of the oil and water system, the attraction that water molecules have for each other is much greater than the attraction that water molecules have for oil molecules (Fig. 7-5).

The liquid phase separation naturally produces a top layer and a lower layer, the order depending on the comparative densities. In the oil-water separation, the water will be the lower phase, since it is slightly denser than oil. If an even denser liquid, such as carbon tetrachloride, were poured first into the tube, we would observe a three phase system (Fig. 7-6).

Some liquids (alcohol or acetone for example) are somewhat miscible with both water and oil. This kind of behavior allows some very useful flexibility in performing separations and extractions in the laboratory. An illustration of this procedure is found in a method for extracting leaf pigments, principally the green chlorophylls, which are not water soluble. Suppose a concentrated solution of these pigments is to be prepared in petroleum ether (which is not miscible with water). The first step would be to grind the leaves in acetone or alcohol, either one being a suitable solvent for the pigments. Using a separatory funnel, petroleum ether is then added. The next step is the addition of water, which is miscible with acetone or alcohol

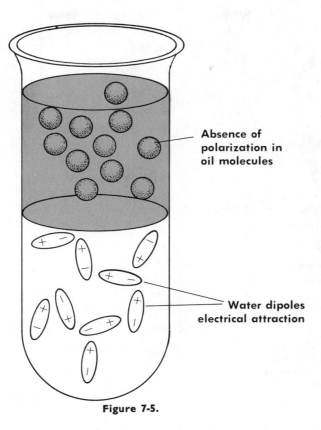

Absence of
polarization in
oil molecules

Water dipoles
electrical attraction

Figure 7-5.

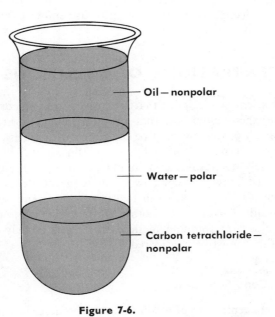

Oil — nonpolar

Water — polar

Carbon tetrachloride —
nonpolar

Figure 7-6.

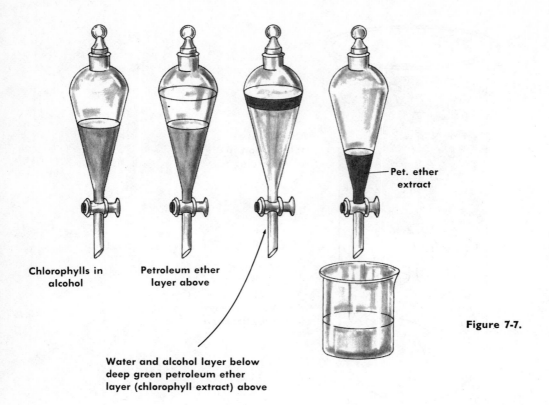

Chlorophylls in
alcohol

Petroleum ether
layer above

Pet. ether
extract

Water and alcohol layer below
deep green petroleum ether
layer (chlorophyll extract) above

Figure 7-7.

but not with the pigments. The net result is the "washing out" of the pigments from the water-loaded phase into the water-free petroleum ether. The actual separation is easily done by opening the stopcock until the water and alcohol (or acetone) solution is removed (Fig. 7-7).

CONCENTRATIONS OF SOLUTIONS

Any scientific discipline that is involved in the recording and communication of experimental data that are measurable and in a form suitable for mathematical processing must be concerned with precision. It is not enough to operate in terms of "weak," "strong," "medium," "small," and "large." These loosely qualitative terms might be adequate for the kitchen, but the chemist must know *precisely* how strong (concentrated) or how weak (dilute). One chemist's "weak" acid, in relation to the metal he is cleaning, might well be deadly "strong" to another chemist working with living organisms. There are several common methods for quantitatively describing solution strengths that must be considered.

Per Cent Solutions

The concentration of a solution may be described in terms of **per cent of solute**. While this method is not the most common or most useful, it is necessary to consider it because it is still encountered in laboratory experience. For example, a 3 per cent solution of iodine in alcohol is a solution containing 3 g of iodine per 97 g of alcohol.*

Commercially prepared chemicals often give us the percentage of solute and the specific gravity of the solution. From this we can calculate the weight of solute in any volume of solution.

Example 1

A label on a bottle of nitric acid reads 60 per cent HNO_3 in water and specific gravity = 1.2. Calculate the weight of HNO_3 in 300 ml of solution.

1. Convert specific gravity to density units: 1.2 g ml^{-1}
2. Multiply by 300 ml: 300 ml (1.2 g ml^{-1}) = 360 g
3. The solution is only 60 per cent HNO_3: 360 g (0.60) = 216 g of HNO_3 in 300 ml of solution.

Example 2

If a solution is prepared by dissolving 4 g of potassium permanganate in 640 g of water, how is this expressed as a per cent solution?

1. The solute is what per cent of total (4 g solute + 640 g solvent = 644 g solution)?

$$4 g = ? \text{ per cent of } 644 g$$

2. $4 g = 644 gX$

$$X = \frac{4}{644} = 0.062$$

3. $0.062 \times 100 = 6.2$ per cent.

Example 3

How does one prepare 200 ml of a 5 per cent table salt solution?

1. On the basis of water density being 1 g ml^{-1}, we know that a 5 per cent solution is 5 g in 95 ml of water or 10 g in 190 g of water.

---

* This example is described as a *w/w* per cent solution. It means that both the solute and solvent are measured by weight. It is sometimes more useful or convenient to prepare solutions in a weight to volume ratio (*w/v*) or volume to volume ratio (*v/v*). However, it is essential that the ratio type used be clearly indicated on the label.

2. Since this total volume is less than 200 ml (about 194 ml), the weight is scaled upwards proportionately:

$$\frac{10 \text{ g salt}}{194 \text{ ml solution}} = \frac{X \text{ g salt}}{200 \text{ ml solution}}$$

$$X = 10.3 \text{ g salt}$$

3. Therefore, place 10.3 g of salt in a volumetric flask and add enough water to bring the volume up to 200 ml.

Molar Solutions

The most useful form of expressing concentrations of solutions for everyday laboratory activity is in terms of moles of solute per liter of solution. Solutions of this type are called **molar**. You will recall from Chapter 2 that a mole of compound is effectively its molecular weight (formula weight) expressed in grams. The tremendous advantage of this approach is that it allows us to know, with considerable precision, the actual number of particles (ions or molecules) per unit of solution volume. This method of expressing concentration is also consistent with the modern chemist's view of the mole as being a central and unifying concept in chemistry. The symbol for molarity is M and the relationship of molarity, mole, and volume can be stated in the following manner:

$$M = nV^{-1} \text{ (molarity = moles per liter)}$$

where M is molarity, n is the number of moles of solute, and V is the volume of solution in liters.

This relationship can be used to calculate the number of moles of solute in a solution by rearranging the first equation.

$$n = MV \text{ (moles = moles } \cancel{\ell}^{-1}(\cancel{\ell}))$$

Example 4

Prepare a liter of 1 M (one molar) sodium hydroxide solution.
1. Calculate the weight of one mole of NaOH:
   Na = atomic weight 23
   O = atomic weight 16
   H = atomic weight 1
        formula weight 40
   40 g = 1 mole NaOH

2. Since molarity = moles per liter ($M = nV^{-1}$) and moles = $\dfrac{\text{g solute}}{\text{g mole}^{-1}}$, we substitute:

$$\text{molarity} = \frac{\text{g}}{\text{g mole}^{-1}} (\ell^{-1})$$

3. To find grams, rearrange the equation:
   g = molarity (g mole⁻¹) (liters)
   g of NaOH = (1 mole X⁻¹) (40 g mole⁻¹)
   g of NaOH = 40

4. Therefore, place 40 g of NaOH in a beaker, add about 800 ml of water and stir until dissolved. Finally, add enough water to bring the volume of solution up to one liter.

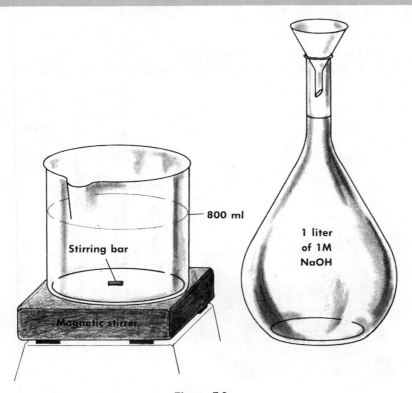

Figure 7-8.

Observe in Example 4 that the relatively simple device for calculating the grams of solute needed to prepare a solution of specific molar concentration is to take the gram-molecular weight (i.e., the weight of one mole) and modify this by the desired molarity and further modify the weight by the volume (in liters) desired.

$$\underset{\text{(solute)}}{g} = \underset{\text{(weight of one mole)}}{\text{molecular weight}} \underset{}{\text{(molarity)}} \underset{\text{(volume)}}{\text{(liters)}}$$

This all-purpose equation, devoid of any mysterious derivation, may be used to solve the host of practical problems that a technician can expect to face.

Example 5

Prepare 250 ml of a 0.2 M solution of NaOH.
1. Molecular weight of NaOH = 40.
2. 250 ml = 0.25 liter.
3. Substitute in the equation:
   g = molecular weight (molarity) (liters)
   g = 40(0.2)(0.25)
   g = 2

4. Dissolve 2 g of NaOH in approximately 200 ml of water.
5. Add enough water to bring the volume of solution to 250 ml.

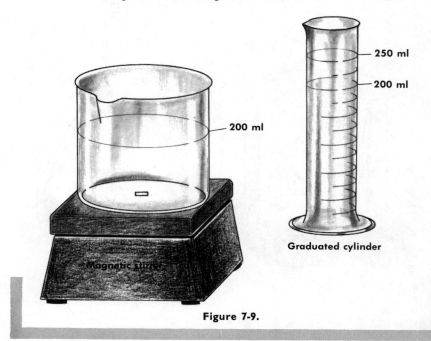

250 ml

200 ml

200 ml

Graduated cylinder

Magnetic stirrer

Figure 7-9.

Example 6

Calculate the number of moles of KCN in 400 ml of 0.5 M solution.
a. Change 400 ml to liters:

$$\frac{400 \text{ ml}}{10^3 \text{ ml } \ell^{-1}} = 0.4 \text{ liters}$$

b. Substitute in the equation:

$$\text{number of moles} = MV$$

$$(0.5 \text{ M})(0.4 \text{ liters}) = 0.2 \text{ moles}$$

Example 7

How many molecules of glucose ($C_6H_{12}O_6$) are there per milliliter in a 2.5 M solution?

a. 2.5 M means 2.5 moles liter^{-1}

b. Use conversion factor 10^3 ml liter^{-1}:

$$\frac{2.5 \text{ moles liter}}{10^3 \text{ ml liter}} = 2.5 \times 10^{-3} \text{ moles ml}^{-1}$$

c. Since there are 6.02×10^{23} molecules mole^{-1}, multiply:

$$2.5 \times 10^{-3} \text{ moles ml}^{-1} (6.02 \times 10^{23} \text{ molecules mole}^{-1})$$
$$= 15 \times 10^{20}, \text{ or } 1.5 \times 10^{21} \text{ molecules ml}^{-1}$$

Example 8

A bottle of sulfuric acid is labeled as 92 per cent by weight and has a specific gravity of 1.83. What is the molarity?

1. Convert specific gravity 1.83 to density-1.83 g ml^{-1}.

2. Use the conversion factor 10^3 ml liter^{-1}:

$$1.83 \text{ g ml}^{-1} (10^3 \text{ ml } \ell^{-1}) = 1830 \text{ g } \ell^{-1}$$

3. Since the acid is 92 per cent by weight:

$$1830 \text{ g } \ell^{-1}(0.92) = 1682 \text{ g } \ell^{-1} H_2SO_4$$

4. Substitute in the equation:

$$M = \frac{g}{\text{g-molec. wt. (liters)}} = \frac{1682}{98(1)} = 17.2 \text{ M}$$

Example 9

Prepare 2 liters of biological saline solution of the following composition:

a. NaCl   0.0120 M
b. $MgSO_4 \cdot 5H_2O$ 0.0034 M
c. KCl   0.0082 M
d. K_2HPO_4 0.0012 M
e. KH_2PO_4 0.0027 M

1. Calculate the weight of a mole of each of the components:
   a. NaCl $(23 + 35.5) = 58.5$ g mole^{-1}
   b. $MgSO_4 \cdot 5H_2O$ $(24 + 32 + 64 + 90) = 210$ g mole^{-1}
   c. KCl $(39 + 35.5) = 74.5$ g mole^{-1}
   d. K_2HPO_4 $(78 + 1 + 31 + 64) = 174$ g mole^{-1}
   e. KH_2PO_4 $(39 + 2 + 31 + 64) = 136$ g mole^{-1}

2. In each case, substitute in the equation:

$$g = g \text{ molec. wt. (M) (liters)}$$

    a. $g = 58.5(0.012 \text{ M}) (2) = 1.40 \text{ g}$
    b. $g = 210(0.0034 \text{ M}) (2) = 1.43 \text{ g}$
    c. $g = 74.5(0.0082 \text{ M}) (2) = 1.22 \text{ g}$
    d. $g = 174(0.0012 \text{ M}) (2) = 0.42 \text{ g}$
    e. $g = 136(0.0027 \text{ M}) (2) = 0.73 \text{ g}$

3. Each of the weights listed in Step 2 are placed in a 2 liter volumetric flask and water is added to the final volume.

The preparation of molar solutions from solid reagents quickly becomes a routine operation, requiring little more than care in substituting data in the following equation:

$$g \text{ (solute)} = g\text{-molec. wt. (M) (V)}$$

Dilution Methods

Many solutions in a research laboratory environment are made from existing stock solutions. For example, the technician may be asked to prepare a certain volume of a dilute concentration of a given solution from a bottle of concentrated stock. A commercial bottle of hydrochloric acid may be 18 M, while the concentration to be used should be 0.1 M. The objective, then, is to solve practical *dilution problems*.

It must be recognized first that a dilution process changes only the volume of solvent while the actual number of molecules, or preferably stated as the number of moles, remains the same. This means that the initial (1) number of moles of solute must equal the final (2) number of moles of solute.

$$\text{number of moles}_1 = \text{number of moles}_2$$

Rearranging the original "all-purpose" equation produces the following equation:

$$\frac{g \text{ (solute)}}{g\text{-molec. wt.}} = MV$$

Remembering that

$$\frac{g \text{ (solute)}}{g\text{-molec. wt.}} = \text{number of moles}$$

substitute:

$$\text{number of moles} = MV$$

Going back to the original equation again, substitute equals for equals. The following equation is the result:

$$M_1 V_1 = M_2 V_2$$

In other words, we are saying that the product of the initial molarity and volume must equal the product of the final molarity and volume. This is exactly the same as saying that it is not the number of moles of solute that is being changed, it is only the amount of water in which the molecules or ions are "swimming" that is being changed.

Example 10

Prepare 500 ml of 0.2 M KCl solution from 1 M stock.
1. How many milliliters of concentrated stock must be diluted with water to give a final volume of 500 ml and a concentration of 0.2 M?

2. Substitute the data in the equation. If the volume in milliliter units is used, the answer will be in milliliters:

$$M_1V_1 = M_2V_2$$
$$(1 \text{ M})(V_1 \text{ ml}) = (0.2 \text{ M})(500 \text{ ml})$$
$$V_1 = 100 \text{ ml}$$

3. Pour 100 ml of 1 M stock into a 500 ml cylinder and add enough water to bring it to the desired volume and concentration. (Note: it is actually preferable to use a volumetric flask rather than a graduated cylinder.)

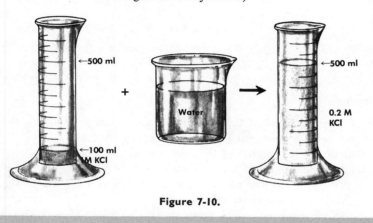

Figure 7-10.

Example 11

An experiment requires the preparation of 300 ml of 0.25 mM (millimolar) KCN solution from 0.2 M stock.
1. Change the concentration units so that they are compatible, i.e., both units molar *or* millimolar:

$$0.2 \text{ mole liter}^{-1} (10^3 \text{ millimoles mole}^{-1})$$
$$200.0 \text{ millimoles liter}^{-1} = 200.0 \text{ mM}$$

2. Substitute in the equations:

$$M_1V_1 = M_2V_2$$
$$(2 \times 10^2 \text{ mM})(V_1) = (2.5 \times 10^{-1} \text{ mM})(300 \text{ ml})$$
$$V_1 = \frac{(2.5 \times 10^{-1} \text{ mM})(300 \text{ ml})}{(2 \times 10^2 \text{ mM})}$$
$$V_1 = 377.5 \times 10^{-3} \text{ ml} = 0.38 \text{ ml}$$

3. To accomplish this preparation, use a 1 ml graduated pipet for the transfer of the stock solution to a 300 ml volumetric flask. Add enough water to bring the solution to 300 ml volume.

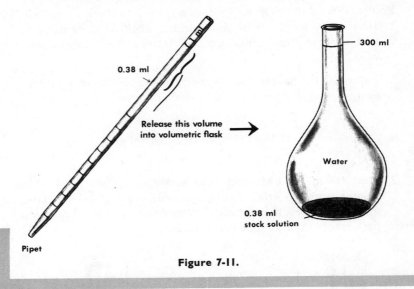

0.38 ml

Release this volume
into volumetric flask

300 ml

Water

0.38 ml
stock solution

Pipet

Figure 7-11.

Example 12

Prepare 500 ml of 0.2 M sulfuric acid from a commercial bottle labeled 96 per cent H_2SO_4, specific gravity 1.84.
1. Calculate the molarity of the commercial acid:
   a. Specific gravity of 1.84 = density of 1.84 g ml^{-1}.
   b. This is the same as 1840 g liter^{-1} multiplied by the conversion factor of 10^3 ml liter^{-1}.
   c. 96 per cent by weight amounts to 1840 gl^{-1} (0.96) = 1764 gl^{-1}.
   d. Using the equation:

$$M = \frac{g \text{ (solute)}}{\text{g-molec. wt. (volume)}} = \frac{1764 \text{ g}}{(98) \text{ g-mole}^{-1}} \text{ (1 liter)}$$

   M = 18 molar (commercial concentration)

2. Now the dilution part:

$$M_1V_1 = M_2V_2$$
$$(18 \text{ M})(V_1 \text{ ml}) = (0.2 \text{ M})(500 \text{ ml})$$
$$V_1 = 5.6 \text{ ml}$$

3. CAUTION! Add 5.6 ml of the 18 M commercial acid to approximately 100 ml of water and then complete the dilution 500 ml. It is important to add the sulfuric acid to water rather than adding water to sulfuric acid because the intense heat produced in the solution process could cause the first drops of water to splatter some acid on the technician.

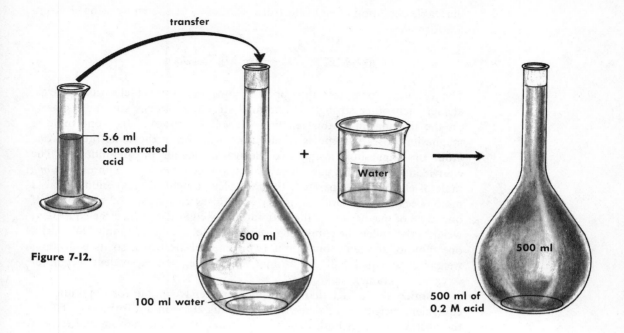

Figure 7-12.

Normal Solutions

There is one more commonly employed method for the preparation of solutions of precise concentration. This third alternative is known as the **normal** solution, usually abbreviated as N. Although the normality method is not as generally useful as molarity, it does have some advantages in the controlling of acid-base neutralization, especially when the stoichiometry is not known. The procedure used in acid-base control is known as **titration**.

Normality may be defined as the **number of equivalent weights**; or, more simply, as the number of **equivalents** of solute per liter of solution

$$N = \text{number of equivalents liter}^{-1}$$

In order to make this concept of equivalents meaningful, further explanation might be in order. One could start by saying that the equivalent weight of any substance is that weight which may gain or lose one mole of electrons.

For example, if magnesium is burned (rapid oxidation) the reaction can be represented by this equation:

$$Mg + \tfrac{1}{2}O_2 \rightarrow MgO$$

This means that 1 mole of magnesium combines with one half mole of oxygen to produce 1 mole of magnesium oxide. Examine the mechanism of electron loss and gain on an atomic level. Observe that magnesium has two valence electrons which may be lost with relative ease (low ionization energy). Oxygen has six valence electrons and a strong tendency (high electron affinity) to gain two more which would result in the formation of the stable condition of eight electrons. This may be shown by using electron-dot formulas

$$\text{Mg}\!\overset{\displaystyle\frown}{:}\; + \;\overset{\circ\circ}{\underset{\circ\circ}{\text{O}}}\!{\circ} \;\longrightarrow\; \text{Mg}^{2+}, \; \left[\overset{\circ\circ}{\underset{\circ\circ}{:\text{O}}}\!{\circ}\right]^{2-}$$

The resulting ions, bearing opposite signs in terms of electrical charge, attract each other strongly in the final structure. Focus still more closely on the mechanism of this reaction. It will be observed that one mole of magnesium loses two moles of electrons—this being the same ratio as one atom of magnesium losing two electrons. Recall the definition: "The equivalent weight of a substance is that weight which may gain or lose *one* mole of electrons." Therefore the equivalent weight of magnesium is half a mole, because one mole of magnesium loses *two* moles of electrons. If one mole of magnesium is the gram-atomic weight (24 g) then the equivalent weight (also called the combining weight) of the metal is 12 grams. Similarly, one mole of oxygen atoms gains two moles of electrons, so its equivalent weight must equal half the atomic weight (i.e., the equivalent weight of oxygen is 8 grams).

Notice also a fact that provides an easier technique for obtaining the equivalent weight of an element than writing electron-dot formulas. This is the observation that both the equivalent weights of magnesium and oxygen are obtained if the atomic weight (i.e., the weight of one mole of an element) is divided by the valence—which is *two* in both cases. This observation also applies to calculating the equivalent weight of the compound MgO, since the decomposition would result in the reverse transfer of the same number of electrons. Consequently, it may be concluded that the equivalent weight of a compound can be obtained by dividing the gram-molecular weight (i.e. the weight of one mole of a compound) by the total valence of either the positive or negative member of the compound. The only exception to this method occurs with oxidation-reduction reactions, which are taken up in a later chapter.

The equivalent weight of MgO then is

$$\frac{\text{g-molec. wt.}}{2}$$

As an arbitrary choice, which hopefully reduces any possible confusion, this rule is usually summarized as

$$\text{Equiv. wt.} = \frac{\text{g-molec. wt.}}{\text{total } \textit{positive} \text{ valence}}$$

Substitute in the equation for MgO (24 + 16):

$$\text{Equiv. wt.} = \frac{40}{2} = 20 \text{ g}$$

These data, derived from the observation of the magnesium and oxygen reaction, are found to hold up consistently when tested by many other measurements on other elements and their compounds.

Although the discussion of acids and bases is taken up in the next chapter, it is expedient to describe a simplified model at this time for the purpose of establishing their relationship to normal concentrations. So, let the definition of an acid be a compound that produces hydrogen ions (H⁺)

in water solution and a base, by contrast, a compound producing hydroxide ions (OH^-) in water solution.

Examples of acids:

$$HCl \xrightarrow{\text{in water}} H^+ + Cl^-$$

$$H_2SO_4 \xrightarrow{\text{in water}} 2H^+ + SO_4^{2-}$$

The formula for an acid is usually identified by the symbol for hydrogen coming first.

Examples of bases:

$$NaOH \xrightarrow{\text{in water}} Na^+ + OH^-$$

$$Ca(OH)_2 \xrightarrow{\text{in water}} Ca^{2+} + 2OH^-$$

The formula for a base may be identified by the symbol for hydroxide coming last in the formula.

In the case of an acid, chemists tend to focus on the hydrogen atom and slightly modify the definition to produce:

$$\text{Equiv. wt. (acid)} = \frac{\text{g-molec. wt.}}{\text{number of replaceable hydrogen atoms}}$$

For example,

HCl (1 + 35.5 = 36.5) (one replaceable hydrogen)

$$\text{Equiv. wt.} = \frac{36.5}{1} = 36.5 \text{ g}$$

H_2SO_4 (2 + 32 + 64 = 98) (two replaceable hydrogens)

$$\text{Equiv. wt.} = \frac{98}{2} = 49 \text{ g}$$

H_3PO_4 (3 + 31 + 64 = 98) (three replaceable hydrogens)

$$\text{Equiv. wt.} = \frac{98}{3} = 32.7 \text{ g}$$

The same idea is used in the case of bases where the denominator is changed to hydroxide units instead of hydrogen atoms.

$$\text{Equiv. wt. (base)} = \frac{\text{g-molec. wt.}}{\text{number of replaceable hydroxide units}}$$

Emphasis is placed on hydrogens and hydroxides since the neutralization process directly concerns itself with the combination of hydrogen ion and hydroxide ion in the formation of water. In practice, neutralization may be usefully defined as the formation of water.

$$H^+ + OH^- \rightarrow H_2O$$

The following are some examples of equivalent weight calculation in the case of bases:

KOH $(39 + 16 + 1)$ (one replaceable hydroxide)

$$\text{Equiv. wt.} = \frac{56}{1} = 56 \text{ g}$$

$Ca(OH)_2$ $(40 + 32 + 2)$ (two replaceable hydroxides)

$$\text{Equiv. wt.} = \frac{74}{2} = 37 \text{ g}$$

Now that the question of equivalent weight has been resolved (hopefully), let's go back to the normal solution.

$$\text{Normality} = \left(\begin{array}{c}\text{number of}\\\text{Equiv. wts.}\end{array}\right) \text{ per liter}$$

or

$$N = \text{number of equiv. liter}^{-1}$$

The number of equivalents is found by dividing the weight of one equivalent into the total weight of solute being used.

$$\text{number of equiv.} = \frac{g}{\text{g-equiv. wt.}}$$

For example, KOH has an equivalent weight of 56 grams. If 112 grams are used, then,

$$\text{number of equivalents} = \frac{112 \text{ g total}}{56 \text{ g per } \textit{one} \text{ equivalent wt.}}$$

The number of equivalents will be 2. Now rewrite the original equation:

a. $N = \text{number of equiv. liters}^{-1}$

b. Substitute $\dfrac{g}{\text{g-equiv. wt.}}$ for the number of equivalents:

$$N = \frac{g}{\text{g-equiv. wt. (liters)}}$$

The next step gives a tremendously useful equation that tells how many grams of some solute are needed to prepare a definite volume of solution of a definite normality.

c. $g = (\text{g-equiv. wt.}) (N) (\text{volume in liters})$

Note the remarkable similarity to the equation involving molarity:

$$g = (\text{g-molec. wt.}) (M) (\text{volume in liters})$$

The difference amounts to the substitution of N for M and equivalent weight for molecular weight.

Example 13

Prepare 150 ml of 0.25 N calcium hydroxide solution.
1. Calculate the equivalent weight of calcium hydroxide from the formula $Ca(OH)_2$ $(40 + 32 + 2 = 74)$:

$$\text{Equiv. wt.} = \frac{\text{g-molec. wt.}}{\text{number of replaceable hydroxides}} = \frac{74}{2} = 37 \text{ g}$$

2. Change the volume to liters:

$$\frac{150 \text{ ml}}{10^3 \text{ ml liters}^{-1}} = 0.15 \text{ liters}$$

3. Substitute in the equation:
   g = equiv. wt. (N) (volume in liters)
   g = (37 g)(0.25 N)(0.15 liters)
   g = 0.14 or 140 mg

4. Place the 140 mg of $Ca(OH)_2$ into a 150 ml volumetric flask and add water to the line marked on the flask.

While Example 13 represents a straightforward method for preparing a solution in normality units of concentration, the fact of the matter is that most normal solutions are prepared from compounds that are already in solution rather than solids like $Ca(OH)_2$. The upshot of this fact in terms of practical application means that most preparations will involve *dilution* calculations. Furthermore, it is often necessary to convert molar concentration to normal. The explanation of the dilution process here is really identical to the explanation for molar solutions. The only difference is that the number of *equivalents* of solute is considered as remaining constant instead of *moles* of solute, since the initial (1) number of equivalents equals the final (2) number of equivalents.

$$\text{number equiv.}_1 = \text{number equiv.}_2$$

Remembering that the number of equivalents = NV, substitute to form:

$$N_1V_1 = N_2V_2$$

This equation states that the initial product of normality and volume must equal the final product of normality and volume.

A simplified method for converting molar concentration to normal may be deduced by considering the example of a 1 M solution of sulfuric acid. Remember that one mole of H_2SO_4 (98 g) is equal to two equivalents since the weight of one equivalent (i.e., the equivalent weight) is 98 ÷ 2 (replaceable hydrogens) or 49 g. Then a 1 M solution (containing one mole per liter) is the same as a 2 N solution (containing two equivalents per liter). This means that the normality may be immediately calculated if the molar concentration is multiplied by the total positive valence or the number of replaceable hydrogens or hydroxides. For H_2SO_4,

$$N = (2 \text{ replaceable hydrogens}) (M)$$

For HCl, $N = (1) (M)$, which means that the normality and molarity are identical. For H_3PO_4, $N = (3) (M)$, meaning that a 1 M solution is equal to 3 N.

Example 14

Prepare 200 ml of 0.33 N H_2SO_4 from an 18 M stock.

1. Convert the 18 M stock to normal concentration:

$$N = (2)(18) = 36 \text{ N}$$

2. Substitute data in the dilution equation:

$$N_1V_1 = N_2V_2$$
$$(36 \text{ N})(V_1) = (0.33 \text{ N})(200 \text{ ml})$$
$$V_1 = 18.3 \text{ ml}$$

3. Observe the caution required in handling H_2SO_4 by pouring the 18.3 ml of concentrated stock into a large volume of water before adding the additional water needed to bring the volume of solution up to 200 ml.

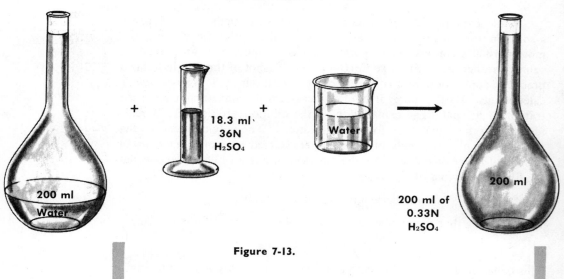

200 ml
Water

+

18.3 ml
36N
H₂SO₄

+

Water

200 ml

200 ml of
0.33N
H₂SO₄

Figure 7-13.

The usefulness of normality as an expression of concentration will be thoroughly investigated in the next chapter. The process of titrating acids against bases in order to determine unknown concentrations and to rigidly control the degree of acidity in a solution is of tremendous practical importance in the laboratory.

PROBLEMS

Set A

1. How would you prepare the following solutions:

a. 35 ml of 2 M $AgNO_3$ from solid $AgNO_3$
b. 160 ml of 0.5 M Na_2SO_4 from solid $Na_2SO_4 \cdot 10H_2O$
c. 450 ml of 0.03 M KOH from solid KOH

2. How do you prepare the following solutions?

a. 200 ml of 10 per cent NaCl using solid NaCl
b. 120 ml of 0.4 N $CaCl_2$ from solid $CaCl_2$
c. 2.5 liters of 0.2 N $MgSO_4$ from solid $MgSO_4 \cdot 5H_2O$

3. What is the weight of copper in 125 ml of 0.1 M $CuSO_4$ solution?

4. What molar concentration is obtained by dissolving 10 g of KOH to a final volume of 200 ml?

5. How do you prepare the following solutions?

a. 500 ml of 0.35 M KCl from 6.0 M stock solution
b. 350 ml of 20 mM KCN from 1.0 M stock solution
c. 50 ml of 0.2 N H_2SO_4 from 1.5 N stock solution

6. Find the molarities and normalities of the following solutions:

a. HNO_3, specific gravity 1.3, 50 per cent by weight
b. H_2SO_4, specific gravity 1.8, 96 per cent by weight
c. $CHCl_3$, specific gravity 0.9, 42 per cent by weight

7. What final concentrations result when 50 ml of 0.2 M $CaCl_2$ is added to 20 ml of 5.0 NaCl?

8. If 20 g of glucose ($C_6H_{12}O_6$) is dissolved in water to a final volume of 400 ml, approximately how many molecules are there in each milliliter?

9. If the density of 6.0 M HCl is 1.1 g ml^{-1}, what is the percentage of HCl by weight?

10. Use the Handbook of Chemistry and Physics to determine the strongest molar concentrations of KIO_3 that can be prepared at 20° C and 100° C.

Set B

11. How would you prepare the following solutions?

a. 50 ml of 0.3 M KCN from solid KCN.
b. 400 ml of 0.04 M $CuSO_4$ from solid $CuSO_4 \cdot 4H_2O$
c. 10 ml of 6.0 M NaOH from solid NaOH

12. How do you prepare the following solutions?

a. 100 ml of 5 per cent KNO_3 using solid KNO_3
b. 750 ml of 1.2 N $Al(NO_3)_3$ from solid $Al(NO_3)_3$
c. 0.025 liters of 40 mM $MgBr_2$ from solid $MgBr_2$

13. What molar concentration results when 4.0 g of KCN is dissolved in water to a final volume of 150 ml?

14. How do you prepare the following solutions?

a. 250 ml of 0.35 M KNO_3 from a 3.2 M stock solution
b. 0.65 liters of 70.0 mM NaOH from a 5.2 M stock solution
c. 3.0 liters of 0.15 N HCl from 6 N stock solution

15. Find the molarities and normalities of the following solutions:

a. HCl, sp. gr. 1.12, 60 per cent by weight
b. H_3PO_4, sp. gr. 1.74, 84 per cent by weight
c. C_2H_5OH, sp. gr. 0.98, 95 per cent by weight

16. What final concentrations result when 120 ml of 0.4 M NaOH is added to 40 ml of 0.8 M KOH?

17. What is the weight of potassium in 50 ml of 0.5 M KCN solution?

18. If 30 g of acetone (C_3H_6O) is dissolved in water to a final volume of 0.5 liters, approximately how many molecules are there in a 10 ml aliquot?

19. If the density of 1.0 M H_2SO_4 is 1.22 g ml^{-1}, what is the percentage of H_2SO_4 by weight?

20. Use the Handbook of Chemistry and Physics to determine the strongest molar concentration of KNO_3 that can be prepared at 100° C.

SUPPLEMENTARY READING

Vanderwerf, C. A.: *Acids, Bases, and the Chemistry of the Covalent Bond.* (paperback) Reinhold Publishing Co., New York, 1961.

A clear presentation of modern acid-base concepts.

Sisler, H. H.: *Chemistry in Non-aqueous Solvents.* (paperback) Reinhold Publishing Co., 1961.

Good discussion of solvents other than water.

Solutions (II) 8

VARIETIES OF SOLVENTS

In the previous chapter, water was used repeatedly as an example of a solvent. Although it is true that water is the most common solvent, it is by no means the only one. The choice of an appropriate solvent depends on the *nature of the solute*. It also depends on what purpose the solubility process is serving. Solvent selection requires consideration of **volatility** (boiling points), **toxicity** (how poisonous they are to handle and breathe), **flammability** (whether heating is accomplished by direct flame, water bath, hot plate, or electric mantle), and the **intended procedure**. "Intended procedure" raises the question of what purpose the solubility process is meant to serve. Are reactants simply put into solution because it is a necessary condition for the reaction to occur, or is the objective to purify a solute by recrystallization? Do we have special adsorption and flow requirements in chromatographic processes? Perhaps we have to effect separations among many solutes in complex systems. Of course, there are still other procedures—enough to discourage a catalogue approach as an answer to the question. The choice of solvent would most likely be indicated in the protocol. If this is not the case, an empirical (trial and error) approach, using very small amounts of solute and solvent, might answer the question. This suggests that milliliter volumes and milligram weights tested under a fume exhaust hood provide small risk. So far as the hazardous aspects of solvents are concerned, a few moments of careful reading of the labels on the containers, as standard procedure, could greatly reduce any risk involved.

From a theoretical point of view, one may think in terms of "like dissolves like." This means a need to consider the geometry, or polarity, of the interacting molecules as discussed in the previous chapter. To summarize the rules of thumb:

1. Polar solutes tend to dissolve in polar solvents (Fig. 8-1).

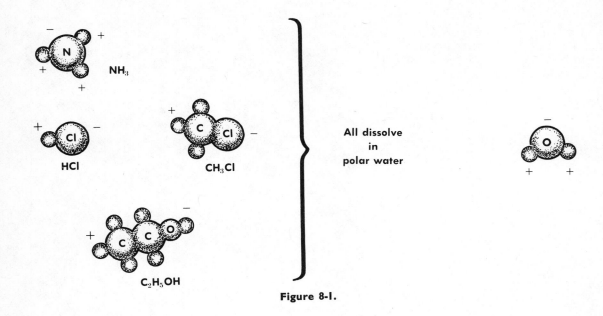

Figure 8-1.

2. Nonpolar solutes tend to dissolve in nonpolar solvents (Fig. 8-2).

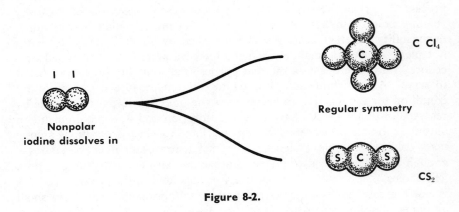

Figure 8-2.

3. Organic compounds tend to dissolve in organic solvents.

There are numerous exceptions to these rules, to be sure. For example, sugar (an organic compound) dissolves extensively in polar water but not in an organic solvent such as carbon tetrachloride. The explanation involves a special case of polarity called **hydrogen bonding**. While this and other exceptions hardly prove the rule, they do not seriously detract from the general usefulness of the "like dissolves like" rule. A list of common solvents might include:

Common Solvents	Boiling Point °C
diethyl ether	35
carbon disulfide	46
acetone	56
chloroform	61
carbon tetrachloride	76
ethanol	78
benzene	80
cyclohexane	81
isopropanol	82
petroleum ether	30–175
toluene	111

Purity of Solvents

When it comes to placing orders for reagent purchase or making a choice for use, economic and practical factors must be considered. Using a very pure and expensive solvent for rough qualitative investigations would be needlessly extravagant. On the other hand, it would be worse to use low grade material for precision work involving sophisticated instruments. Chemical suppliers help by indicating the degree of purity in their catalogues and on their labels (Fig. 8-3). If a spectrophotometer is to be used for some critical analysis, the appropriate label would probably read "spectro quality."

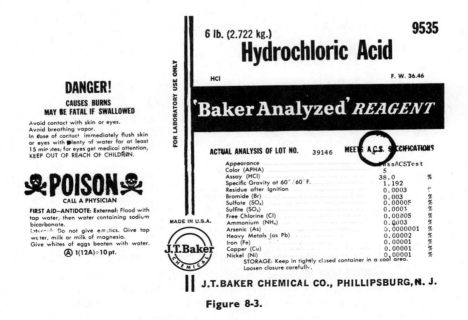

Figure 8-3.

The common grading system in decreasing order of purity is listed.

Abbreviation	Grade
Spectro	Spectroquality—specially purified and controlled for minimal absorption.
Anal. Reagent	Reagent—high purity for analytical work. Per cent impurities listed.
A.C.S.	American Chemical Society—meets A.C.S. high standards for purity.
Baker	J. T. Baker Chemical Co.—high standard for purity. Suitable for most research and general lab use.
C.P.	Chemically Pure—suitable for routine lab work. No analysis of impurities.
U.S.P.	United States Pharmacopeia—satisfactory for routine work.
N.F.	National Formulary—used for routine work.
Practical	Purified low grade organic compounds—suitable for syntheses.
Purified	A step above lowest grade—not suitable for analyses.
Tech.	Technical grade—used mostly for industrial operations. Not suitable for lab use.

ACIDS AND BASES

It is often satisfactory to think of acids and bases as producers of hydrogen ions and hydroxide ions respectively. The formulas of compounds described as acids and bases usually indicate their nature by the positions of H and OH. Here are some examples of common acids.

$\underline{H}Cl$ (hydrochloric acid)

$\underline{H}NO_3$ (nitric acid)

$\underline{H}_2SO_4$ (sulfuric acid)

The following are some examples of bases:

$Na\underline{OH}$ (sodium hydroxide)

$K\underline{OH}$ (potassium hydroxide)

$Ca(\underline{OH})_2$ (calcium hydroxide)

Acids may be described as sour tasting, corrosive to metals, and as compounds that change blue litmus paper to red. Bases are said to be bitter in taste, slippery to touch, and are compounds causing red litmus paper to turn blue. These descriptions are fine—most of the time. But what does the chemist mean when he calls cyanide ion a strong base, aluminum ion an acid, or hydrofluoric acid (which notoriously attacks glass or skin) a *weak* acid? These random examples suggest a broader definition of acids and

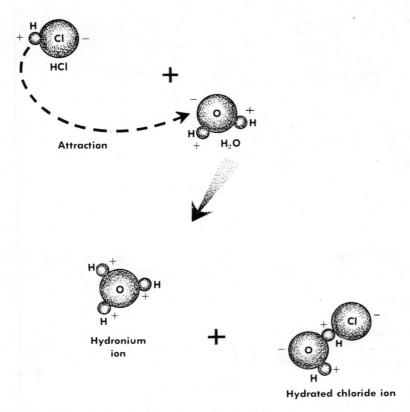

Figure 8-4.

bases, and that is precisely the case. Consider hydrogen chloride (HCl) with regard to how this "classical acid" leads us to the broader definition (Fig. 8-4).

The hydrogen ion, being strongly attracted by the negatively charged part of a water molecule, actually transfers from the HCl molecule to the water, forming a structure called a **hydronium ion**. The hydrogen ion, being "donated" by the HCl molecule, is a **proton** because the only difference between a hydrogen atom and a proton is the single electron. By virtue of this mechanism, an acid may be defined as a proton **donor**. Conversely, a base is defined as a proton acceptor. In Figure 8-4, water is acting as a base.

Expanding this newer and broader definition, the tendency of negatively charged particles to function as proton (a unit of positive charge) acceptors can be understood. To answer the question of why cyanide ion is a base, note that a negative ion predictably attracts a proton.

$$CN^- \leftarrow H^+ = HCN \text{ (hydrogen cyanide gas)}$$

The result is the formation of very slightly soluble hydrogen cyanide gas. Note: Do not attempt to verify this reaction experimentally because HCN is a very deadly gas. The vigor with which this reaction occurs attests to the strong proton affinity of the cyanide ion. This is the reason why CN^- is described as a *strong* base.

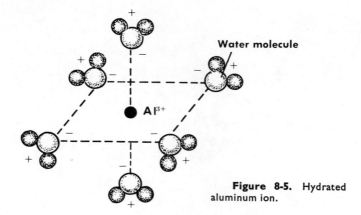

Figure 8-5. Hydrated aluminum ion.

An aluminum ion in water solution has a strong attraction for water molecules because of its considerable positive charge.

$$Al \rightarrow Al^{3+} + 3e^-$$

Studies on the structure of a *hydrated* aluminum ion suggest a geometry that looks like that shown in Figure 8-5, resulting in the common formula for hydrated aluminum ion:

$$Al(H_2O)_6^{3+}$$

If sodium carbonate, for example, were added to hydrated aluminum ion, the first step of the reaction would be

a proton transfer

$$\underset{\substack{\text{proton}\\\text{donor}}}{Al(H_2O)_6^{3+}} + \underset{\substack{\text{proton}\\\text{acceptor}}}{CO_3^{2-}} \rightarrow Al(H_2O)_5(OH)^{2+} + HCO_3^-$$

By definition, then, a hydrated aluminum ion has to be classified as an acid and carbonate ion as a base. This concept of acids and bases not only broadens the general classification, but also serves to give meaning to directions in a lab that call for the addition of cyanide ion to render a solution basic.

Chart of Some Acids and Bases

Increasing acid strength	Acid	Base	Increasing basic strength
	$HClO_4$	ClO_4^-	
	HI	I^-	
	HCl	Cl^-	
	HNO_3	NO_3	
	H_2SO_4	HSO_4	
	H_3O^+	H_2O	
	H_3PO_4	H_2PO_4	
	$Al(H_2O)^{3+}$	CO_3^{2-}	
	H_2S	S^{2-}	
	H_2O	OH^-	

Acid-Base Titrations

The process called titration is a laboratory method that enables the chemist to balance volumes of acids and bases to a desired degree of acidity or alkalinity or to find out an unknown concentration of one of them as it is compared to a standardized concentration of the other.

While the former purpose has more practical application in the research lab, it would be useful to discuss the latter purpose first.

There is a bottle of sulfuric acid on the shelf, and its concentration is unknown. How can a titration be performed to discover what its molar concentration is? Initially, the reaction with a common base, such as sodium hydroxide, may be proposed.

$$H_2SO_4 + 2NaOH \rightarrow Na_2SO_4 + H_2O$$

The essential part of this reaction is the fact that the acid-base reaction is the combination of hydrogen ion from H_2SO_4 with the hydroxide ion from NaOH to form water. **Neutralization** is the term applied to equal numbers of H^+ and OH^- combining to form water.

$$H^+ + OH^- \rightarrow H_2O$$

Titration I. Therefore, what is needed is to set up an apparatus that will control the flow and measure the volumes of acids and bases combined. When the equivalent numbers of H^+ and OH^- have reacted, it is known that the number of equivalents of acid and base are equal. If a known normality of NaOH is used, the unknown normality of H_2SO_4 can be calculated by knowledge of the volume required. For example,

number of equivalents acid = number of equivalents base

Since

Number of equivalents = $NV_{(normality)(volume\ in\ liters)}$

we can say

$$N_{acid} \times V_{acid} = N_{base} \times V_{base}$$

Although a pH meter is used in the following illustrations, it should be mentioned that color indicators may also be employed. A table of appropriate color indicators is included in Appendix A.

Example I

Experimentally determine the normal concentration of an unlabeled bottle of hydrochloric acid (Fig. 8-6).
1. Prepare 100 ml of 1 N NaOH as the standard base.
2. Fill a buret with the NaOH solution and open the stopcock in order to fill the buret to the tip, closing it when the volume reads 50 ml.
3. Perform the same operation with the unknown acid in a second buret.

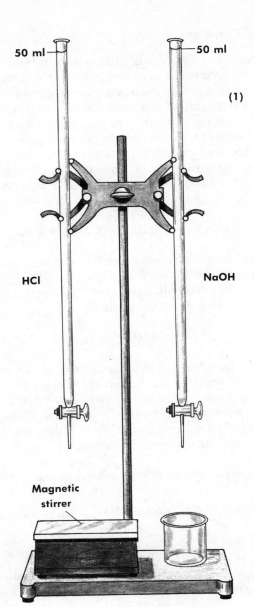

50 ml   50 ml

(1)

HCl   NaOH

Magnetic
stirrer

Figure 8-6.

4. Release 20 ml of NaOH into a beaker containing a stirring bar and place it on a magnetic stirrer. Adjust the rate of stirring so that splattering does not occur.
5. Insert the electrodes, or combination electrode of a pH meter that has been adjusted with standard buffer at pH 7.0.
6. Open the stopcock of the acid buret so that a steady dropwise flow is obtained.
7. Close the stopcock when the pH meter reads 7.0 and record the volume. A pH value of 7.0 is the neutralization point.
8. Suppose 5.2 ml of acid was required to produce the pH 7.0;

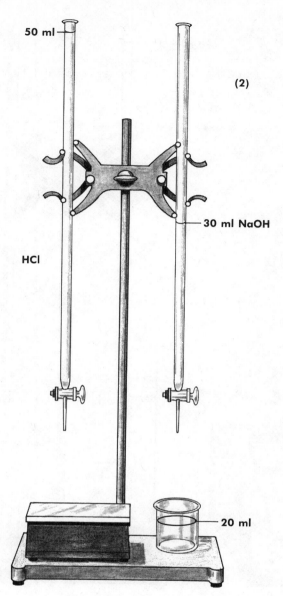

50 ml

(2)

30 ml NaOH

HCl

20 ml

Figure 8-6 continued.

calculate the normality of the acid:

$$N_{acid} \times V_{acid} = N_{base} \times V_{base}$$
$$N_{acid}(5.2 \text{ ml}) = 1.0 \text{ N} (20 \text{ ml})$$
$$N_{acid} = \frac{(1.0 \text{ N})(20 \text{ ml})}{5.2 \text{ ml}}$$
$$N_{acid} = 3.85 \text{ N}$$

If too much acid were added, e.g., the pH shifted too quickly to 5.0, then it is necessary to **back-titrate**. This presents no real problem if you record the additional volume of NaOH used and incorporate this value in the equation.

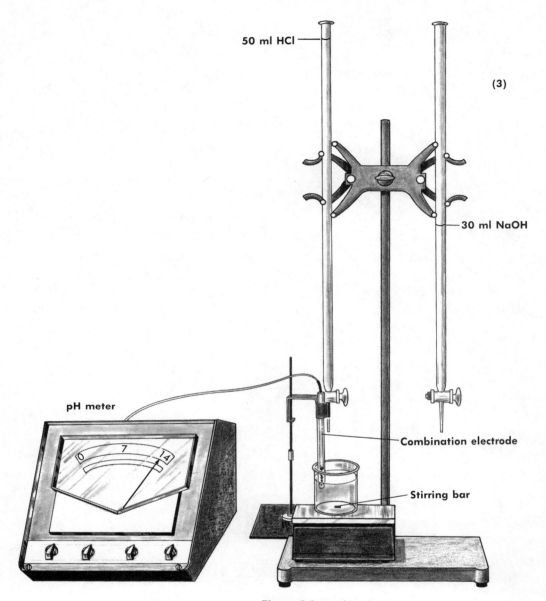

Figure 8-6 continued.

The Concept of pH

pH (the "strength" of the hydrogen ion) is a convenient method for determining the relative acidity or alkalinity of a solution. It does not tell anything about the strength of the acid or base, however. Equal pH values can be obtained by using a strong acid (vigorous proton donor) like HCl or a weak acid such as acetic acid (definitely not a vigorous proton donor).

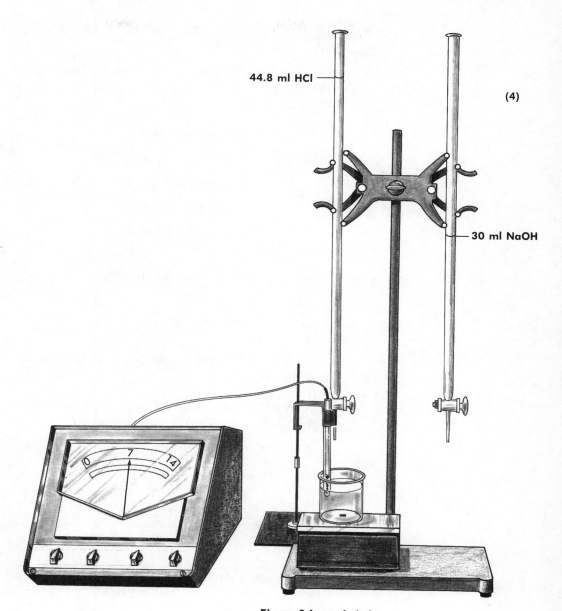

44.8 ml HCl

(4)

30 ml NaOH

Figure 8-6 concluded.

Of course, much more acetic acid would be needed to produce the same hydrogen ion concentration.

pH is defined as the negative logarithm of the H⁺ ion concentration in moles per liter.

$$pH = -\log [H^+]$$

The convenience of the pH method is that it expresses hydrogen ion concentrations in easily manageable small positive numbers.

Example 2

Calculate the pH of a solution having a H$^+$ ion concentration of 1×10^{-3} M.

a. Write the equation:

$$pH = -\log (H^+)$$

b. Substitute:

$$pH = -\log 1 + \log 10^{-3}$$
$$pH = -[0 + (-3)]$$
$$pH = -(0 - 3)$$
$$pH = 3$$

Example 3

Calculate the pH of a solution having a much smaller [H$^+$] (i.e., less acidic) of 3.2×10^{-10} M.

a. $pH = -\log (3.2 + 10^{-10})$

b. The log of 3.2 from the L scale of the slide rule is 0.52.

c. $pH = -(0.52 + (-10))$

$$pH = -(0.52 - 10)$$
$$pH = -(-9.48)$$
$$pH = 9.48$$

Now consider pure water as the standard of neutrality so that the pH values in the above examples, 3.0 and 9.48, may be properly interpreted as acidic or basic.

The [H$^+$] of pure water is very low because water molecules are extremely stable and, therefore, do not tend to ionize to any appreciable degree. In fact, the concentration of H$^+$ ions in pure water is 1×10^{-7} moles per liter. The pH of water then is

$$pH = -\log (1 + 10^{-7})$$
$$pH = -(0 - 7)$$
$$pH = 7$$

The hydroxide ion concentration of pure water is also 1×10^{-7} M since

$$H_2O \rightleftharpoons H^+ + OH^-$$
$$\text{1 mole} \quad \text{1 mole} \quad \text{1 mole}$$

One hydroxide ion forms for each hydrogen ion produced. The product of the ion concentrations in pure water is then

$$(H^+)(OH^-) = \text{ion product}$$

$$(1 \times 10^{-7})(1 \times 10^{-7}) = 1 \times 10^{-14}$$

If the H^+ ion concentration became maximal by the addition of an acid, then the OH^- ion concentration would be correspondingly minimal. The limit would be

$$(H^+)(OH^-) = 1 \times 10^{-14}$$

$$(1)(1 \times 10^{-14}) = 1 \times 10^{-14}$$

The pH of a H^+ ion concentration of 1 M is

$$pH = -\log (0)$$

$$pH = 0$$

On the other hand, if the H^+ ion concentration were reduced to a minimum by the addition of a base, the pH value would swing to the other extreme and, theoretically, achieve a value of 14.

The range of 0 to 14 then comprises the extreme limits of the pH scale (Fig. 8-7).

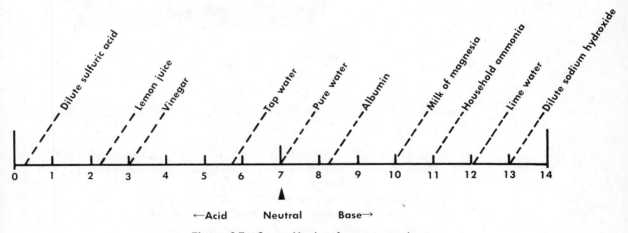

Figure 8-7. Some pH values for common substances.

Any reading on the pH meter below 7 indicates an acidic medium while readings above 7 are alkaline. The pH meter, as a measuring instrument, will be discussed in a later chapter.

"Bucket" titration. A typical titration procedure in the lab involves the deliberate control of pH without regard to unknown concentrations of acid or base. For example, living organisms or tissues have very critical pH limits—a few tenths of a pH unit too low or too high might destroy biological activity. Preparations of various solutions might well dictate to a technician that the pH of a salt solution would have to be adjusted to a pH of 6.8 to 7.2. The procedure would be called **titration**, but a pipet is used rather than a buret to control the change.

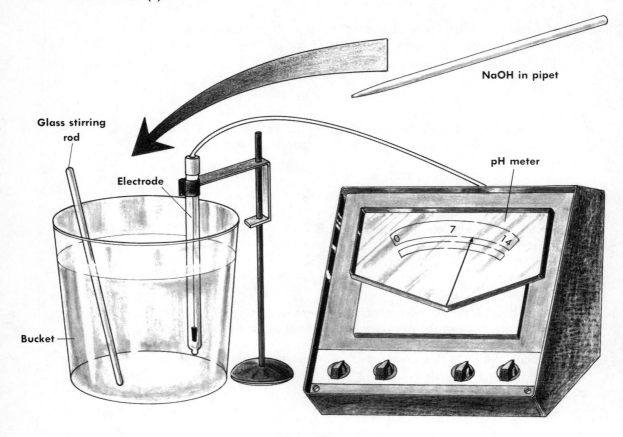

Figure 8-8.

Example 4

Given a few liters of minced muscle tissue in salt water where the fatty acids released by the minced tissue has lowered the pH to 5.8, how can the pH be raised to 7.4?

1. Prepare a solution of NaOH or KOH to a concentration of 1 M to 2 M. (Higher concentrations of OH⁻ may cause a pH shift so rapid and extreme that control is difficult).
2. Use a pipet to add the hydroxide solution carefully (possibly dropwise) while the pH meter electrode is immersed in the bucket (plastic or glass). Stir constantly with a large glass stirring rod.
3. As soon as the pH reaches 7.4, the titration is complete.

Buffers and pH

A **buffering agent** is any substance that prevents sharp changes in pH. If an acid is added to a buffered solution, the buffering agent acts as a proton acceptor, "binding" them so that the H^+ ion concentration does not appreciably increase. The buffering agent may also act as a proton donor if a base is added, forming water and thus keeping the OH^- ion concentration in check.

Buffered systems are absolutely essential in any biological experimentation where the pH range is critical. For example, human blood cannot vary more than a few tenths of a pH unit if the body is to remain alive. Blood is described as having a high buffer capacity, and if this were not the case, ordinary acidic breakfast fruit juice could be fatal. Most experiments involving living tissue require high buffer capacities that fix the pH between 6.8 and 7.2. A typical buffered medium is a 0.1 M phosphate buffer.

The two usual phosphates are K_2HPO_4 (potassium monohydrogen phosphate) and KH_2PO_4 (potassium dihydrogen phosphate). The essential mechanism involves hydrogen phosphate ions while the potassium ion is simply called a "spectator ion." Consider the $H_2PO_4^-$ ion first. In water $H_2PO_4^-$ tends to be a proton donor. Remember that the H_3O^+ (hydronium

proton transfer

$$H_2PO_4^- + H_2O \rightleftharpoons HPO_4^{2-} + H_3O^+$$

ion) may be thought of as a soggy hydrogen ion; therefore, it can be expected that the pH will drop into the acid range. In fact, the pH does drop to about 4.2. The KH_2PO_4 is often called potassium acid phosphate.

On the other hand, K_2HPO_4 tends to be a proton acceptor and it is often called potassium basic phosphate. Once again, the essential base is the monohydrogen phosphate ion while the potassium ion is still relegated

proton transfer

$$HPO_4^{2-} + H_2O \rightleftharpoons H_2PO_4^- + OH^-$$

to the "spectator" category. The production of OH^- ion in water predictably shifts the pH into the basic range, reaching about 8.4.

Example 5

Prepare a liter of 0.1 M phosphate buffer of pH 7.0 to suspend some homogenized rat liver tissue.
1. Prepare about a liter each of 0.1 M K_2HPO_4 and 0.1 M KH_2PO_4 to be used as stock solutions.
2. Pour about 400 ml of the acid phosphate into a liter beaker containing a magnetic stirring bar and adjust the pH electrodes just under the surface.
3. While the magnetic stirrer is running, carefully pour the basic phosphate into the beaker until the desired pH of 7.0 is reached.
4. Since both molarities are 0.1, the final molarity will remain unchanged.

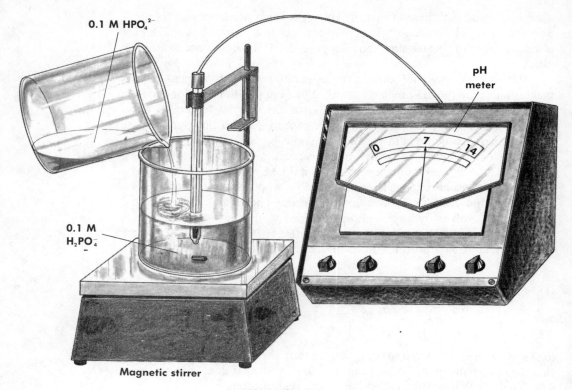

0.1 M HPO₄²⁻

0.1 M
H₂PO₄⁻

pH
meter

0
7
14

Magnetic stirrer

Figure 8-9.

5. If the volume is less than a liter at pH 7.0, add more acid phosphate and titrate as before.
   Alternative method
1. Prepare stock solutions of 1.0 M concentration.
2. Titrate the buffers in exactly the same way, except that the final volume will be 100 ml.
3. Use distilled water to dilute the 100 ml of 1.0 M phosphate to a liter volume, which will then be 0.1 M.

An investigation into what makes a pH meter work, i.e., a little theory about the electrochemistry involved, will be taken up in Chapter 10.

CHROMATOGRAPHY

Chromatography is a laboratory method of separating, identifying, and possibly purifying a mixture of complex organic solutes (plant pigments, proteins, fats) that are dissolved in a common solvent. When the solvent flows through some **adsorbing** (a kind of superficial absorption) substance, the various components of the solute mixture flow along at different rates depending on how tenaciously the various molecules hold to the adsorbing material. Usually the strength of attraction depends on the distribution of electrical charges on the molecule, which may also be described as the **polarity** of the molecules. Other factors in this complex solute-solvent-adsorbent relationship are sizes and shapes of molecules, molecular weights, and the electrical charge on the adsorbent material.

Sometimes chromatographic separation is accomplished by using two immiscible solvents, one of which saturates the filter paper or column packing material, while the other solvent flows over the stationary one. The movement of solute molecules depends on its comparative solubilities in the stationary and moving solvents. This phenomenon is called **partition chromatography**.

The third explanation of chromatography, after adsorption and partition, is **ion-exchange**. Ion-exchange chromatography is often carried out in columns of resins called ion-exchange resins. The resins, which are capable of exchanging anions or cations, form a network that strongly attracts polar molecules or ions. Those molecules having the lowest degree of polarity will move most rapidly through the column and be eluted first. Solutions that have percolated through a column are called **eluates**.

Adsorbents used for adsorption chromatography usually include charcoal, sodium silicate, aluminum oxide, and sucrose; partitioning solids may be cellulose, filter paper, diatomaceous earths, or a huge variety of commercial preparations. The technician's concern with experimenting with different media would be good procedure if time is available and if the existing protocols outline a procedure that seems inefficient.

Paper Chromatography

This kind of chromatographic separation is obviously carried out on paper. Sheets of Whatman No. 1 or No. 3 filter paper are usually satisfactory. Sometimes the paper is specially treated beforehand, but this question would be related to a specific direction in a protocol.

The sample. The sample to be investigated is first prepared as a concentrated extract of plant or animal tissues. Consider a specific example that illustrates the separation of leaf pigments. Green leaves (spinach is good) may be ground in a mortar with sand and acetone or chopped in a high speed blender with acetone for a few seconds. The green suspension is then poured through several layers of cheesecloth to strain out the solid debris. Allow the acetone suspension to evaporate until a few milliliters of concentrated extract remain.

Spotting. Cut a rectangular piece of filter paper to a size that fits conveniently in a glass jar covered with a square glass plate. Draw a pencil line parallel to one end, about two centimeters from it, and use a micropipet or capillary tube to place small (no more than 5 mm in diameter) dots of extract on the line. Allow the spots to dry before making an addition. Keep spotting until the dots are dark green and then dry thoroughly (Fig. 8-10). Clip the paper to a bar or frame that fits in the glass jar.*

---

* Note: The filter paper should not be handled directly since any oil on the technician's hands may introduce an unexpected and puzzling spot.

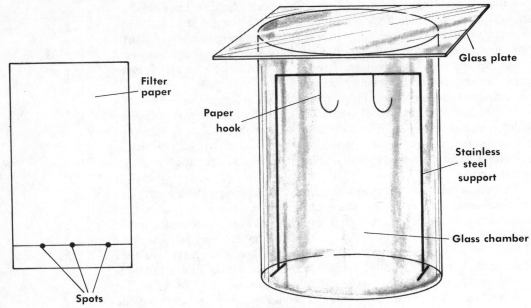

Filter paper

Paper hook

Glass plate

Stainless steel support

Glass chamber

Spots

Figure 8-10.

Solvent. The flow solvent depends on the nature of the sample. For leaf pigments, a solvent of 5 per cent acetone and 95 per cent petroleum ether may be used. Pour the solvent into the glass jar to a depth of a couple of centimeters. Cover the jar in order to allow the air to be saturated with solvent vapor. Clip the spotted paper to the jar at the top and adjust the height so that the spots are about one centimeter above the solvent (Fig. 8-11).

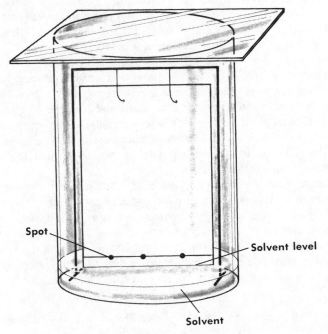

Spot

Solvent level

Solvent

Figure 8-11.

Development.   As the solvent rises, it will be observed that several colored blotches are rising at different rates. When a good separation has been made, remove the chromatogram and outline the blotches with a pencil. Sometimes the compounds will be invisible, in which case methods such as ultraviolet fluorescence, chemical sprays or dips, or Geiger counters (for radioactive samples) may be used to reveal the positions of the components.

R_f values.   A method of identification of a substance is to measure the distance it flows compared to the distance the solvent flows. A reference text will contain R_f values for various substances in given solvents. For the green pigment separation the picture might be as designated in Figure 8-12.

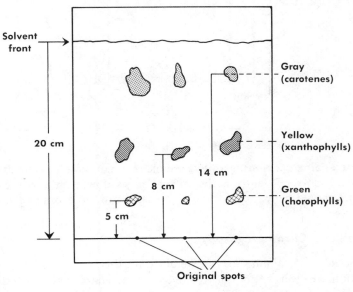

Figure 8-12.

For chlorophyll, the R_f value would be:

$$R_f = \frac{\text{solute rate flow}}{\text{solvent rate flow}}$$

Distance may be substituted for *rate*, since the time factor is constant.

$$R_f = \frac{5 \text{ cm}}{20 \text{ cm}} = 0.25$$

For the xanthophylls:

$$R_f = \frac{8 \text{ cm}}{20 \text{ cm}} = 0.4$$

For the carotenes:

$$R_f = \frac{14 \text{ cm}}{20 \text{ cm}} = 0.7$$

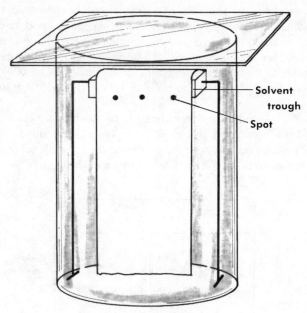

Figure 8-13.

Descending chromatography. Some chromatographic separations are performed more efficiently if the solvent flows downward. However, the R_f values should be expected to be in the same ratio to ascending R_f values.

Column Chromatography

Special glass columns may be obtained for this type of chromatographic separation. However, any glass tubing can be adapted with a little glass wool and rubber stoppers. The usual column looks like the one in Figure 8-14.

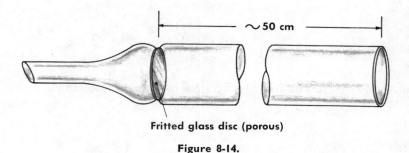

Fritted glass disc (porous)

Figure 8-14.

The sample. For comparison's sake, leaf pigments are again a good example. As before, sand or a blender may be used to break up the leaf structure and obtain the green pigmented extract. Filtration through cheese-cloth follows. At this point, the acetone suspension is treated with petroleum ether in a separatory funnel as described in the previous chapter. The petroleum ether extract is **lyophilized** (freeze-dried) by rotating a flask of the

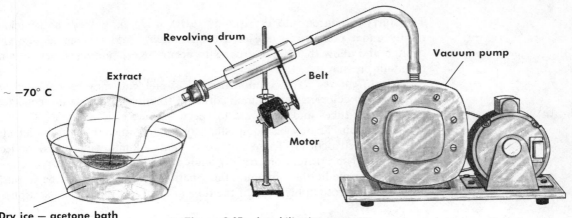

Figure 8-15. Lyophilization apparatus.

extract in a dry ice-acetone bath under low pressure obtained with a vacuum pump. The dry residue in the flask is suspended in a few milliliters of 2,2,4-trimethylpentane (isooctane). This small volume of highly concentrated solution is placed on the top of the column.

The column. Finely ground sucrose is the adsorbing medium. The powdered sugar is mixed with isooctane to make a fluid mush called a **slurry**.

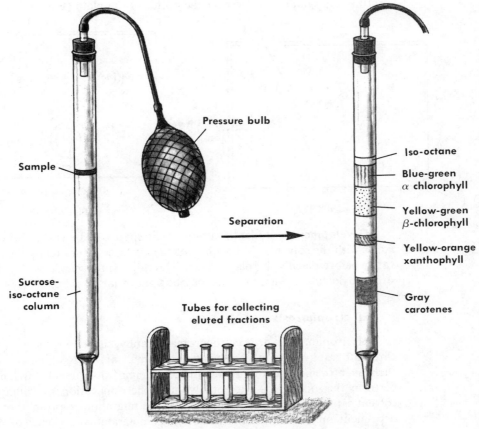

Figure 8-16.

The slurry is poured into the column, extra isooctane added, and packed tightly under pressure by use of a rubber bulb. Add the pigment to the column and allow the **elution** process to occur. Keep adding isooctane and maintain pressure.

Identification. The components of leaf pigments may be identified by color, or if a colorless compound is desired, fractions of eluate are collected in small test tubes and subjected to spectrophotometric analysis. For example, vitamin K or plastoquinone may be identified by ultraviolet absorption spectroscopy. There are instruments in many labs that collect fractions and spectrophotometrically analyze them automatically.

Another method of obtaining the separated pigments would be to push the packed sucrose "plug" out of the tube and cut out the desired sections.

Thin Layer Chromatography (TLC)

When paper (cellulose) is not the most suitable medium, other material, such as silica gel, can be used by being made up as a thin layer on a glass plate. Today, such ready-made plates can be commercially obtained. TLC has the advantage of permitting separations of very small samples, since the diffusion of spots does not occur as it does on paper.

Sample and solvent. Concentrated solutions are made as usual for spotting purposes. The solvent used should be of a similar polarity to the substances being separated. Selection of a solvent can be made empirically. Spotting the sample is done near the center of the plate (Fig. 8-17).

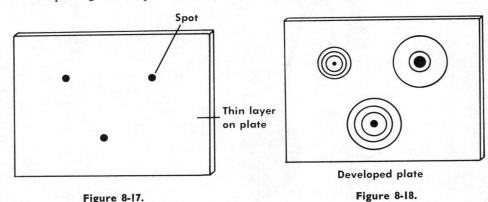

Figure 8-17.   Figure 8-18.

Development and identification. The spotted plate is placed in a container with the solvent. Allow time for the air to be saturated by the solvent vapor before placing the plate inside. Development is very rapid, and heat, strong oxidizing agents, or acids may be used to locate the rings (Fig. 8-18).

Electrophoresis

Electrophoresis, or electrochromatography, is a variation of paper chromatography that takes advantage of small amounts of polar samples such as proteins. Again, there are commercial instruments in labs that perform this operation. Essentially, it is the application of voltage to a solvent saturated chamber which induces the migration of proteins at a rate dependent on their degree of polarity. A diagrammatic sketch of such a system is found in Figure 8-19.

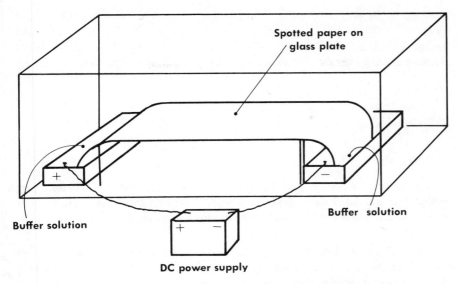

Figure 8-19.   Diagrammatic representation of electrophoresis apparatus.

Dialysis

Dialysis is a technique using a membrane (cellophane tubing, animal intestine, and some plastic tubing) that has pores large enough to permit small molecules through but not larger molecules. This selective movement in and out of the tube is called **semipermeability**.

Any concentration of molecules tends to diffuse—to spread out and become random in its arrangement. By pouring a mixture of molecules

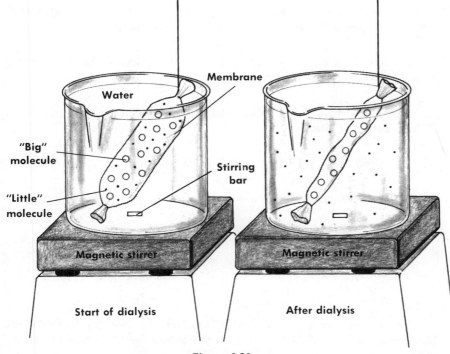

Figure 8-20.

(call them "bigs" and "littles") in a length of seamless dialysis tubing and placing the filled tube in a large beaker of water, the "little" molecules will gradually diffuse out into the water, leaving the "big" molecules behind. Thus, a separation is made. The dialysis process may be speeded by stirring.

PROBLEMS

Set A

1. Use a reference text to determine the geometry of the following molecules, and label them polar or nonpolar on that basis: a. NH_3 b. HBr c. cis-dichloroethane d. trans-dichloroethane e. carbon disulfide.

2. Sulfuric acid is needed for (a) cleaning pipets, (b) drying a moist salt in a desiccator, and (c) acidifying a solution before colorimetric analysis. What grade or grades should be used? Why?

3. Complete and balance the following acid-base reactions:

a. $HCl + CO_3^{2-} \rightleftharpoons$
b. $H_3PO_4 + H_2O \rightleftharpoons$
c. $S^{2-} + H_2SO_4 \rightleftharpoons$
d. $Zn(H_2O)_4^{2+} + H_2O \rightleftharpoons$
e. $HPO_4^{2-} + HNO_3 \rightleftharpoons$

4. If 23.2 ml of HCl is required to neutralize 147.0 ml of 0.5 N KOH, what is the normality of the acid? What is the molarity?

5. How many milliliters of 6 N H_2SO_4 are required to neutralize 0.06 g of $Ca(OH)_2$?

6. What is the pH of a solution having a hydrogen ion concentration of 0.0036 moles per liter?

7. Calculate the pH of a solution where the $[H^+]$ is 7.7×10^{-9} M.

8. If a solution of HNO_2 has a pH of 3.42, what is its molarity?

9. Define the following terms:

a. proton donor
b. hydronium ion
c. hydrated

d. equivalent weight
e. buffer

10. If a certain protein moved 8.7 cm from its spot while the solvent front moved a total of 11.6 cm on a paper chromatogram, calculate the R_f value.

Set B

11. Fluorine has a very strong attraction for electrons. How do you explain the fact that HF is polar while CF_4 is nonpolar?

12. Complete and balance the following reactions:

a. $HNO_3 + H_2O \rightleftharpoons$
b. $Co(H_2O)_6^{3+} + H_2O \rightleftharpoons$
c. $CN^- + H_2S \rightleftharpoons$

13. Complete and balance the following reactions:

a. $HClO_4 + CO_3^{2-} \rightleftharpoons$
b. $H_3O^+ + OH^- \rightleftharpoons$
c. $OH^- + HPO_4^{2-} \rightleftharpoons$

14. What is the normal concentration of H_2SO_4 if 15.2 ml is required to neutralize 85.5 ml of 0.3 M NaOH? What is the molarity of the H_2SO_4?

15. What volume of 18 N HCl is needed to neutralize 4.0 g of NaOH?

16. What is the pH of a solution in which the hydrogen ion concentration is 0.00025 M?

17. Calculate the pH of a solution in which the $[H^+]$ is 4.5×10^{-4} M.

18. What is the molar concentration of a solution of HCl which has a pH of 4.7?

19. Define or explain the following terms:

a. titration
b. eluate
c. slurry
d. lyophilize
e. semipermeable

20. Calculate the R_f value on a paper chromatogram if a particular spot moved 2.3 cm while the solvent front moved a total of 17.5 cm.

SUPPLEMENTARY READING

Abbott, D., and Andrews, R. S.: *An Introduction to Chromatography.* (paperback) Houghton Mifflin Co., Boston, 1965.

A concise introduction to the theory and practice of chromatography.

Jaski, T.: Electronic pH Measurement. *Electronics World*, September, 1960, p. 44.

Very good discussion of electronic pH measurement for laboratory technicians.

Christensen, H. N.: *pH and Dissociation.* (programmed instruction) W. B. Saunders Company, Philadelphia, 1963.

An excellent self-study text for understanding acids and bases.

Heftman, E. (ed.): *Chromatography*, 2nd ed. Reinhold Publishing Corporation, New York, 1967.

A very thorough text that compiles chapters on chromatography contributed by outstanding scientists.

9 Oxidation-Reduction

The type of chemical reaction described as oxidation-reduction, often contracted to **redox**, is a topic of major importance in chemistry and biology. The tendency of substances to lose and gain electrons in their drive toward stability and lower energy content has been described in Chapter 3. In living organisms, there is also the use of intricate chains of redox reactions to absorb and store the available energy for life functions.

An understanding of the principles of oxidation-reduction provides you with the tools to understand some of the mysteries of chemistry encountered in the useful laboratory activities a research technician is likely to perform. Why does an apparently harmless solution of copper (II) sulfate "attack" and "eat up" a piece of zinc? How does a redox reaction produce electricity? What is a redox titration, and what is it used for? These are just a few of many questions to be investigated.

Any person employed in the investigation of redox phenomena has a grave responsibility to be knowledgeable. The chronicles of scientific activity are filled with stories of explosions, damage to skin and eyes, fires, billions of dollars of loss in corrosion, and gruesome accounts of what surgical metals have done to teeth and bones. On the other hand, knowledge has permitted great strides forward in the development of electrochemical cells, safety procedures in laboratories, modern dental and surgical methods, and a growing understanding of the use and production of energy for life.

REDOX PRINCIPLES REVIEWED

Oxidation of a substance occurs when there is an increase in its oxidation number. This is accomplished by a *loss* of electrons. **Reduction** results in a decrease in the oxidation number. This is achieved by a *gain* of electrons. Any substance which promotes the loss of electrons, because of its own tendency to gain them, is called an **oxidizing agent**. The substance which

readily gives up its electrons is called the **reducing agent**. Tables 9-1 and 9-2
list some of the most prominent oxidizing and reducing agents.

TABLE 9-1. COMMON OXIDIZING AGENTS.

Oxidizing Agents	Number of Electrons Gained	Product
F_2 (g)	2	$2F^-$ (aq)
MnO_4^- (aq) (in acid)	5	Mn^{2+} (aq)
ClO_4^- (aq) (in acid)	8	Cl^- (aq)
Cl_2 (g)	2	$2Cl^-$ (aq)
$Cr_2O_7^{2-}$ (aq) (in acid)	6	$2Cr^{3+}$ (aq)
O_2 (g) (in acid)	4	$2H_2O$ (l)
Br_2 (l)	2	$2Br^-$ (aq)
NO_3^- (aq) (in acid)	3	NO (g)

TABLE 9-2. COMMON REDUCING AGENTS.

Reducing Agent	Number of Electrons Lost	Product
Li (s)	1	Li^+ (aq)
Ca (s)	2	Ca^{2+} (aq)
Na (s)	1	Na^+ (aq)
Zn (s)	2	Zn^{2+} (aq)
H_2 (g)	2	$2H^+$ (aq)
Cu (s)	2	Cu^{2+} (aq)

Consider now the question raised earlier about the reaction between
zinc metal and copper (II) sulfate solution. When the zinc is placed in a
tube of the blue solution, it darkens immediately as reddish clumps of solid
copper form. The zinc gradually disappears, and the solution turns from blue
to colorless (Fig. 9-1). This is oxidation-reduction. The zinc is a good
reducing agent, which is another way of saying that zinc becomes oxidized
quite easily. The copper (II) ion has changed to plain copper metal, which is
a feat accomplished by gaining electrons (i.e., undergoing reduction). The
equation for the entire redox reaction is the sum of the oxidation part and
the reduction part. The number of electrons lost must equal the number of

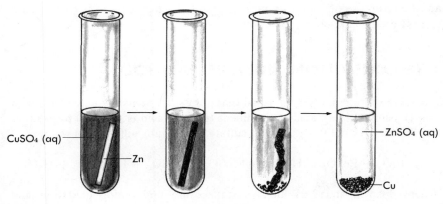

$CuSO_4$ (aq)

Zn

$ZnSO_4$ (aq)

Cu

Figure 9-1.

electrons gained, since elecrtic charge is also subject to the law of conservation.

$$\overset{\frown\text{oxidation number}}{Zn^0(s)} \xrightarrow{\hspace{3cm}} \overset{\frown\text{oxidation number}}{Zn^{2+}(aq)} + 2e^-$$

metallic zinc   zinc ion   two electrons lost

The rise in the oxidation number from zero to positive two, fits the definition of oxidation.

$$Cu^{2+}(aq) + 2e^- \rightarrow Cu^0(s)$$

copper   two   metallic copper
ion   electrons
from   gained
$CuSO_4$
solution

The generally accepted convention is to transpose the $+2e^-$ from the left hand side of the equation to the right for purposes of uniformity and simplicity.

$$Cu^{2+}(aq) \rightarrow Cu^0 - 2e^-$$

When the two half-reactions (half oxidation and half reduction) are added, the net redox equation is obtained.

$$Zn^0 \rightarrow Zn^{2+} + 2e^-$$
$$Cu^{2+} \rightarrow Cu^0 - 2e^-$$
$$\overline{\hspace{5cm}}$$
$$Zn^0 + Cu^{2+} \rightarrow Zn^{2+} + Cu^0$$

zinc + copper (II) ion $\rightarrow$ zinc ion + copper

The resulting reaction explains the observations. The formation of reddish clumps is the appearance of copper. The loss of blue color occurs as the copper (II) ion disappears, and the zinc is "eaten away" as it changes to zinc ion.

Most redox reactions that are encountered in a research laboratory are not as simple as the zinc and copper (II) ion example. When the technician needs a balanced equation for stoichiometric calculations, guidance and practice are required to give him the skill to do the job. There are two common methods suited to the task of completing and balancing redox reactions: (1) the oxidation number method for molecular equations, and (2) the ion-electron method for ionic equations. The first method is easier, and therefore a good starting point, while the latter method is more useful since it omits "spectator" ions.

THE OXIDATION NUMBER METHOD

It is experimentally established that potassium permanganate reacts with iron (II) sulfate in an acid environment (sulfuric acid is suitable) to produce manganese (II) sulfate, iron (III) sulfate, potassium sulfate, and water.

$$KMnO_4 + FeSO_4 + H_2SO_4 \rightarrow MnSO_4 + K_2SO_4 + Fe_2(SO_4)_3 + H_2O$$

Where do you start? Which substances are undergoing oxidation and reduction?

Redox reactions should *not* be balanced by inspection. The numerical coefficients that represent the mole ratios are so varied that a "trial and error" method usually results in acute frustration.

One method is to look for common oxidizing and reducing agents. $KMnO_4$ is quickly recognized as an oxidizing agent, but the iron (II) in $FeSO_4$ is not so easily identified. There is an alternative method based on three clues.

1. When an element is in the *middle* of a ternary (three element) compound on one side of the equation but not on the other side. For example,

$$KMnO_4 \rightarrow MnSO_4$$

in the middle of   not in the middle
a ternary compound

2. When an element is in a *compound* on one side of the equation but is in the *free* (uncombined) state on the other side.

$$CuSO_4 \rightarrow Cu$$

combined with   free (uncombined)
sulfate ion

3. When a metal ion exhibits one oxidation number on one side of the equation and a *different oxidation number* on the other side.

$$FeSO_4 \rightarrow Fe_2(SO_4)_3$$

iron (II)   iron III

Using the original equation, the clues may be applied to find the elements undergoing oxidation and reduction.

$$KMnO_4 + FeSO_4 + H_2SO_4 \rightarrow MnSO_4 + Fe_2(SO_4)_3 + K_2SO_4 + H_2O$$

iron II ──────────→ iron III
change in oxidation number (clue #3)

middle of   not in the
ternary compound ──────────→ middle (clue #1)

The next steps are to write the oxidation-reduction half-reactions apart from the equation, determine the electron loss and gain, balance the half-reactions, and finally replace the data in the equation. The electron loss and gain require preliminary calculations.

1. Calculate the oxidation number of manganese in the compound $KMnO_4$.
    a. The formula indicates the oxidation number of potassium is $1+$ (K^+, MnO_4^-) from the table of common oxidation numbers.
    b. The oxidation number of oxide is $2-$. There are a total of four oxides. Therefore, the *total* oxidation value for the oxide is minus eight.

$$4 \text{ times } (2-) = 8- \qquad \overset{1+}{\underset{K_1^+}{\nearrow}} \qquad \overset{8-}{\underset{MnO_4^x}{2-}}$$

c. Since the sum of the oxidation numbers of any compound must equal zero, the value for the Mn in the formula has to be 7+.

$$\begin{array}{cc} 1+ & 8- \\ & (7+) \\ K_1^+ & MnO_4^{2-} \end{array}$$

1+ and 7+ and 8− = zero

2. The oxidation number of manganese in $MnSO_4$ is 2+, and this is clearly indicated by the name of the compound, manganese (II) sulfate.

$$Mn^{2+}(SO_4)^{2-}$$

3. Iron (II) sulfate and iron (III) sulfate also indicate the oxidation numbers by their names.

Now, write the half-reactions:

$$Mn^{7+} \text{ is reduced to } Mn^{2+}$$

This requires a *gain* of five electrons (negative charges):

$$Mn^{7+} + \text{ ? negative charges} \rightarrow Mn^{2+}$$
$$Mn^{7+} + 5e^- \rightarrow Mn^{2+}$$

Transpose the $+5e^-$:

$$Mn^{7+} \rightarrow Mn^{2+} - 5e^- \text{ (reduction)}$$

Fe^{2+} is oxidized to Fe^{3+} which means a *loss* of one electron per atom. Since two units of Fe^{3+} are produced, you must start with two units of Fe^{2+}. Two units of Fe^{2+} undergoing oxidation will require a two electron loss.

$$2Fe^{2+} \rightarrow Fe_2^{3+} + 2e^- \text{ (oxidation)}$$

Write the two half-reactions together for addition and balancing.

$$Mn^{7+} \rightarrow Mn^{2+} - 5e^-$$

$$2Fe^{2+} \rightarrow Fe_2^{3+} + 2e^-$$

The law of conservation requires a balanced loss and gain of electrons. To accomplish this, the half-reactions must be multiplied to obtain the lowest common denominator. This means that 10 moles of Fe^{2+} will lose 10 moles of electrons to produce a total of 10 moles of Fe^{3+}. Two moles of Mn^{7+} will gain 10 moles of electrons to produce 10 moles of Mn^{2+}.

$$2Mn^{7+} \rightarrow 2Mn^{2+} - 10e^-$$
add $$\underline{10Fe^{2+} \rightarrow 5Fe_2^{3-} + 10e^-}$$

$$\underline{2}Mn^{7+} + \underline{10}Fe^{2+} \rightarrow \underline{2}Mn^{2+} + \underline{5}Fe_2^{3+}$$

2 moles + 10 moles → 2 moles + 5 moles

Replace these coefficients in the original equation.

$$2KMnO_4 + 10FeSO_4 + H_2SO_4 \rightarrow 2MnSO_4 + 5Fe_2(SO_4)_3 + K_2SO_4 + H_2O$$

The completion of the balancing is done by inspection. The following order of inspection is recommended:
1. Balance the metals.
2. Balance the nonmetals, including the radicals.
3. Balance the hydrogen.
4. Balance the oxygen.

$8H_2SO_4$ gives a total of 16 hydrogens on the left. The final balancing requires $8H_2O$ to produce 16 hydrogens on the right. The total number of oxygens are also balanced at this point.

The balanced equation is

$$2KMnO_4 + 10FeSO_4 + 8H_2SO_4 \rightarrow 2MnSO_4 + 5Fe_2(SO_4)_3 + K_2SO_4 + 8H_2O$$

Another example, more concisely developed, should reinforce the previous rules and principles.

$$MnO_2 + HCl \rightarrow MnCl_2 + Cl_2 + H_2O$$

1. Identify the redox participants.
   a. Chloride changes to free chlorine gas (clue #2).
   b. Manganese changes from Mn^{4+} in MnO_2 (manganese (IV) oxide) to Mn_2^+ in $MnCl_2$ (manganese (II) chloride) (clue #3).
2. Write the half reactions.
   a. Balance the chlorine:

$$2Cl^- \rightarrow Cl_2^0$$

   Going from negative one *up* to zero is oxidation. One mole of chloride ion loses one mole of electrons. Two moles of chloride ion lose two moles of electrons:

$$2Cl^- \rightarrow Cl_2^0 + 2e^-$$

   b. One mole of manganese (IV) ion gains two moles of electrons in decreasing its oxidation number to manganese (II).

$$Mn^{4+} \rightarrow Mn^{2+} - 2e^-$$

3. Combine and add the half-reactions:

$$\begin{array}{l} 2Cl^- \rightarrow Cl_2^0 + \cancel{2}e^- \text{ (oxidation)} \\ Mn^{4+} \rightarrow Mn^{2+} - \cancel{2}e^- \text{ (reduction)} \end{array}$$

add

$$\overline{Mn^{4+} + 2Cl^- \rightarrow Cl_2^0 + Mn^{2+}}$$

4. Replace the coefficients in the original equation.

$$MnO_2 + \underline{2}HCl \rightarrow MnCl_2 + Cl_2 + H_2O$$

5. Balance the equation by adjusting the number of chlorine units.

$$MnO_2 + 4HCl \rightarrow MnCl_2 + Cl_2 + H_2O$$

2 chlorines

4 chlorines

+

2 chlorines

4 chlorines

6. Balance the hydrogen and check the total number of oxygens.

2 oxygen   2 oxygen

$$MnO_2 + 4HCl \rightarrow MnCl_2 + Cl_2 + 2H_2O$$

4 hydrogen   4 hydrogen

Here is a summary of the six steps in the oxidation number method of balancing molecular equations.
1. Identify the elements involved in the redox activity by use of the three clues.
2. Write the half-reactions, showing the loss and gain of electrons.
3. Balance the number of atoms undergoing oxidation or reduction if subscripts are in the formulas.
4. Add the balanced half-reactions and transfer the coefficients to the original equation.
5. Balance the metals, nonmetals and radicals, and then hydrogen—in that order.
6. Check the oxygens for final balancing.

THE ION-ELECTRON METHOD FOR BALANCING IONIC REDOX REACTIONS

The ion-electron method is concerned with conservation of charge rather than individual oxidation numbers of elements. The same reaction previously investigated looks different when stripped of "spectator" ions.

$$KMnO_4 + FeSO_4 + H_2SO_4 \rightarrow MnSO_4 + Fe_2(SO_4)_3 + K_2SO_4 + H_2O$$

becomes

$$MnO_4^-(aq) + Fe^{2+}(aq) + H^+(aq) \rightarrow Mn^{2+}(aq) + Fe^{3+}(aq) + H_2O$$

The oxidizing and reducing agents and their products can be immediately identified. As familiarity is gained, an equation can be completed when only the reactants are given.

1. MnO_4^- (permanganate ion) in acid medium ($H^+[aq]$) is identified as an oxidizing agent that becomes reduced to manganese (II) ion.

$$MnO_4^- \rightarrow Mn^{2+}$$

2. The 4 units of oxygen on the left are balanced by the oxygen contained in 4 units of water on the right.

$$MnO_4^- \rightarrow Mn^{2+} + 4H_2O$$

3. Since the reaction occurs in acid medium, the 8 hydrogens on the right are balanced by $8H^+(aq)$ (hydrogen ion in water solution) on the left.

$$MnO_4^- + 8H^+ \rightarrow Mn^{2+} + 4H_2O$$

4. The total electronic charge on the left is $7+$ (one negative permanganate unit added to 8 positive hydrogen ion units). The total electronic charge on the right is $2+$, from the single unit of Mn^{2+}.

$$7+ \rightarrow 2+$$

This imbalance is adjusted by adding 5 electrons to the left or, as convention prefers, subtracting 5 electrons from the right.

$$MnO_4^- + 8H^+ \rightarrow Mn^{2+} + 4H_2O - 5e^-$$

5. Fe^{2+} (iron [II] ion) is a reducing agent that becomes oxidized to Fe^{3+} (iron [III] ion).

$$Fe^{2+} \rightarrow Fe^{3+}$$

6. Balancing the electronic charge is accomplished by adding one electron to the right side.

$$Fe^{2+} \rightarrow Fe^{3+} + e^-$$

7. Add the two half-reactions after balancing the loss and gain of electrons.

$$MnO_4^- + 8H^+ \rightarrow Mn^{2+} + 4H_2O - \cancel{5}e^-$$
add $$\underline{5Fe^{2+} \rightarrow 5Fe^{3+} + \cancel{5}e^-}$$

$$MnO_4^- + 5Fe^{2+} + 8H^+ \rightarrow Mn^{2+} + 5Fe^{3+} + 4H_2O$$

In conclusion, it can be seen that the ratio of 5 moles of Fe^{2+} reacting with 1 mole of MnO_4^- is exactly the same ratio as the stoichiometry of the molecular equation

$$2KMnO_4 \quad + \quad 10\,FeSO_4$$
$$2 \quad : \quad 10 \quad = \quad 1:5$$

Another example of this method may be seen in the reaction where ethanol (ethyl alcohol) is oxidized to acetic acid (ethanoic acid). In more familiar terms, this is what happens when wine turns to vinegar. The laboratory reaction is observed when potassium dichomate (oxidizing agent) is added to ethanol in an acidic medium provided by sulfuric acid. The ionic equation is

$$\underset{\text{ethanol}}{C_2H_5OH} + Cr_2O_7^{2-} + H^+ \rightarrow \underset{\text{acetic acid}}{C_2H_4O_2} + Cr^{3+} + H_2O$$

1. The dichromate ion is the oxidizing agent which becomes reduced to chromium (III) ion.

$$Cr_2O_7^{2-} \rightarrow Cr^{3+}$$

2. The 7 units of oxygen on the left are balanced by 7 units of water on the right.

$$Cr_2O_7^{2-} \rightarrow Cr^{3+} + 7H_2O$$

3. Since the reaction occurs in acid medium, the 14 hydrogens on the right are balanced by $14H^+(aq)$ on the left.

$$Cr_2O_7^{2-} + 14H^+ \rightarrow Cr^{3+} + 7H_2O$$

4. The 2 chromium units on the left are balanced by placing the co-efficient 2 in front of Cr^{3+}.

$$Cr_2O_7^{2-} + 14H^+ \rightarrow 2Cr^{3+} + 7H_2O$$

5. The total electronic charge on the left is $12+$ (2 negative dichromate and $14+$ hydrogen ion units). The total electronic charge on the right is $6+$ from the two units of chromium (III) ion.

$$12+ \rightarrow 6+$$

The imbalance is adjusted by subtracting 6 electrons from the right side.

$$Cr_2O_7^{2-} + 14H^+ \rightarrow 2Cr^{3+} + 7H_2O - 6e^-$$

6. The oxidation of ethanol to acetic acid half-reaction is balanced by adding water to the left, and hydrogen ion on the right.

$$C_2H_5OH + H_2O \rightarrow C_2H_4O_2 + 4H^+$$

7. The total electronic charge on the left is zero, and the total of $4+$ (due to $4H^+$) on the right is balanced by adding 4 electrons to that side.

$$0 = (4+) + (4e^-)$$

$$0 = 0$$

8. Balance the loss and gain of electrons and add the two half-reactions

multiply by 2 $(Cr_2O_7^{2-} + 14H^+ \rightarrow 2Cr^{3+} + 7H_2O - 6e^-)$

multiply by 3 $(C_2H_5OH + H_2O \rightarrow C_2H_4O_2 + 4H^+ + 4e^-)$

$$2Cr_2O_7^{2-} + 28H^+ \rightarrow 4Cr^{3+} + 14H_2O - \cancel{12}e^-$$

add $3C_2H_5OH + 3H_2O \rightarrow 3C_2H_4O_2 + 12H^+ + \cancel{12}e^-$

$$\overline{2Cr_2O_7^{2-} + 3C_2H_5OH + 28H^+ + 3H_2O \rightarrow 4Cr^{3+} + 3C_2H_4O_2 + 12H^+ + 14H_2O}$$

9. The equation should be simplified by subtracting $3H_2O$ and $12H^+$ from both sides.

$$2Cr_2O_7^{2-} + 3CrH_5OH + 28H^+ + 3H_2O\,4Cr^{3+} + 3C_2H_4O_2 + 12H^+ + 14H_2O$$
$$- 12H^+ - 3H_2O \qquad\qquad - 12H^+ - 3H_2O$$
$$2Cr_2O_7^{2-} + 3C_2H_5OH + 16H^+ \rightarrow 4Cr^3 + 3C_2H_4O_2 + 11H_2O$$

The previous examples of oxidation-reduction were in acidic media. An example of redox in a basic medium involves some modification.

Permanganate ion reacts with cyanide ion (a strong base since it is a vigorous proton acceptor) to produce manganese (IV) oxide and cyanate ion. Because of the CN^-, the reaction occurs in a basic environment.

$$MnO_4^- + CN^- \rightarrow MnO_2 + OCN^-$$

1. The MnO_4^- undergoes reduction to MnO_2.

$$MnO_4^- \rightarrow MnO_2$$

2. The 4 units of oxygen on the left compared to the 2 units on the right require water for balancing.

$$MnO_4^- \rightarrow MnO_2 + 2H_2O$$

3. The 4 hydrogens on the right are balanced by $4H^+$—just the same as in an acidic medium reaction!

$$MnO_4^- + 4H^+ \rightarrow MnO_2 + 2H_2O$$

4. Now, add $4\,OH^-$ units to both sides of the equation in order to convert the $4H^+$ to water

$$4H^+ + 4\,OH^- \rightarrow 4H_2O$$
$$MnO_4^- + 4H_2O \rightarrow MnO_2 + 2H_2O + 4\,OH^-$$

5. Simplify the half-reaction by subtracting $2H_2O$ from both sides.

$$MnO_4^- + 2H_2O \rightarrow MnO_2 + 4\,OH^-$$

6. Balance the electronic charge since the total on the left is 1 negative and the total on the right side is 4 negatives. Subtract 3 electrons from the right side.

$$1- = (4-) - (3e^-)$$
$$1- = 1-$$
$$MnO_4^- + 2H_2O \rightarrow MnO_2 + 4OH^- - 3e^-$$

7. The oxidation of cyanide to cyanate follows the same pattern.

$$CN^- \rightarrow OCN^-$$

8. Add water on the left to balance the oxygen and add H^+ on the right to balance the hydrogen in H_2O.

$$CN^- + H_2O \rightarrow OCN^- + 2H^+$$

9. Add $2OH^-$ to both sides of the equation in order to convert the H^+ to water.

$$CN^- + H_2O + 2OH^- \rightarrow OCN^- + 2H_2O$$

10. Simplify the half-reaction by subtracting one H_2O unit from both sides of the equation.

$$CN^- + 2OH^- \rightarrow OCN^- + H_2O$$

11. Balance the electronic charge by adding 2 electrons to the right side.

$$\underbrace{CN^- + 2OH^-}_{3 \text{ negatives}} \rightarrow \underbrace{OCN^- + 2e^-}_{1 \text{ negative} + 2 \text{ negatives}}$$

12. The two half-reactions are then balanced with regard to loss and gain of electrons, and finally added.

multiply by 2 $(MnO_4^- + 2H_2O \rightarrow MnO_2 + 4OH^- - 3e^-)$

multiply by 3 $(CN^- + 2OH^- \rightarrow OCN^- + H_2O + 2e^-)$

$$2MnO_4^- + 4H_2O \rightarrow 2MnO_2 + 8OH^- - \cancel{6}e^-$$

add $3CN^- + 6OH^- \rightarrow 3OCN^- + 3H_2O + \cancel{6}e^-$

$$\overline{2MnO_4^- + 3CN^- + 6OH^- + 4H_2O \rightarrow 2MnO_2 + 3OCN^- + 8OH^- + 3H_2O}$$

13. Simplify the final equation by subtracting $6OH^-$ units and $3H_2O$ units from both sides.

$$2MnO_4^- + 3CN^- + 6\cancel{OH^-} + 4H_2O \rightarrow 2MnO_2 + 3OCN^- + 8OH^- + 3\cancel{H_2O}$$
$$-\cancel{6}OH^- - 3H_2O \qquad\qquad -6OH^- -\cancel{3}H_2O$$

$$\overline{2MnO_4^- + 3CN^- + H_2O \rightarrow 2MnO_2 + 3OCN^- + 2OH^-}$$

Summary. The following checks should be applied to the balancing of redox equations when the ion-electron method is used.

1. The actual number of each species of atom must be balanced. H^+ and H_2O will have to be added in acidic medium and OH^- and water in basic medium.
2. Electrons must be added or subtracted on the right side of the half-reactions to achieve electrical neutrality.
3. The loss and gain of electrons must be balanced before the half-reactions are added.
4. The final equation must be simplified by removing excess H^+, OH^-, and H_2O.

REDOX STOICHIOMETRY

When an oxidation-reduction reaction is known to be a complete conversion of reactants to products (stoichiometric), the methods for performing useful calculations are essentially the same as those employed for nonredox reactions.

Gravimetric Problems

Whenever weights of solids are involved, the measurements required are described as gravimetric.

Example 1

Calculate the weight of sodium iodide required for the production of 5.0 g of iodine in the following reaction:

$$NaI + H_2SO_4 \rightarrow Na_2SO_4 + I_2 + H_2S + H_2O$$

1. CAUTION: The reaction should be carried out under a fume hood since the H_2S gas produced is both foul smelling and toxic.
2. Balance the equation to obtain the stoichiometry:

$$NaI + H_2SO_4 \rightarrow Na_2SO_4 + I_2 + H_2S + H_2O$$

a. Sulfur and iodine fit the clues (1 and 2) used to identify the redox participants:

$$2I^- \rightarrow I_2^0 + 2e^- \text{ (oxidation)}$$

$$2 + \boxed{?}\ \ 8- = 0 \quad \boxed{6+}$$

$$H_2^+SO_4^{2-} = H_2\ \ SO_4^{2-} = S$$

$$S^{6+} \rightarrow S^{2-} - 8e^- \text{ (reduction)}$$

b. Balancing the half-reactions and replacing the coefficients in the equation:

$$8I^- \rightarrow 4I_2^0 + \cancel{8e^-}$$
$$S^{6+} \rightarrow S^{2-} - \cancel{8e^-}$$
$$\overline{8I^- + S^{6+} \rightarrow 4I_2^0 + S^{2-}}$$

$$8NaI + H_2SO_4 \rightarrow Na_2SO_4 + 4I_2 + H_2S + H_2O$$

c. Completing the balancing by inspection:

$$8NaI + 5H_2SO_4 \rightarrow 4Na_2SO_4 + 4I_2 + H_2S + 4H_2O$$
8 moles ⸻ produce ⸻→ 4 moles

More simply, 2 moles of NaI will produce 1 mole of I_2.

3. Organize the data:

$$2 \text{ moles NaI} = 2 \times \text{g-molec. wt.} = 2 \times 150 \text{ g} = 300 \text{ g}$$

NaI
↙   ↘
$$23 + 127 = 150 \text{ g mole}^{-1}$$

$$1 \text{ mole } I_2 = 2 \times \text{g-atomic wt.} = 2 \times 127 \text{ g} = 254 \text{ g}$$

4. Solving by the proportion method:

$$\begin{array}{cc} X\text{ g} & 5\text{ g} \\ \text{NaI} \rightarrow & I_2 \end{array}$$

$$300 \text{ g} \quad\quad 254 \text{ g}$$

$$\frac{X}{300} = \frac{5}{254}$$

$$X = 5.9 \text{ g of NaI required}$$

5. Solving by the unit-factoring method:

mole fraction

$$\text{g of NaI} = \frac{5 \cancel{g}}{254 \cancel{g}} \times 300 \text{ g}$$

$$\text{g of NaI} = 5.9$$

Example 2

How many grams of solid potassium dichromate must be added to oxalic acid in acid medium to produce 500 ml of carbon dioxide measured at STP? The partial equation is

$$H_2C_2O_4 + Cr_2O_7^{2-} \text{ (acidic)} \rightarrow$$

1. Complete the reaction with the knowledge (or reference to a table of redox half-reactions) that $Cr_2O_7^{2-}$ becomes reduced to Cr^{3+} in acidic medium:

$$Cr_2O_7^{2-} \rightarrow 2Cr^{3+}$$

$H_2C_2O_4$ becomes oxidized to CO_2.

2. Balance the half-reactions:

$$Cr_2O_7^{2-} + 14H^+ \rightarrow 2Cr^{3+} + 7H_2O - 6e^-$$

multiply by 3 $(H_2C_2O_4 \rightarrow 2CO_2 + 2H^+ + 2e^-)$

$$Cr_2O_7^{2-} + 14H^+ \rightarrow 2Cr^{3+} + 7H_2O - \cancel{6}e^-$$

add $\qquad 3H_2C_2O_4 \rightarrow 6CO_2 + 6H^+ + \cancel{6}e^-$

$$Cr_2O_7^{2-} + 3H_2C_2O_4 + 8H^+ \rightarrow 2Cr^{3+} + 6CO_2 + 7H_2O$$

3. The pertinent stoichiometry is

$$1 \text{ mole } Cr_2O_7^{2-} \cdots \boxed{\text{produces}} \rightarrow 6 \text{ moles } CO_2$$

or $\qquad$ $1 \text{ mole } K_2Cr_2O_7 \rightarrow \boxed{\text{produces}} \rightarrow 6 \text{ moles } CO_2$

4. Organize the data:

1 mole $K_2Cr_2O_7$ = g-molec. wt. = 294 g

$2 \times 39 \qquad 7 \times 16$

$\qquad 2 \times 52$

1 mole of gas occupies $22.4 \ \ell$ at STP

6 moles occupy $6 \times 22.4 \ \ell = 134.4 \ \ell$

500 ml of $CO_2 = 0.5 \ \ell$

5. Solving the problem by the proportion method:

$$\frac{Xg}{294 \text{ g}} = \frac{0.5 \ \ell}{134.4 \ \ell}$$

$$X = 1.1 \text{ g of } K_2Cr_2O_7$$

Volumetric Problems

In a manner analogous to acid-base reactions proceeding toward a completion called "neutralization," redox reactions move toward an **end point** where electrons are no longer gained and lost. Volumetric problems are probably more useful than gravimetric, since most reactions occur in solutions, where the experimenter's concern is with volumes and concentrations. Two approaches toward solving practical problems will be discussed.

Solution stoichiometry

Example 3

What is the molar concentration of nitric acid if 20 ml is required to completely oxidize 4 g of copper to copper (II) nitrate in the reaction $Cu + HNO_3 \rightarrow Cu(NO_3)_2 + NO_2 + H_2O$?

1. The first and most critical step is to balance the equation:

$$Cu + HNO_3 \rightarrow Cu(NO_3)_2 + NO_2 + H_2O$$

$$1 + \overset{\uparrow}{\boxed{5+}} 6- = O \qquad Cu^0 \rightarrow Cu^{2+} + 2e^-$$

$$H^+NO_3^{2-} \qquad\qquad 2N^{5+} \rightarrow 2N^{4+} - 2e^-$$

$$Cu^0 + 2N^{5+} \rightarrow Cu^{2+} + 2N^{4+}$$

$$Cu + 4HNO_3 \rightarrow Cu(NO_3)_2 + 2NO_2 + 2H_2O$$

$$1 \text{ mole} + 4 \text{ moles} \rightarrow \text{completion}$$

2. Organize the data:

$$1 \text{ mole Cu} = \text{g-atomic wt.} = 63.5 \text{ g mole}^{-1}$$

$$\text{moles of } HNO_3(aq) = MV$$
$$\text{(molarity)(volume in liters)}$$

$$20 \text{ ml} = 0.02 \ \ell$$

3. Using the proportion method:

$$\begin{array}{cc} 4 \text{ g} & X \text{ moles} \\ Cu + & 4HNO_3 \rightarrow \\ 63.5 \text{ g} & 4 \text{ moles} \end{array}$$

Since moles = MV

$$\begin{array}{cc} 4 \text{ g} & M(0.02 \ \ell) \\ Cu = & HNO_3 \\ 63.5 \text{ g} & 4 \text{ moles} \end{array}$$

$$\frac{4}{63.5} = \frac{M(0.02)}{4}$$

cross multiply

$$0.02(63.5)M = 4(4)$$

$$M = \frac{16}{0.02(63.5)} = \frac{16}{1.27} = 12.4$$

20 ml of 12.4 M HNO_3 are needed.

Example 4

What molar concentration of MnO_4^-(aq) is used if 40 ml is needed to completely convert 25 ml of 0.2 M Fe^{2+}(aq) to Fe^{3+}(aq) in the reaction

$$MnO_4^- + 5\ Fe^{2+} + 8H^+ \rightarrow Mn^{2+} + 5Fe^{3+} + 4H_2O$$

1. The stoichiometry states that 1 mole of MnO_4^- reacts with 5 moles of Fe^{2+}.

2. Using the proportion method and remembering that moles = MV:

$$MV = MV$$
$$\text{(moles)} \quad \text{(moles)}$$

$$MnO_4^- = 5\ Fe^{2+}$$
$$\text{1 mole} \quad \text{5 moles}$$

3. Substituting the data:

$$\frac{M(0.04\ \ell)}{\text{1 mole}} \quad \frac{(0.025\ \ell)(0.2\ M)}{\text{5 moles}}$$

4. Cross multiply:

$$M(0.04)(5) = (0.025)(0.2)$$

$$M = \frac{0.025(0.2)}{0.04(5)} = 0.025$$

40 ml of 0.025 M MnO_4^-(aq) used.

Example 5

For versatility, convert the 0.025 M concentration to normal concentration.

1. Recall that the first chapter on solutions developed the equation for interconverting molarity and normality:

$$N = \text{(valence) (M)}$$
$$\text{or}$$
$$N = \text{(number of replaceable } H^+ \text{ or } OH^-\text{) (M)}$$

2. Extending the original reasoning to the specific example of MnO_4^-(aq), it is observed that 1 mole of MnO_4^- gains five moles of electrons in the redox reaction. Therefore $\frac{1}{5}$ mole of MnO_4^- gains one mole of electrons, which means that $\frac{1}{5}$ mole is the equivalent weight.

3. Since normality is based on equivalent weights, one molar MnO_4^- is equal to 5 normal.

4. A new equation is obtained for interconverting molarity and normality in the case of oxidizing and reducing agents:

$$N = \left(\begin{array}{c}\text{number of moles of electrons} \\ \text{gained or lost}\end{array}\right)(M)$$

5. The solution to the problem is

$$N = (0.025 \text{ M})(5 \text{ moles of electrons})$$
$$N = 0.125$$

Redox titration. The principal advantage of using normal concentrations for oxidation-reduction titrations is that the stoichiometry is not required for determining unknown concentrations. Just as neutralization in acid-base titrations is described as the point when equivalent numbers of H^+(aq) and OH^-(aq) have formed water, redox titrations are concerned with balancing the number of equivalents of oxidizer and reducer. The titration end point in this case is marked by a stoppage in the electron transfer from the reducing agent to the oxidizing agent.

The end point of a redox titration can sometimes be seen as a sudden and sharp color change. If this advantage is not present, a **potentiometric titration** is necessary. A pH meter employs the principles of potentiometric titration in acid-base reactions. This subject will be discussed in the next chapter.

Consider the MnO_4^-(aq) and Fe^{2+}(aq) redox system again. This time a laboratory titration will be described. A color change is employed in this system as an end point indicator. MnO_4^-(aq) is a deep purple color. When MnO_4^-(aq) has been completely reduced to Mn^{2+}(aq) the purple color disappears. This marks the end point of the titration.

$$MnO_4^-\text{(aq)} \rightarrow Mn^{2+}\text{(aq)}$$

$$\text{purple} \qquad\qquad \text{colorless}$$

Example 6

Pour 40 ml of a solution of an unknown concentration of $KMnO_4$ into a beaker, add a stirring bar and a few drops of concentrated H_2SO_4, and place the beaker on a magnetic stirrer. Fill a 50 ml buret with 0.2 N $FeSO_4$ solution and add it carefully to the $KMnO_4$ solution. Stop immediately as the purple color disappears. Calculate the normality of the $KMnO_4$ from the data.

1. Set up the apparatus (Fig. 9-2).

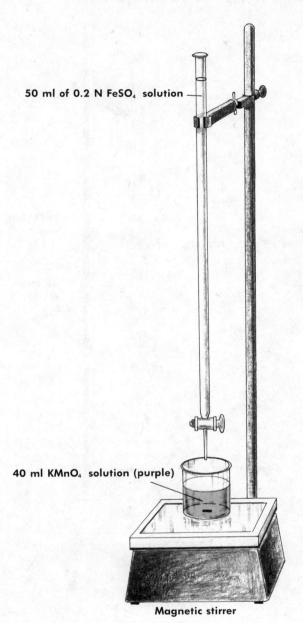

50 ml of 0.2 N FeSO₄ solution

40 ml KMnO₄ solution (purple)

Magnetic stirrer

Figure 9-2.

2. Perform the experiment and record results.

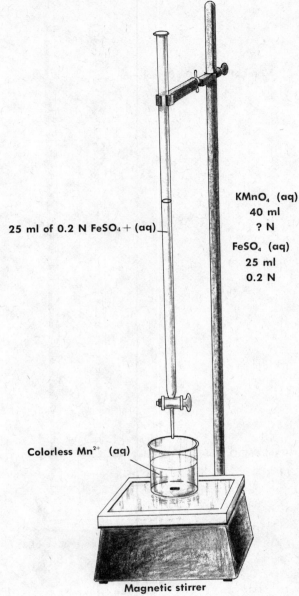

25 ml of 0.2 N FeSO₄ + (aq)

KMnO₄ (aq)
40 ml
? N

FeSO₄ (aq)
25 ml
0.2 N

Colorless Mn²⁺ (aq)

Magnetic stirrer

Figure 9-3.

3. At the end point:

number of equivalents = number of equivalents
of oxidizing agent   of reducing agent

Number of equivalents = NV (liters)

Number of milliequivalents = NV (milliliters)

Therefore $NV_{(ox)} = NV_{(red)}$

4. Substitute the data:

$$(40 \text{ ml})(X \text{ N}) = (25 \text{ ml})(0.2 \text{ N})$$

$$N = \frac{25(0.2)}{40} = 0.125$$

The $KMnO_4$ is 0.125 N

This answer agrees with the result obtained in the solution stoichiometry approach.

Example 7

A solution of oxalic acid contains 0.75 g of $H_2C_2O_4 \cdot 2H_2O$. In a titration, 20.0 ml of $KMnO_4(aq)$ was required to complete the redox reaction. Calculate the normal and molar concentration of the $KMnO_4$ solution.

1. Calculate the number of equivalents of oxalic acid from the data.

   a. $H_2C_2O_4 \cdot 2H_2O = 2 + 24 + 64 + 36 = 126$ g mole^{-1}

   $2 \times 1 \quad 4 \times 16 \quad 2 \times 18$

   2×12

   b. The half-reaction:

   $$H_2C_2O_4 \rightarrow 2CO_2 + 2H^+ + 2e^-$$

   This indicates 2 moles of electrons lost per mole of $H_2C_2O_4$ oxidized.

   c. Therefore, the equivalent weight $= \dfrac{\text{g-molec. wt.}}{2}$

   $$\text{Equiv. wt.} = \frac{126 \text{ g}}{2} = 63 \text{ g}$$

   d. The number of equivalents is calculated from the equation,

   $$\text{Number of equiv.} = \frac{g}{\text{g-equiv. wt.}}$$

   $$\text{Number of equiv.} = \frac{0.75 \text{ g}}{63 \text{ g}} = 0.012$$

2. In a titration, the number of equivalents of oxidizer equals the number of equivalents of reducer. For $KMnO_4$, the number of equiv. = NV (liters); for $H_2C_2O_4 \cdot 2H_2O$, the number of equiv. = 0.012.

Therefore,

$$NV_{(ox)} = 0.012_{(red)}$$

$$N = \frac{0.012}{0.020\ell} = 0.60$$

The $KMnO_4$ is 0.60 N.

3. The half-reaction for MnO_4^- reduction is

$$MnO_4^- + 8H^+ \rightarrow Mn^{2+} + 4H_2O - 5e^-$$

Converting normality to molarity,

$$M = \frac{N}{\text{(moles of electrons gained or lost)}} = \frac{0.6}{5} = 0.12 \text{ M}$$

It has been repeatedly emphasized that oxidation-reduction is a matter of electrons being transferred between oxidizing and reducing agents. A movement of electric charge is what electricity is all about. The next chapter will explore the relationship between oxidation-reduction and electricity in various interesting and practical ways.

PROBLEMS

Set A

1. Define the following terms:

a. oxidizing agent
b. oxidation
c. stoichiometric
d. half-reaction
e. ternary compound

2. Label the following changes as oxidation or reduction:

a. MnO_4^- becomes Mn^{2+}
b. SO_4^{2-} becomes SO_3^{2-}
c. Cr^{3+} becomes $Cr_2O_7^{2-}$
d. Zn^{2+} becomes Zn°
e. Cl^- becomes Cl_2°

3. Calculate the oxidation number of the underlined elements:

a. $H\underline{Cl}O_3$
b. $\underline{S}O_2$
c. $\underline{N}O_3^-$
d. $K\underline{I}O_3$
e. $Na_2\underline{S}O_4$

4. Balance the following equations:

a. $Bi_2S_3 + HNO_3 \rightarrow Bi(NO_3)_3 + S + NO + H_2O$
b. $MnO + PbO_2 + HNO_3 \rightarrow HMnO_4 + Pb(NO_3)_2 + H_2O$
c. $K_2Cr_2O_7 + HCl + FeCl_2 \rightarrow CrCl_3 + FeCl_3 + KCl + H_2O$
d. $PbCrO_4 + KI + HCl \rightarrow PbCl_2 + CrCl_3 + KCl + I_2 + H_2O$
e. $AuCl_3 + H_2C_2O_4 \rightarrow Au_2 + HCl + CO_2$
f. $KMnO_4 + HBr \rightarrow MnBr_2 + KBr + Br_2 + H_2O$

5. Balance the following equations by the ion-electron method.

a. $Ag + NO_3^- \rightarrow Ag^+ + NO$ (acidic medium)
b. $Mn^{2+} + Br_2 \rightarrow MnO_2 + Br^-$ (basic medium)
c. $CrO_4^{2-} + HSnO_2^- \rightarrow HSnO_3^- + CrO_2^-$ (basic medium)
d. $Cr_2O_7^{2-} + I^- \rightarrow Cr^{3+} + I_2$ (acidic medium)
e. $PbO_2 + Cl^- \rightarrow Pb^{2+} + Cl_2$ (acidic medium)

6. How many milligrams of sulfur can be obtained by reacting 50.0 mg of zinc with sulfuric acid? The unbalanced equation is

$$Zn + H_2SO_4 \rightarrow ZnSO_4 + S + H_2O$$

7. What volume (milliliters) of 0.3 M nitric acid is required to produce 300 ml of NO gas in the following unbalanced reaction equation:

$$PbS + HNO_3 \rightarrow PbSO_4 + NO + H_2O$$

8. What is the molar concentration of MnO_4^-(aq) if 75 ml is required to completely oxidize 10.0 g of pure methanol (methyl alcohol) to formaldehyde? The unbalanced equation is

$$MnO_4^- + CH_3OH + H^+ \rightarrow MnO_2 + CH_2O + H_2O$$
methanol $\qquad\qquad$ formaldehyde

9. What is the normality of a solution of $KMnO_4$ if 60.0 ml is titrated to the end point with 45.0 ml of 0.5 N $Na_2C_2O_4$ in acid medium?

10. What is the normality and molarity of $KMnO_4$ if 40 ml is titrated to the end point with 0.4 g of KCN dissolved in water? The unbalanced equation is $MnO_4^- + CN^- + H_2O \rightarrow Mn^{2+} + OCN^- + OH^-$.

Set B

11. Define the following terms:

a. redox
b. reducing agent
c. redox titration
d. titration end point
e. equivalent weight (in oxidation-reduction)

12. Label the following changes as oxidation or reduction:

a. Mn^{2+} becomes MnO_2
b. IO_3^- becomes I_2
c. I^- becomes I_2
d. SO_4^{2-} becomes S^{2-}
e. $S_2O_3^{2-}$ becomes S°

13. Calculate the oxidation number of the underlined elements:

a. $\underline{Sn}SO_4$
b. $\underline{N}H_4^+$
c. $H\underline{Cl}O_4$
d. $\underline{As}O_4^{3-}$
e. $\underline{S}_2O_3^{2-}$

14. Balance the following equations:

a. $HgS + HNO_3 + HCl \rightarrow HgCl_2 + S + NO + H_2O$
b. $CuS + HNO_3 \rightarrow Cu(NO_3)_2 + S + NO + H_2O$
c. $KMnO_4 + HCl \rightarrow MnCl_2 + KCl + Cl_2 + H_2O$
d. $BiCl_3 + Na_2SnO_2 + NaOH \rightarrow Bi + Na_2SnO_3 + NaCl + H_2O$
e. $K_2Cr_2O_7 + H_2S + HCl \rightarrow CrCl_3 + KCl + S + H_2O$
f. $NaI + H_2SO_4 \rightarrow Na_2SO_4 + I_2 + H_2S + H_2O$

15. Balance the following equations by the ion-electron method.

a. $MnO_4^- + H_2S \rightarrow Mn^{2+} + S$ (acidic medium)
b. $Cr_2O_7^{2-} + HNO_2 \rightarrow Cr^{3+} + NO_3^-$ (acidic medium)
c. $HS^- + MnO_4^- \rightarrow HSO^- + MnO_2$ (basic medium)
d. $MnO_4^- + NO_2^- \rightarrow MnO_2 + NO_3^-$ (basic medium)
e. $H_2C_2O_4 + MnO_2 \rightarrow Mn^{2+} + CO_2$ (acidic medium)

16. How many milliliters (at STP) of NO_2 gas will be liberated when 60 mg of copper is added to an excess of concentrated nitric acid? The unbalanced equation is

$$Cu + HNO_3 \rightarrow Cu(NO_3)_2 + NO_2 + H_2O$$

17. What is the molar concentration of a KI solution if 42 ml are used in the formation of 10 mg of iodine in the following unbalanced reaction equation?

$$KI + K_2Cr_2O_7 + H_2SO_4 \rightarrow Cr_2(SO_4)_3 + K_2SO_4 + I_2 + H_2O$$

18. How many grams of copper are needed to completely reduce the silver ion in 50 ml of 0.4 N $AgNO_3$ (aq)?

19. What is the normality of an iodine solution if 2.5 ml is titrated to the end point (iodine color disappears) with 26.2 ml of 0.05 N $S_2O_3^{2-}$ (aq) (thiosulfate ion)?

20. Calculate the normality and molarity of a $K_2C_2O_4$ solution if 35.0 ml is titrated to the end point with 0.46 g of $KMnO_4$ dissolved in water. The unbalanced equation is

$$MnO_4^- + C_2O_4^{2-} + H^+ \rightarrow Mn^{2+} + CO_2 + H_2O$$

SUPPLEMENTARY READING

Masterton, W. L., and Slowinski, E. J.: *Chemical Principles*, 2nd ed. W. B. Saunders Company, Philadelphia, 1969.
  Excellent chapters on oxidation-reduction.

Kirschbaum, J.: Biological Oxidations and Energy Conservation. *J. Chem. Educ.*, January, 1968.
  A fine paper on biological oxidations.

Taube, H.: How Do Redox Reactions Occur? *Chemistry*, March, 1965.
  Answers to a fundamental question indicated by the title of the article.

Electrochemistry 10

Electricity is commonly defined as the flow of electrons or ions. A metal wire usually provides a pathway for electron movement while a liquid medium allows for the mobility of ions. When an oxidation-reduction reaction is controlled by the reactants that are selected or the physical apparatus constructed, the gained and lost electrons may become electric current. The chemical system producing an electron source is called a **voltaic cell** or **electrochemical cell.** The voltages produced may be harnessed to perform useful work such as that obtained from dry cells (batteries) and fuel cells, which are common examples of electrochemical cells. Electron flow may be measured by galvanometers, and voltages may be measured by sensitive voltmeters. These data help us to understand why some redox reactions occur spontaneously while others will not. The energy absorbed or lost in a chemical reaction may be determined from the electrical measurements. This increases man's knowledge of chemical events in living organisms. The movement of ions also permits the application of gold, silver, platinum, and chromium coatings to cheaper metals.

COULOMBS, FARADAYS, AND THE MOLE

The mystical appearance of Avogadro's number early in the text can be profitably investigated at this time. If 12 grams of carbon 12 are really composed of 6.02×10^{23} atoms, then a mole (or the gram-atomic weight, 107.87 grams) of silver should also be composed of Avogadro's number of atoms.

Since the reduction of silver ion is represented by the equation $Ag^+(aq) + e^- \rightarrow Ag^0(s)$, the stoichiometry is one mole of silver ion gains one mole of electrons in becoming one mole of silver metal. The experimental technique is to "count" the number of electrons gained in the process of depositing 107.87 grams of silver. Of course, a lesser amount of silver could be deposited and calculations could then be adjusted by the arithmetic of proportionality. The depositing of silver is accomplished by establishing a **cathode** (negative electrode), which is a metal plate connected to the negative

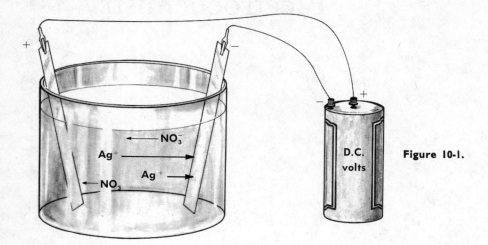

Figure 10-1.

terminal of a D.C. power supply. The positive silver ions will migrate toward the cathode, gain electrons, and stick to the cathode as a silver coating (Fig. 10-1).

In order to "count" the electrons, the relationships between current, time, and units of electrical charge must be established. The unit of electrical charge is called a **coulomb**. A coulomb is the amount of electrical charge that crosses a surface in one second at a current of one ampere. In the early

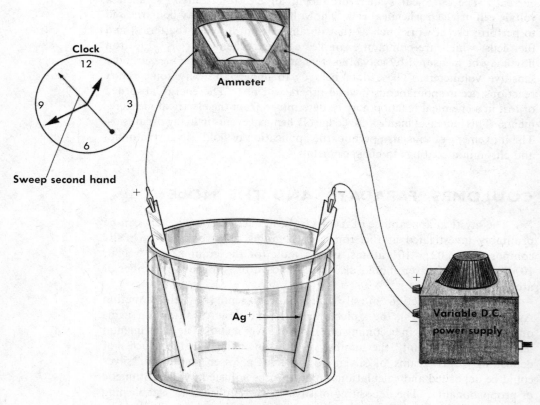

Figure 10-2.

part of this century, an American scientist, R. A. Millikan, performed an experiment (known as the "oil drop experiment") in which he found the charge on a single electron to be 1.6×10^{-19} coulomb. A very small bit of electricity indeed. The measurement of Avogadro's number requires an ammeter and a stopwatch. The calculation of the number of coulombs needed to deposit 107.87 grams of silver is achieved by multiplying the amperage by the number of seconds required. The apparatus is shown in Figure 10-2 The relationship is expressed in this equation:

$$Q = It$$
coulombs amps seconds

If the experiment is performed at a constant current for many seconds, the answer obtained is 9.65×10^4 coulombs needed to deposit 107.87 grams of silver. The value 9.65×10^4 coulombs is called a **faraday** and is symbolized by the letter $\mathscr{F}$. Calculating Avogadro's number:

$$\frac{9.65 \times 10^4 \text{ coul. mole}^{-1}}{1.6 \times 10^{-19} \text{ coul. (per electron)}} = 6.02 \times 10^{23} \text{ electrons per mole}$$

Example I

If a current of 2.0 amps is passed through molten $AlCl_3$ for ten minutes, how many grams of aluminum will be deposited on the cathode?

1. Write the half-reaction for the reduction of Al^{3+} ion.

$$Al^{3+} + 3e^- \rightarrow Al^0$$

1 mole of aluminum gains 3 moles of electrons
Therefore, $\frac{1}{3}$ mole of aluminum gains 1 mole of electrons.

2. Organize the data:

amps   2.0
time   10 min $= 6.00 \times 10^2$ seconds

3. Substitute in the equation:

$$Q = It = 2.0(6.00 \times 10^2)$$

$$Q = 1.2 \times 10^3 \text{ coulombs}$$

4. Reasoning that one faraday would allow 9 grams of aluminum ($\frac{1}{3}$ mole) to be deposited, the weight deposited by 1.2×10^3 coulombs is

$$\frac{1.20 \times 10^3 \text{ coul.}}{9.65 \times 10^4 \text{ coul.}} \times 9 \text{ g} = 1.12 \times 10^{-1} \text{ g} = 0.112 \text{ g of aluminum}$$

Coulometry

Coulometry is an analytical method in chemistry which determines an amount of chemical change by using the fact that a faraday is the number of coulombs in one mole of electrons. It is one mole of electrons that is gained or lost by the gram-equivalent weight of a substance in the processes of oxidation-reduction, electrolysis, and electrodeposition.

Electrolysis is the forced breakdown of a substance by an external current. For example, water can be decomposed into hydrogen and oxygen by electrolysis (Fig. 10-3).

Water is forced to ionize into $H^+(aq)$ and $OH^-(aq)$:

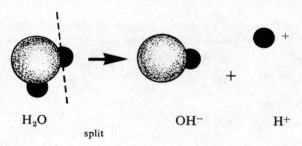

H_2O	OH^-	H^+

split

and the reaction at the cathode is

$$2H^+ \rightarrow H_2 - 2e^-$$

while the anodic oxidation is

$$2OH^- \rightarrow \tfrac{1}{2}O_2 + H_2O + 2e^-$$

The overall reaction may be written

$$H_2O \rightarrow H_2 + \tfrac{1}{2}O_2$$

Another example is found in the separation of sodium and chlorine from molten salt (Fig. 10-4). In this case, sodium ions become reduced to free sodium by reduction at the cathode.

$$Na^+ \rightarrow Na^0 - e^-$$

The chloride ion is oxidized to chlorine gas at the anode.

$$2Cl^- \rightarrow Cl_2 + 2e^-$$

Electrodeposition is the process linked to electrolysis when the oxidation or reduction of some species results in the coating or plating of an element on a selected electrode. One practical application of this, other than routine operations such as silver plating spoons, is the **platinizing** of electrodes for measuring other electrochemical events in solution. The platinizing of an electrode consists of using the platinum ion in solution (this solution is commercially available) to coat a smooth piece of platinum metal with a dark, granular coating of platinum grains. This coating has the effect of

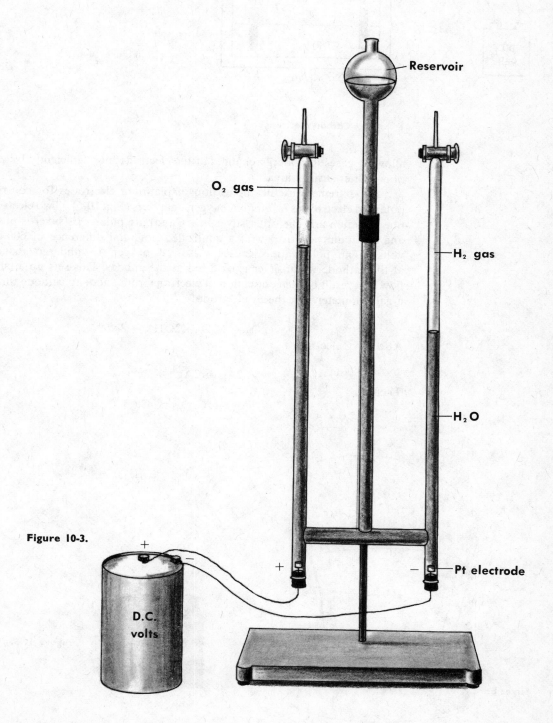

Reservoir

O₂ gas

H₂ gas

H₂O

Figure 10-3.

+

−

+

−   **Pt electrode**

D.C. volts

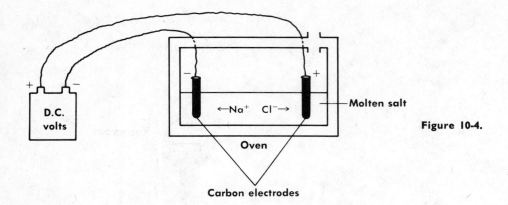

Figure 10-4.

allowing gases to undergo changes at the electrode more efficiently by virtue of a greater surface area.

An extremely useful application of platinum electrodes for controlled potential electrolysis is found in oxygen analyzers (Fig. 10-6). Two electrodes, one platinum and the other silver in KCl(aq) are polarized (silver is positive and platinum negative) with a controlled potential difference of 600 millivolts. The polarization causes dissolved oxygen to undergo reduction at the cathode while silver is oxidized at the anode. This sets up a current flow as a result of the conduction of electrons from anode to cathode through a galvanometer. At the Pt electrode:

$$\tfrac{1}{2}O_2 + H_2O \rightarrow 2OH^- - 2e^-$$

At the Ag electrode:

$$2Ag \rightarrow 2Ag^+ + 2e^-$$

Then

$$Ag^+ + Cl^- \rightarrow AgCl(s)$$

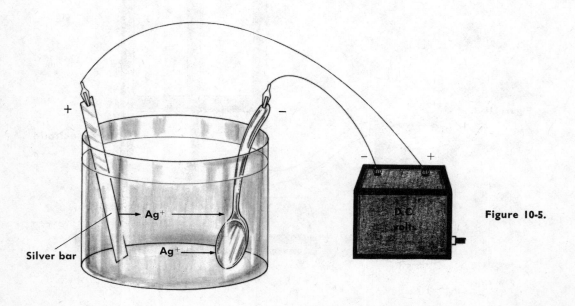

Figure 10-5.

Figure 10-6. A biological oxygen monitor which gives Warburg-type data in minutes.

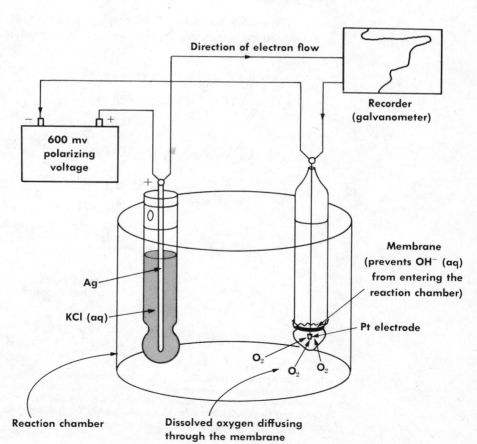

Direction of electron flow

Recorder (galvanometer)

− 600 mv polarizing voltage +

Membrane (prevents OH⁻ (aq) from entering the reaction chamber)

Ag

KCl (aq)

Pt electrode

O_2

O_2 O_2

Figure 10-7.

Reaction chamber

Dissolved oxygen diffusing through the membrane

This device is used in biochemical analyses to determine the rate and extent of oxygen uptake by living cells. As living cells use up the oxygen, there is less available for reduction at the electrode and the current drops. By comparing the results of experimental samples with established standards, the galvanometer provides investigators with accurate measurements of respiratory rates. This example of oxygen determination may be considered as a practical example of a polarographic analysis.*

Another use of platinum electrodes is for the determination of standard electrode potentials of redox half-reactions. This application will be discussed later in this chapter.

The relationship between chemical change and coulometry is summarized in this equation:

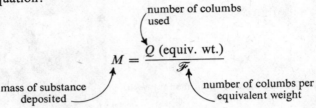

$$M = \frac{Q \text{ (equiv. wt.)}}{\mathscr{F}}$$

number of columbs used

mass of substance deposited

number of columbs per equivalent weight

Example 2

What weight of iron from a solution of iron (II) sulfate will be deposited on a cathode by a current of 5 amps after 3 minutes?

1. Calculate Q:

$$3 \text{ min} = 180 \text{ sec}$$

$$Q = It = 5 \text{ A} \, (180 \text{ sec})$$

$$Q = 900 \text{ coulombs}$$

2. The equivalent weight of iron in $FeSO_4$ is

$$Fe = 55.85 \text{ g mole}^{-1}$$

$$Fe^{2+} \rightarrow Fe^0 - 2e^-$$

$$\text{The equiv. wt.} = \frac{\text{mass of a mole}}{N}$$

numbers of moles of electrons lost or gained

$$\text{equiv. wt.} = \frac{55.85}{2} = 27.92$$

---

* Many polarographic analyses are performed by using a mercury-dropping electrode coupled with a standard calomel (Hg_2Cl_2) reference electrode. The tiny drops of mercury coming from a capillary tube have two significant advantages: (1) each drop of mercury is fresh and therefore cannot be poisoned, and (2) the increasing surface of an emerging drop compensates for any decrease in current.

3. Develop the equation:

$$\text{Mass} = \frac{Q}{\mathscr{F}} \times \text{Equiv. wt.} = \frac{9 \times 10^2}{9.65 \times 10^4} \times 27.92$$

Mass deposited $= 0.26$ g

Conductimetry

There is a great deal of research in chemistry and biology requiring water that is free of ions. When an investigator wants to observe the effect of added metal ions in a reaction, the experimental results would be worthless if the water contained a variety of metal ions before any additions were made. While water may be triple-distilled in quartz distillation apparatus to assure its being ion-free, a method called **conductimetry** could check this assumption. Many industries using water would be expected to perform conductimetric determinations of ion concentrations.

Conductimetry is based on the observation that ions in water solution are able to conduct electricity. Any substance that forms mobile ions in solution is called an **electrolyte**. A qualitative example of this effect is provided by an electrolyte tester. This is a simple circuit using household current (A.C. at 60 cycles per second). If an electrolyte is in solution, the circuit is completed and the bulb lights up. A weak electrolyte, which does not form ions extensively, allows the bulb to shine dimly. And a nonelectrolyte, such as ion-free water, does not complete the circuit and no light appears (Fig. 10-8). When the fundamental principle of electrolyte conductance is rigidly controlled, precise conductimetric measurements can be made.

The apparatus consists basically of two platinum electrodes placed in the solution to be tested. Alternating current is used to avoid the effects of polarization which were discussed in coulometry. The frequency may be varied, and it usually is. A frequency of 1000 Hz (cycles per second) seems to operate more satisfactorily than the standard 50 Hz. The current that flows is affected by the applied voltage, the resistance of the solution between the electrodes, the concentration of ions, and the area of the platinum. It is a matter of Ohm's law which says that

$$\overset{\text{volts}}{\underset{\text{amps}}{I} = \underset{\text{ohms}}{V/R}}$$

Conductance, which is a measure of the efficiency with which the ions conduct electricity, is described as the **reciprocal of the resistance**, $1/R$, or ohms^{-1}. A special name and symbol have been coined for ohms^{-1}, which is ohm spelled backwards—**mho**. The symbol for the mho is L. The equation can be rewritten to read

$$\underset{\text{amps}}{I} = \underset{\text{volts}}{V} \quad \underset{\text{mhos}}{L}$$

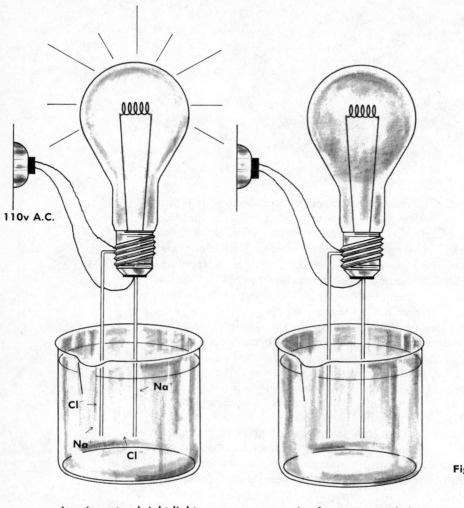

110v A.C.

Na⁺

Cl⁻ →

Na⁺

Cl⁻

Figure 10-8.

Ions in water, bright light   Ion-free water, no light

When nearly pure water is measured for conductance, as an evaluation of its ion-free condition, the value for conductance may be described in micromhos (μL).

ELECTROCHEMICAL CELLS

In the beginning of Chapter 9, an oxidation-reduction reaction between zinc metal and copper (II) ion was described in which electrons were transferred from zinc metal to the copper (II) ion. The electrochemical or **voltaic** cell is based on the physical separation of the oxidizing and reducing agents. An electrolyte bridge links the separated sections so that ions can flow even though the electron transfer cannot occur directly. The developed voltage acts as a driving force for the movement of the electrons through an external wire. This flow of electrons may be put to useful service as it is in the case of any "battery."

The tendency of atoms and ions to undergo oxidation or reduction is a chemical property peculiar to structure of the particular atom or ion. When a substance that stands to lose electrons easily (oxidation) is coupled with one that strongly attracts electrons (reduction) the components of a good voltaic cell are at hand. The separation of the zinc metal from the copper (II) ion is an example of a voltaic cell (Fig. 10-9). In Figure 10-10, the zinc metal (anode) undergoes oxidation as it ionizes.

$$Zn \rightarrow Zn^{2+} + 2e^-$$

The electrons pile up on the zinc metal. At the copper electrode, the copper (II) ion becomes reduced to copper metal as it removes electrons (Fig. 10-11). The salt bridge permits the oxidation to occur because of the fact that the two electrodes are "joined," or the circuit is completed by the mobility of ions. The ions that compose the salt bridge do not enter into the oxidation-reduction since the other ions in the system are more easily oxidized and reduced. Since electrons are building up on the zinc at the same time as they are being depleted from the copper, a potential difference is developed. The connection of the two electrodes by an external wire (a good electron conductor) provides a pathway for electron flow. The voltmeter measures the developed voltage. Under standard conditions of 25° C and ion concentration of 1 molar, the observed voltage is 1.10 volts. This is usually written as $\mathscr{E}^0_{cell} = +1.10$ v. The superscript of zero to the upper right of the symbol ($\mathscr{E}^0$) means standard conditions. The reaction continues until the oxidation of zinc and the reduction of copper (II) ion is completed. When this happens the electrochemical cell no longer functions. When the redox reaction in a flashlight battery is complete, it is said that the battery is "dead."

It is interesting to note that while copper is the site of the reduction of copper (II) ion when zinc is in the system, it may assume an opposite role with some other metal ion that is more easily reduced. An example is found

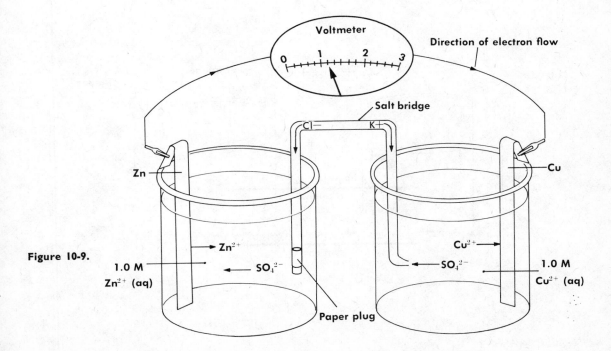

Figure 10-9.

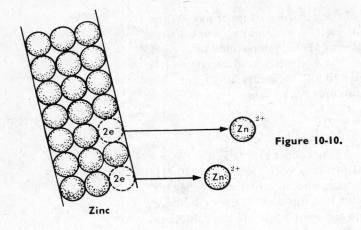

Figure 10-10.

Zinc

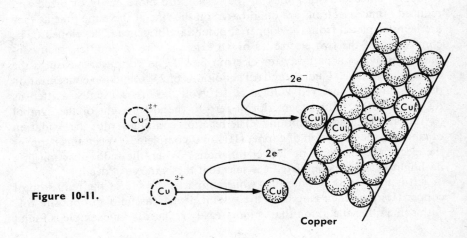

Figure 10-11.

Copper

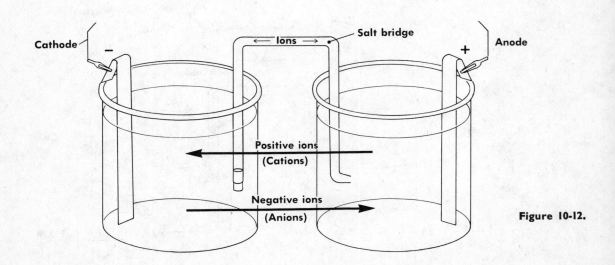

Figure 10-12.

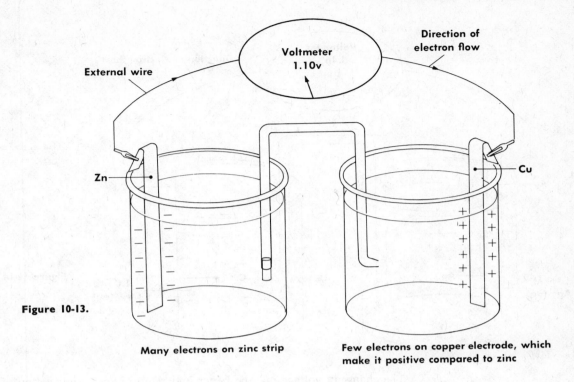

Figure 10-13.

Many electrons on zinc strip   **Few electrons on copper electrode, which make it positive compared to zinc**

in a copper–copper (II) ion and silver ion–silver system. Here, the silver ion undergoes reduction while the copper metal is oxidized to copper (II) ion. The voltage produced is read on the voltmeter as +0.46 volts (Fig. 10-14). The reaction $Cu^0 \rightarrow Cu^{2+} + 2e^-$ causes an electron build-up for the copper while the reaction $2Ag^+ \rightarrow 2Ag^0 - 2e^-$ results in a removal of electrons from the silver. Thus, a voltage is developed:

$$\mathscr{E}^0_{cell} = +0.46 \text{ v}$$

In the examples described, standard conditions were specified. A question now presents itself. What effect will nonstandard conditions have on the voltage developed? A qualitative answer is appropriate here because the mathematics involved in the development of an equation that incorporates the nonstandard parameters (known as the Nernst equation) requires information beyond the scope of this text. The Nerst equation is

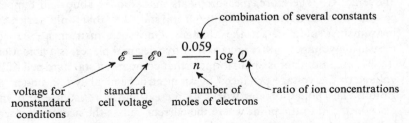

combination of several constants

$$\mathscr{E} = \mathscr{E}^0 - \frac{0.059}{n} \log Q$$

voltage for nonstandard conditions   standard cell voltage   number of moles of electrons   ratio of ion concentrations

But the use of an equation without the understanding of how it was developed is intellectually frustrating. It is also not germane to the needs of a laboratory technician.

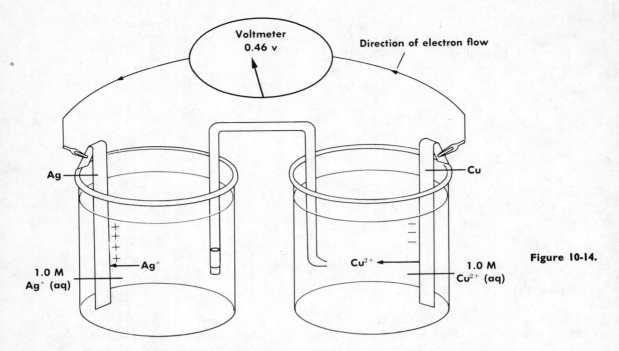

Figure 10-14.

The change in voltage may be better understood by an appreciation of **Le Chatelier's principle**. Le Chatelier collected empirical evidence to support the principle which says, in effect, that when a "stress" is applied to a system that is in a balanced state (equilibrium), the concentrations of reactants and products will shift in such a way so as to restore the balance.

Consider the standard condition of one molar ion concentration in the voltaic cell. If the ion concentration of just the zinc ion were decreased, would this increase or decrease the cell voltage? Le Chatelier's principle indicates that the system would change in a way that would restore the balance. In effect, the rate of zinc metal oxidation to zinc ion would increase. The result would then be an increase in the cell voltage because of the increased build-up of electrons on the zinc electrode (Fig. 10-15). Instead of $\mathscr{E}^0_{cell} = +1.10$ volts, the $\mathscr{E}_{cell}$ which is 0.1 M Zn^{2+}(aq) is increased to $+1.13$ volts.

Compared to the effect that concentration changes have on standard cell voltage, the variation of temperature from 25° C is often negligible. The cell voltage only increases by about 0.1 millivolt for each degree rise above 25° C. Therefore, measurements made under abnormal temperature conditions would vary so slightly from standard that only very sensitive instruments could detect the change. One such instrument, capable of measuring voltages with a precision of five decimal places, is a **potentiometer**.

A potentiometer is an instrument that compares a standard cell of known voltage to the voltage produced by an electrochemical system under investigation. The potentiometer works by increasing the resistance of a variable resistance coil to the point where the current causes the needle on a sensitive galvanometer to read "zero." Then the switch is "flipped" to the unknown cell. The current produced by the unknown cell's voltage causes a change in the galvanometer reading. The calibrated dial on the potentiometer is turned until the galvanometer reading returns to zero (see Fig. 10-16). This

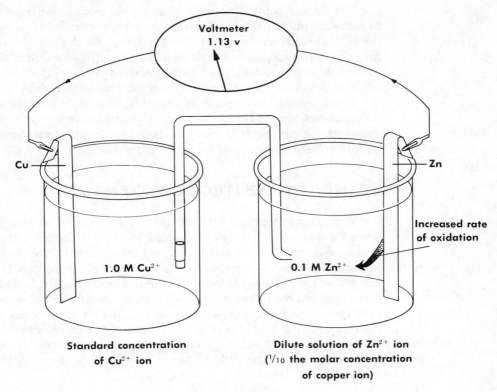

Figure 10-15.

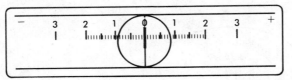

Galvanometer scale

Figure 10-16.

Figure 10-17.

is called the **null point**. The calibrations on the potentiometer are designed to read volts rather than the ohms that are actually changed. In other words, the potentiometer automatically converts R to V for the operator instead of his using paper and pencil to calculate $V = RI$. The direction of the needle shift on the galvanometer, which may be marked $+$ and $-$, tells the operator which electrodes of the unknown cell are the anode and cathode.

Other instruments capable of recording very slight changes in voltage are **electrometers** and pH meters. An electrometer is a very sensitive electronic voltmeter. A pH meter is really a specially adapted electrometer with a built in voltage compensator for temperature effects.

STANDARD ELECTRODE POTENTIALS

A standard electrode potential of a half-cell is the voltage produced when the particular half-cell is coupled with a hydrogen electrode. The hydrogen electrode is arbitrarily assigned standard oxidation or reduction potentials of **zero**. If the direction of the electron flow is from the second half-cell to the hydrogen half-cell, it is an indication that the second half-cell is undergoing an oxidation reaction while the hydrogen half-cell is undergoing reduction. The voltage produced would be due directly to the oxidation reaction since the hydrogen is assigned a value of 0.000 volts. For example, the electrode potential for a zinc–zinc ion half-cell is observed according to the diagram in Figure 10-18. In the voltaic cell in that diagram, zinc is oxidized to Zn^{2+} and $H^+(aq)$ is reduced to $H_2(g)$.

$$Zn^0 \rightarrow Zn^{2+} + 2e^- \quad \mathscr{E}^0_{ox} = +0.76 \text{ v}$$

$$2H^+ \rightarrow H_2 - 2e^- \quad \mathscr{E}^0_{red} = 0.00 \text{ v}$$

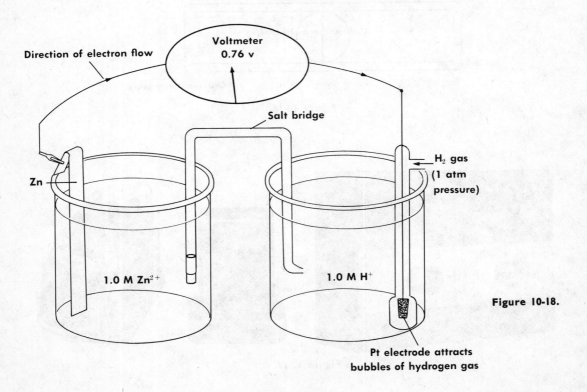

Direction of electron flow

Voltmeter 0.76 v

Salt bridge

H_2 gas (1 atm pressure)

Zn

1.0 M Zn^{2+}

1.0 M H^+

Figure 10-18.

Pt electrode attracts bubbles of hydrogen gas

The $\mathscr{E}^0_{cell}$ is the sum of the standard oxidation and reduction potentials. Since the $\mathscr{E}^0_{cell}$ is due only to the zinc–zinc ion half-cell, this value is recorded as the standard oxidation potential $\mathscr{E}^0_{oxid}$. Because Zn is oxidized, the sign of $\mathscr{E}^0_{ox}$ is positive. If zinc ion were reduced, a negative sign would indicate the reduction potential, -0.76 v.

An example of a negative oxidation potential is provided by a $Cu^0/Cu^{2+}(aq)$ half-cell where reduction takes place more easily than in the $H_2/2H^+(aq)$ half-cell (Fig. 10-19). In this cell the $Cu^{2+}(aq)$ is reduced to free copper and the $H_2(g)$ is oxidized to $2H^+(aq)$:

$$Cu^{2+} \rightarrow Cu^0 - 2e^- \quad \mathscr{E}^0_{red} = +0.34 \text{ v}$$

$$H_2 \rightarrow 2H^+ + 2e^- \quad \mathscr{E}^0_{ox} = 0.00 \text{ v}$$

Since the voltage is developed by the reduction of $Cu^{2+}(aq)$ to Cu^0, the cell voltage is known as the standard reduction potential of the Cu^0/Cu^{2+} half-cell.

Consider again, the voltaic cell made of a Zn^0/Zn^{2+} half-cell and a Cu^0/Cu^{2+} half-cell that produced a cell voltage of 1.10 volts. The tendency of zinc to undergo oxidation coupled with the tendency of Cu^{2+} to undergo reduction made a cell that vigorously and spontaneously produced a fairly high voltage. If the sum of the $\mathscr{E}^0_{ox}$ of zinc half-cell is *added* to the $\mathscr{E}^0_{red}$ of the copper half-cell, the actual $\mathscr{E}^0_{cell}$ is observed.

$$Zn^0 \rightarrow Zn^{2+} + 2e^- \quad \mathscr{E}^0_{ox} + 0.76 \text{ v}$$

add $\qquad\dfrac{Cu^{2+} \rightarrow Cu^0 - 2e^- \quad \mathscr{E}^0_{red} + 0.34 \text{ v}}{Zn^0 + Cu^{2+} \rightarrow Cu^0 + Zn^{2+} \quad \mathscr{E}^0_{cell} = +1.10 \text{ v}}$

Whenever a redox reaction occurs spontaneously, or if there is a *tendency* for the reaction to occur, the cell voltage will have a *positive* value. If two half-cells (one oxidation and one reduction) are added and produce a negative

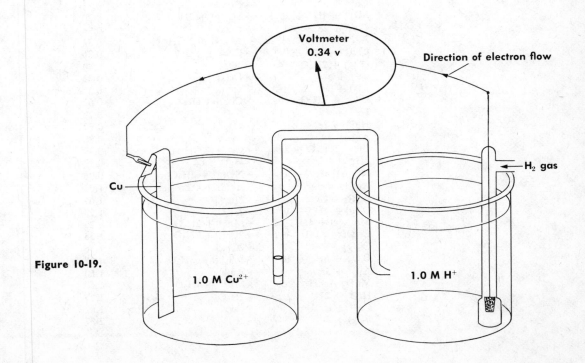

Figure 10-19.

Voltmeter
0.34 v

Direction of electron flow

H_2 gas

Cu

1.0 M Cu^{2+}

1.0 M H^+

TABLE 10-1. STANDARD REDUCTION POTENTIALS

Reduction Half Reaction	Standard Reduction Potential (volts)
$Li^+ + e^- \rightarrow Li(s)$	-3.05
$K^+ + e^- \rightarrow K(s)$	-2.93
$Ba^{2+} + 2e^- \rightarrow Ba(s)$	-2.90
$Ca^{2+} + 2e^- \rightarrow Ca(s)$	-2.87
$Na^+ + e^- \rightarrow Na(s)$	-2.71
$Mg^{2+} + 2e^- \rightarrow Mg(s)$	-2.37
$Al^{3+} + 3e^- \rightarrow Al(s)$	-1.66
$Zn(OH)_4^{2-} + 2e^- \rightarrow Zn(s) + 4\,OH^-$	-1.22
$Mn^{2+} + 2e^- \rightarrow Mn(s)$	-1.18
$Fe(OH)_2(s) + 2e^- \rightarrow Fe(s) + 2\,OH^-$	-0.88
$2H_2O + 2e^- \rightarrow H_2(g) + 2\,OH^-$	-0.83
$Zn^{2+} + 2e^- \rightarrow Zn(s)$	-0.76
$Cr^{3+} + 3e^- \rightarrow Cr(s)$	-0.74
$Fe(OH)_3(s) + e^- \rightarrow Fe(OH)_2(s) + OH^-$	-0.56
$S(s) + 2e^- \rightarrow S^{2-}$	-0.48
$Fe^{2+} + 2e^- \rightarrow Fe(s)$	-0.44
$Cr^{3+} + e^- \rightarrow Cr^{2+}$	-0.41
$Cd^{2+} + 2e^- \rightarrow Cd(s)$	-0.40
$PbSO_4(s) + 2e^- \rightarrow Pb(s) + SO_4^{2-}$	-0.36
$Cu(OH)_2(s) + 2e^- \rightarrow Cu(s) + 2\,OH^-$	-0.36
$Tl^+ + e^- \rightarrow Tl(s)$	-0.34
$Co^{2+} + 2e^- \rightarrow Co(s)$	-0.28
$Ni^{2+} + 2e^- \rightarrow Ni(s)$	-0.25
$AgI(s) + e^- \rightarrow Ag(s) + I^-$	-0.15
$Sn^{2+} + 2e^- \rightarrow Sn(s)$	-0.14
$Pb^{2+} + 2e^- \rightarrow Pb(s)$	-0.13
$CrO_4^{2-} + 4H_2O + 3e^- \rightarrow Cr(OH)_3(s) + 5\,OH^-$	-0.12
$2H^+ + 2e^- \rightarrow H_2(g)$	0.00
$NO_3^- + H_2O + 2e^- \rightarrow NO_2^- + 2\,OH^-$	0.01
$AgBr(s) + e^- \rightarrow Ag(s) + Br^-$	0.10
$S(s) + 2H^+ + 2e^- \rightarrow H_2S$	0.14
$Sn^{4+} + 2e^- \rightarrow Sn^{2+}$	0.15
$Cu^{2+} + e^- \rightarrow Cu^+$	0.15
$SO_4^{2-} + 4H^+ + 2e^- \rightarrow SO_2(g) + 2H_2O$	0.20
$Cu^{2+} + 2e^- \rightarrow Cu(s)$	0.34
$Ag_2O(s) + H_2O + 2e^- \rightarrow 2Ag(s) + 2\,OH^-$	0.34
$ClO_4^- + H_2O + 2e^- \rightarrow ClO_3^- + 2\,OH^-$	0.36
$O_2(g) + 2H_2O + 4e^- \rightarrow 4OH^-$	0.40
$Cu^+ + e^- \rightarrow Cu(s)$	0.52
$I_2(s) + 2e^- \rightarrow 2I^-$	0.53
$ClO_3^- + 3H_2O + 6e^- \rightarrow Cl^- + 6\,OH^-$	0.62
$Fe^{3+} + e^- \rightarrow Fe^{2+}$	0.77
$Hg_2^{2+} + 2e^- \rightarrow 2Hg(l)$	0.79
$Ag^+ + e^- \rightarrow Ag(s)$	0.80
$ClO^- + H_2O + 2e^- \rightarrow Cl^- + 2\,OH^-$	0.89
$2Hg^{2+} + 2e^- \rightarrow Hg_2^{2+}$	0.92
$NO_3^- + 4H^+ + 3e^- \rightarrow NO(g) + 2H_2O$	0.96
$AuCl_4^- + 3e^- \rightarrow Au(s) + 4Cl^-$	1.00
$Br_2(l) + 2e^- \rightarrow 2Br^-$	1.07
$O_2(g) + 4H^+ + 4e^- \rightarrow 2H_2O$	1.23
$MnO_2(s) + 4H^+ + 2e^- \rightarrow Mn^{2+} + 2H_2O$	1.23
$Cr_2O_7^{2-} + 14H^+ + 6e^- \rightarrow 2Cr^{3+} + 7H_2O$	1.33
$Cl_2(g) + 2e^- \rightarrow 2Cl^-$	1.36
$ClO_3^- + 6H^+ + 5e^- \rightarrow \frac{1}{2}Cl_2(g) + 3H_2O$	1.47
$Au^{3+} + 3e^- \rightarrow Au(s)$	1.50
$MnO_4^- + 8H^+ + 5e^- \rightarrow Mn^{2+} + 4H_2O$	1.52
$H_2O_2 + 2H^+ + 2e^- \rightarrow 2H_2O$	1.77
$Co^{3+} + e^- \rightarrow Co^{2+}$	1.82
$F_2(g) + 2e^- \rightarrow 2F^-$	2.87

$\mathscr{E}^0_{cell}$, the prediction can be made that this reaction will not have a tendency to proceed spontaneously. This does not mean that the reaction cannot be forced to proceed. There is such a phenomenon known as an endothermic (energy absorbing) reaction. Many oxidation-reduction reactions vital to the energy-absorbing needs of living organisms have negative cell potentials, but they do occur. The use of standard electrode potentials allows predictions of tendencies, but this is not at all the same as considering them as inflexible laws.

Predicting Tendencies

Given a pair of half-cell reactions, it is possible to predict the reaction that is most likely to occur by arranging them so that the sum of the standard electrode potentials is a positive value. These data can be obtained from Table 10-1, a table of standard reduction potentials. The table could certainly be arranged as one of standard oxidation potentials, but most of the world's scientific community prefers the former arrangement.

Example 3

Given two reduction half-reactions, write the equation having the greatest tendency to proceed spontaneously.

$$Cr^{3+} \rightarrow Cr^0 - 3e^- \qquad \mathscr{E}^0_{red} = -0.74 \text{ v}$$

$$Sn^{4+} \rightarrow Sn^{2+} - 2e^- \qquad \mathscr{E}^0_{red} = +0.15 \text{ v}$$

1. The only half-reactions that can possibly yield a positive value for the sum of the electrode potentials is the oxidation of Cr^0 to Cr^{3+} and the easy reduction of Sn^{4+} to Sn^{2+}. The ease of reduction is indicated by the positive sign of the standard reduction potential.

2. Arrange the appropriate half-cells for addition:

$$Cr^0 \rightarrow Cr^{3+} + 3e^- \qquad \mathscr{E}^0_{ox} = +0.74 \text{ v}$$

Note: The voltage is the same, only the sign is reversed.

$$Sn^{4+} \rightarrow Sn^{2+} - 2e^- \qquad \mathscr{E}^0_{red} = +0.15 \text{ v}$$

3. The loss and gain of electrons must be balanced:

$$2(Cr^0 \rightarrow Cr^{3+} + 3e^-)$$

$$3(Sn^4 \rightarrow Sn^{2+} - 2e^-)$$

This becomes

$$2Cr^0 \rightarrow 2Cr^{3+} + 6e^- \qquad \mathscr{E}^0_{ox} = +0.74 \text{ v}$$

add $\quad \dfrac{3Sn^{4+} \rightarrow 3Sn^{2+} - 6e^- \qquad \mathscr{E}^0_{red} = +0.15 \text{ v}}{2Cr^0 + 3Sn^{4+} \rightarrow 2Cr^{3+} + 3Sn^{2+} \qquad \mathscr{E}^0_{cell} = +0.89 \text{ v}}$

Note: The standard electrode potentials are *not* multiplied as the half-reactions are. This is due to the fact that the potential is not a function of the number of electrons, but it is rather a measure of the vigor with which oxidation and reduction occurs.

Example 4

Would the following reaction be expected to occur spontaneously under standard conditions?

$$5Fe^{2+} + MnO_4^- + 8H^+ \rightarrow Mn^{2+} + 5Fe^{3+} + 4H_2O$$

1. Arrange the reaction into oxidation and reduction half-cells and add the electrode potentials:

$$5Fe^{2+} \rightarrow 5Fe^{3+} + 5e^- \qquad \mathscr{E}^0_{ox} = -0.77 \text{ v}$$

$$MnO_4^- + 8H^+ \rightarrow Mn^{2+} + 4H_2O - 5e^- \qquad \mathscr{E}^0_{red} = +1.52 \text{ v}$$

$$\mathscr{E}^0_{cell} = +0.75 \text{ v}$$

2. The cell potential is positive which results in a prediction that the reaction does have a tendency to occur spontaneously.

Example 5

Predict the probable tendency for the following reaction to occur,

$$2Au^0 + Cr_2O_7^{2-} + 14H^+ \rightarrow 2Au^{3+} + 2Cr^{3+} + 7H_2O$$

1. Add the two indicated half-cell reaction potentials:

$$2Au^0 \rightarrow 2Au^{3+} + 6e^- \qquad \mathscr{E}^0_{ox} = -1.50 \text{ v}$$

$$Cr_2O_7^{2-} + 14H^+ \rightarrow 2Cr^{3+} + 7H_2O - 6e^- \qquad \mathscr{E}^0_{red} = +1.33 \text{ v}$$

$$\mathscr{E}^0_{cell} = -0.17 \text{ v}$$

2. The conclusion drawn from the negative cell voltage is that the dichromate oxidation of gold is *not* likely to occur spontaneously.

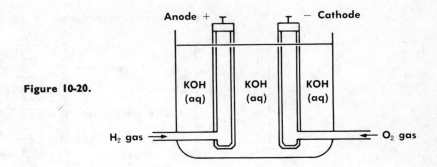

Figure 10-20.

FUEL CELLS

Fuel cells are electrochemical systems designed to efficiently convert chemical energy into useful electricity. Fuel cells are a relatively recent practical application of electrochemistry. It is a current area of scientific investigation with the hope that simpler and more efficient cells will be developed.

An example of a fuel cell is one in which hydrogen and oxygen are used. Porous carbon cylinders, containing catalysts, act as the electrodes. A diagrammatic cell is illustrated in Figure 10-20. The reactions occurring are

$$H_2(g) + 2OH^-(aq) \rightarrow 2H_2O + 2e^- \qquad \mathscr{E}^0_{ox} + 0.83 \text{ v}$$

add
$$\tfrac{1}{2}O_2(g) + H_2O \rightarrow 2OH^- - 2e^- \qquad \mathscr{E}^0_{red} + 0.40 \text{ v}$$

$$H_2 + \tfrac{1}{2}O_2 \rightarrow H_2O \qquad \mathscr{E}^0_{cell} = 1.13 \text{ v}$$

THE pH METER

A pH meter is a sensitive electrometer that is specifically designed to measure the hydrogen ion concentration in a liquid medium. It performs the measurement by indicating the voltage developed between two electrodes. One electrode is a standard reference which is not affected by $H^+(aq)$, and the other electrode is glass. The potential of the cell depends on the difference between the $H^+(aq)$ outside the glass electrode and a standard electrolyte solution inside the bulb-shaped tip. A tenfold change in $H^+(aq)$ concentration produces a change of 59 millivolts. In other words, when the pH changes by 1 unit, this corresponds to a change of one order of magnitude in the $H^+(aq)$ concentration. For example, a change from $H^+(aq) = 1 \times 10^{-5}$ M to 1×10^{-4} M, is a pH change from 5 to 4. This reflects a cell potential change of 0.059 volts. A pH meter is often equipped with a millivolt scale so that potentiometric measurements involving ions other than hydrogen may be performed. The pH meter may be used with a variety of electrodes appropriate for measurements and titrations with other ions. Such titrations are known as potentiometric. However, a scale calibrated in pH units is used for $H^+(aq)$, in addition to the millivolt scale. A diagrammatic illustration of the electrodes is given in Figure 10-21.

The glass bulb at the tip of the glass electrode is permeable only to hydrogen ions. The amount of current amplified by the pH meter is directly proportional to the $H^+(aq)$ concentration. The reference calomel (Hg_2Cl_2

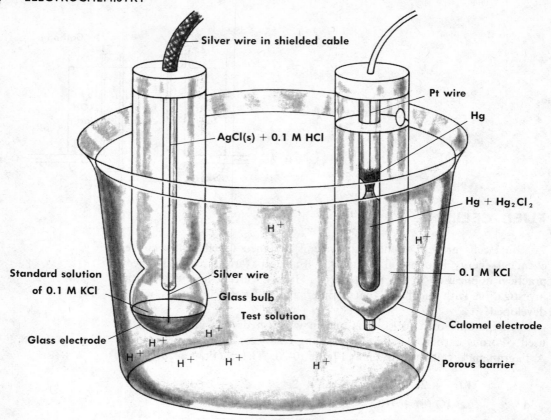

Silver wire in shielded cable

Pt wire

Hg

$AgCl(s) + 0.1\ M\ HCl$

$Hg + Hg_2Cl_2$

H^+

H^+

Standard solution
of 0.1 M KCl

Silver wire

0.1 M KCl

Glass bulb

Test solution

Glass electrode

H^+

H^+

Calomel electrode

H^+

H^+

H^+

H^+

H^+

Porous barrier

Figure 10-21.

is mercury [I] chloride) electrode completes the cell but it does not depend on the hydrogen ion.

Today most pH meters are used with a single combination electrode, i.e., both the calomel and glass electrodes are in the same length of glass tubing, although physically separated. This is a great convenience when measuring the pH of small volumes in beakers too small to accommodate two separate electrodes.

The rules for correct use of any model pH meter are few, but important.

1. Allow several minutes for vacuum tube meters to warm up.

2. Temperature compensator must be adjusted to the temperature of the test sample. Temperature fluctuations of the test sample need to be avoided.

3. Constant stirring is essential for even distribution of the $H^+(aq)$, especially during titrations.

4. Electrodes must be washed thoroughly by squirting with distilled water. Glass electrodes are easily contaminated, which leads to false readings.

5. The pH meter readings must be controlled by use of commercially prepared standard buffers. In this way you can adjust the pH value so that it corresponds exactly to a known pH (the standardizing buffer).

FREE ENERGY CHANGE AND VOLTAGE

When scientists talk about a free energy change, they are describing the amount of energy available for the performance of useful work. Free energy change, symbolized ΔG, is a measure of the tendency of a reaction to occur spontaneously. The tendency for a reaction to "go" is also indicated by a positive cell voltage in an electrochemical cell. This indicates a definite relationship between cell voltages and free energy change. The striking difference is in the sign, positive or negative, applied to the particular measurement.

When energy is available as a result of a chemical change, it means that this energy must have come primarily from the potential (stored) energy of the reactants. Since the products of the reaction are "energy poor" as a result of the change, their loss of energy is described by a negative sign for the ΔG value. In other words, a $-\Delta G$ means that a reaction is feasible. Just how feasible depends on the magnitude of the negative free energy change. For the sake of accuracy, it should be mentioned that the energy content of the reactants and products must be modified by the entropy changes that accompany the chemical change. Entropy has been described before as a tendency toward the disordered arrangement of matter. This procession toward randomness often necessitates an absorption of some energy so that it no longer becomes available for the performance of useful work. The result may be a total $-\Delta G$ value that is less than expected.

It is interesting to note that many energy-storing reactions in living organisms have $+\Delta G$ values. Due to the remarkable activity of biological catalysts, called enzymes, many unfeasible reactions do absorb energy and will occur. Our bodies can break down food in the process of releasing energy for life ($-\Delta G$) and at the same time "package" this energy ($+\Delta G$) for distribution to living cells. The conversion of sugar to CO_2 and H_2O yields a high negative free energy change while the process of adding inorganic phosphate to a large molecule converts adenosine diphosphate (ADP) to adenosine triphosphate (ATP) in an efficient energy absorbing reaction. The question to be answered is, how can a cell voltage measurement be converted to the common units of energy measurement, which are calories?

The solution may be accomplished in steps, the first of which is to convert the cell voltage to energy units. An energy unit discussed earlier is the **joule**. Joules are the product of volts and coulombs, and a joule may be called a volt-coulomb as a synonym. Remember also that a faraday is a mole of electrons, which equals 9.65×10^4 coulombs of electrical change. Now, if two moles of electrons are lost by a mole of a reducing agent, this is equivalent to two faradays. Symbolically, multiplying $(n)(\mathscr{F})(\mathscr{E})$ results in energy units.

The relationship is expressed in the equation

$$-\Delta G^\circ = \quad n \quad\quad \mathscr{F} \quad\quad \mathscr{E}^0$$

free energy	moles of electrons	faraday	cell voltage
change	per mole of	(in coulombs)	
	reducing agent		

The superscript of zero to the upper right of the free energy symbol denotes standard temperature and pressure conditions (25° C and 1 atmosphere).*

---

* Although the equation represents a thermodynamic relationship, the discussion presented in this text has sidestepped the topic of **thermodynamics** because a systematic and thorough study of the laws governing heat and energy relationships is very time-consuming and not especially relevant to the needs of the laboratory technician.

Supplying units to the equation, it becomes simplified:

$$-\Delta G^\circ = n\mathscr{F}\mathscr{E}^0$$

$$-\Delta G^\circ = (\text{moles})(9.65 \times 10^4 \text{ coulombs mole}^{-1})(\text{volts})$$

$$-\Delta G^\circ = (\text{coulombs})(\text{volts})$$

Since a joule = (volt)(coulomb)

$$-\Delta G^\circ = \text{joules}$$

Joules may be converted easily to calories by use of the conversion factor 0.239 cal joule^{-1}

$$-\Delta G^\circ = \text{joules } (0.239 \text{ cal joule}^{-1})$$

$$-\Delta G^\circ = \text{calories}$$

Example 6

Calculate the free energy change of a voltaic cell in which the $\mathscr{E}^0_{\text{cell}} = +1.10$ volts and one mole of zinc loses 2 moles of electrons as it becomes oxidized. $n = 2$ because

$$\underset{\substack{\text{1 mole of} \\ \text{reducing agent}}}{\text{Zn}^0} \rightarrow \text{Zn}^{2+} + \underset{\substack{\text{2 moles of} \\ \text{electrons}}}{2e^-}$$

1. Write the equation relating the data organized below.

$$-\Delta G^\circ = n\mathscr{F}\mathscr{E}^0$$

$-\Delta G^\circ$	?
n	2
$\mathscr{E}^0$	1.10 volts
$\mathscr{F}$	9.65×10^4 coulombs mole^{-1}

2. It is useful to convert the faraday to units incorporating kilocalories by using the conversion factor, 4.18 joules per calorie:

$$\mathscr{F} = \frac{9.65 \times 10^4 \text{ coulombs mole}^{-1}}{\underset{\text{(joules)}}{4.18 \text{ volt-coulombs cal}^{-1}}} = 2.31 \times 10^4 \text{ cal volt}^{-1} \text{ mole}^{-1}$$

Change calories to kilocalories, and

$$\mathscr{F} = 23.1 \text{ kcal volt}^{-1} \text{ mole}^{-1}$$

3. $-\Delta G° = (2 \text{ moles})(23.1 \text{ kcal volt}^{-1} \text{ mole}^{-1})(1.10 \text{ volts})$

$-\Delta G° = 50.8 \text{ kcal}$

4. The free energy change is -50.8 kcal. The indicated tendency for the reaction to occur spontaneously agrees with the cell voltage of $+1.10$ volts.

Example 7

In a catalyzed biochemical reaction the cell potential obtained from the addition of the standard electrode potentials was $\mathscr{E}^0_{\text{cell}} = -0.046$ volts. Calculate the free energy change. Was the reaction an energy producing or an energy storing change? The value for $n = 3$.

1. Establish the equation and organize the data:

$$-\Delta G° = n\mathscr{F}\mathscr{E}^0$$

$-\Delta G°$	?
n	3 moles
$\mathscr{F}$	23.1 kcal volt^{-1} mole^{-1}
$\mathscr{E}^0$	-0.046 volts

2. Develop the equation:

$-\Delta G° = (3 \text{ moles})(23.1 \text{ kcal volt}^{-1} \text{ mole}^{-1})(-0.046 \text{ volts})$

$-\Delta G° = -3.19 \text{ kcal}$

$\Delta G° = +3.19 \text{ kcal}$

3. This is an energy-storing reaction which accounts for need of a catalyst. The reaction that could possibly occur spontaneously, would be the reverse of the one described.

Electrochemistry is an area of scientific inquiry which has many questions still unanswered. The basic facts yet to be discovered and applications that are still waiting to find their way into a man's imagination constitute the fascination of the subject. As science moves even closer toward instrumentation, some knowledge of electronics and electrochemistry becomes essential.

PROBLEMS

Set A

1. Define or explain the following terms:

a. voltaic cell

b. coulomb

c. cathode

d. electrolysis

e. conductiometry

f. electrolyte

g. mho

h. Le Chatelier's principle

2. How many coulombs are required to electroplate 12.0 grams of magnesium?

3. How many hours are needed to reduce 26.0 grams of chromium (III) ion to chromium metal if a current of 500 milliamps is flowing?

4. What is the rationale for the usual practice of welding zinc plates to the steel hulls of ocean going ships?

5. Draw a diagrammatic set-up for the silver plating of a copper spoon. Label the electrodes and indicate the direction of the ion flow. How much silver would be deposited by a current of 2.5 amps flowing for 10 hours?

6. Explain why the standard reduction potential for the $2H^+(aq) \rightarrow H_2 \rightarrow 2e^-$ half-cell is zero.

7. A pair of platinum electrodes extend from a 3.0 volt dry cell into a weak electrolyte solution. A 1000 ohm resistor is in the circuit. If the galvanometer reads 0.0028 amps, what is the conductance (L) of the sample?

8. A voltaic cell is composed of one half-cell having a strip of nickel in a 1 M solution of $Ni^{2+}(aq)$ and the other half-cell having a strip of lead in a solution of $Pb^{2+}(aq)$, also 1 molar.

a. Write the equation for the cell reaction.

b. Diagram and label the cell.

c. Calculate the voltage of the cell.

9. If sulfide ion were added to the $Ni^{2+}(aq)$ in the cell in Problem 8, causing the decrease of Ni^{2+} concentration, would the cell voltage increase or decrease? Explain.

What would happen if the sulfide ion were added to reduce the $Pb^{2+}(aq)$ instead?

10. Given the following half-cell reactions (standard conditions) write the equations for the reactions most likely to proceed spontaneously. Calculate the cell voltages:

a. $Co^{2+} \rightarrow Co(s) - 2e^-$ $\mathscr{E}^0_{red} - 0.28$ v

$SO_4^{2-} + 4H^+ \rightarrow SO_2(g) + 2H_2O - 2e^-$ $\mathscr{E}^0_{red} = +0.20$ v

b. $CrO_4^{2-} + 4H_2O \rightarrow Cr(OH)_3(s) + 5\,OH^- - 3e^-$ $\mathscr{E}^0_{red} = -0.12$ v

$Cu^{2+} \rightarrow Cu^0(s) - 2e^-$ $\mathscr{E}^0_{red} = +0.34$ v

11. Predict the tendency of the following reactions to occur spontaneously.

a. $MnO_4^- + 2H^+ + \frac{1}{2}Cl_2 \rightarrow ClO_3^- + Mn^{2+} + H_2O$

b. $Zn + Ag_2O + 2OH^- \rightarrow Zn(OH)_4^{2-} + 2Ag + H_2O$

12. Calculate the $\Delta G°$ for the following reactions. Comment on the feasibility of the reaction.

a. $2MnO_4^- + 10Cl^- + 16H^+ \rightarrow 2Mn^{2+} + 5Cl_2 + 8H_2O$

b. $2Au + 8Cl^- + 3Sn^{4+} \rightarrow 2AuCl_4^- + 3Sn^{2+}$

Set B

13. Define or explain the following terms:

a. faraday

b. electrode

c. coulometry

d. electrodeposition

e. standard reduction potential

f. fuel cell

g. anode

h. calomel electrode

14. What current is needed for the electrodeposition of 50 mg of silver in 10 minutes?

15. What reaction takes place at the electrodes during the electrolysis of the following solutions:

a. Na_2SO_4

b. $Zn(NO_3)_2$

c. $MgCl_2$

16. What will be the final molar concentration of 500 ml of 0.4 M $CuSO_4$ solution if it is electrolyzed by a 2 amp current for three hours?

17. What volume of oxygen gas (at STP) could be liberated by 0.1 faraday of electricity?

18. Use standard electrode potentials to explain the inadvisability of a surgeon using gold screws to implant a chromium plate at the site of a bone fracture.

19. If the conductance of a weak electrolyte is 4×10^{-4} mhos, what is the galvanometer reading in amps? A 25 volt dry cell is used in the system.

20. A voltaic cell uses the reaction

$$2Al + 3Cu^{2+}(1.0 \text{ M}) \rightarrow 2Al^{3+}(1.0 \text{ M}) + 3Cu$$

a. Diagram and label the cell and salt bridge.

b. Calculate the cell voltage.

21. If additional Al^{3+} were added to the aluminum half-cell described in Problem 20, what effect would be predicted for the cell voltage? Explain. Would an addition of Cu^{2+} to the copper cell have the same effect? Why?

22. Calculate the cell voltages for the most feasible reactions, given the following standard half-cell reactions:

a. $NO_3^- + H_2O \rightarrow NO_2^- + 2OH^- - 2e^-$ $\mathscr{E}^0_{red} = +0.01$ v
   $Zn(OH)_4^{2-} \rightarrow Zn + 4OH^- - 2e^-$ $\mathscr{E}^0_{red} = -1.22$ v

b. $AgBr(s) \rightarrow Ag(s) + Br^- - e^-$ $\mathscr{E}^0_{red} = +0.10$ v
   $MnO_2(s) + 4H^+ \rightarrow Mn^{2+} + 2H_2O - 2e^-$ $\mathscr{E}^0_{red} = +1.23$ v

23. Predict the tendency of the following to occur spontaneously.

a. $3Sn^{2+} + 2Cr^{3+} \rightarrow 3Sn^{4+} + 2Cr(s)$

b. $F_2 + Cu \rightarrow 2F^- + Cu^{2+}$

24. Calculate the standard cell potential for a voltaic cell if 2 moles of electrons are lost by one mole of a reducing agent in producing a $-\Delta G°$ of 50.7 kcal.

SUPPLEMENTARY READING

Angrist, S. W., and Helper, L. G.: *Order and Chaos.* Basic Books, Inc., New York, 1967.
   An imaginative and non-mathematical discussion of thermodynamics.

McGlashan, M. L.: The Use and Misuse of Thermodynamics. *J. Chem. Educ.*, May, 1966.
   A very thought-provoking article. Should be read.

Zemansky, M. W.: *Temperatures Very Low and Very High*, Van Nostrand Co., Inc., Princeton, 1964.
   Excellent chapters on entropy and absolute zero.

Wyatt, P. A. H.: *Energy and Entropy in Chemistry.* Macmillan, London, 1967.
   A short introductory text on thermodynamics using little mathematics.

Organic Chemistry 11

"Organic chemistry nowadays drives one mad. To me it appears like a primeval tropical forest full of the most remarkable things, a dreadful endless jungle into which one dares not enter, for there seems no way out." When the German chemist, Friedrich Wöhler, wrote this in 1835, he effectively described the vastness of this particular branch of chemistry. Organic chemistry is no longer a "dreadful jungle." It is now a highly organized discipline which is constantly expanding in a purposefully controlled manner.

While organic chemistry is commonly and accurately called the chemistry of carbon, it is possible to be more descriptive. It can be said that organic chemistry is the chemistry of life. It is the chemistry of fuels, medicines, drugs, fabrics, plastics, and food. Organic chemistry deals with the analysis, isolation, and purification of natural products—vitamins, antibiotics, drugs, waxes, oils, hormones, and dyes from plants and animals. It deals with the synthesis of complex substances which often perform the same functions as natural products, and often more efficiently. Organic chemists have synthesized the polyethylene and vinyl plastics, the Orlon and Dacron fabrics, and many of the newer lifesaving pharmaceuticals that we have come to know so well in times of need.

There are many thousands of organic compounds available today. The carbon atom, which has the chemical property of forming covalent bonds with other carbon atoms and various other atoms, forms incredibly complex chains and rings which may lead to the formation of infinite varieties of compounds.

It is the purpose of this chapter to familiarize the student with the nomenclature, variety, usefulness, and most significant properties of selected organic compounds. The synthesis of new compounds and the study of the mechanisms of organic reactions is fascinating. However, the needs of the laboratory technician are more closely allied to the development of competence in the reading, writing, and identification of organic compounds in addition to appreciating the uses and hazards involved. An essential working vocabulary is indicated also.

STRUCTURAL FORMULAS AND ISOMERISM

Since carbon has four electrons in its valence shell, it commonly tends to form four covalent bonds. While this is not an invariable rule, it is an accurate description in most cases. The energy requirement for the transfer of four electrons in the formation of a carbon ion, C^{4+} or C^{4-} is far in excess of available energy in ordinary chemical reactions. The bonds formed by carbon may be between carbon atoms in a chain-like structure as well as with other atoms such as halogens, oxygen, nitrogen, and most notably hydrogen.

The geometry of carbon compounds depends on the number of atoms attached to the carbon atoms. Physical conditions, such as temperature, pressure, catalysts, and the chemical properties of the reactants will determine the formation of single, double, or triple bonds. The types of bonds present in a molecule and their geometrical arrangement are very accurately described because of the instrumental methods of analysis available to the chemist today. The types of bonds are revealed by infrared spectrophotometry, and the geometry of the molecules may be determined by x-ray diffraction.

Carbon forms its most stable compounds through a process known as **orbital hybridization**. Orbital hybridization results from the absorption of sufficient energy to permit a different arrangement of carbon's valence electrons. The arrangement that is most favorable for the formation of four single bonds, and the resulting molecular stability, finds the hybrid probability distribution of the four electrons at the four corners of a regular tetrahedron (Fig. 11-1).

An example of the **tetragonal** structure in carbon compounds may be found in methane and in chains of carbon atoms having only single bonds (Fig. 11-2).

Double bonds result from another kind of hybridization of the orbitals. Infrared spectroscopy indicates that double bonds are shorter and have higher vibrational frequencies. The more reactive compounds containing double bonds are characterized by a **trigonal** geometry. Ethylene is a classic example (Fig. 11-3).

The formation of triple bonds between carbon atoms requires another variation of the orbital hybrids. In the case of acetylene, for example, the resulting geometry of the very short triple bond is a linear structure for the molecule (Fig. 11-4).

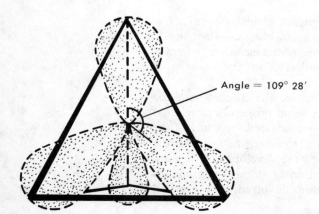

Angle = 109° 28′

Figure 11-1.

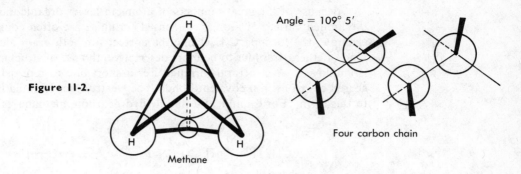

Figure 11-2.

Angle = 109° 5'

Methane

Four carbon chain

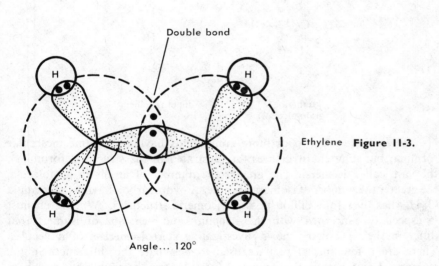

Double bond

Ethylene   Figure 11-3.

Angle... 120°

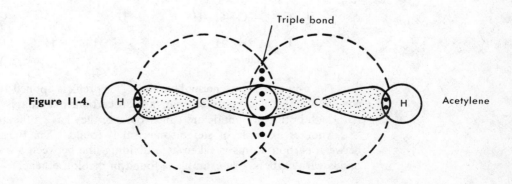

Triple bond

Figure 11-4.   H   C   C   H   Acetylene

Because of the great numbers of atoms in larger organic molecules and the limited variety of atoms, condensed formulas are often confusing. For example, the formula C_2H_6O may be correct for both grain alcohol and a kind of ether. In order to be more descriptive, the use of **structural formulas** is employed. A structural formula uses a short line to represent a pair of shared electrons in a covalent bond. The electron-dot formula is converted to this form. For example, the simple hydrocarbon, methane, is shown:

$$CH_4$$

condensed

electron-dot

structural

The formula C_2H_6O can be similarly illustrated.

ethanol
(grain alcohol)

dimethyl ether

When there are two or more compounds having the same molecular formula, but different structures (as illustrated in the structural formulas), they are called **isomers**. The greater the number of atoms in a molecule, the greater the number of isomers possible. A hydrocarbon having the formula $C_{30}H_{62}$ has over four billion possible isomeric structures. When a chemist is especially concerned with the structure and behavior of isomers that differ in their geometry, he is investigating **stereoisomerism** of molecules. There are a few special prefixes used to designate the differences in the isomers. For example, the structural formulas for the simple hydrocarbon, butane, are presented:

n-butane

isobutane

The n-butane means **normal** butane. This prefix is applied to continuous chain molecules. The prefix **iso** means a hydrocarbon chain that is "branched." Although both compounds are called butane, they have different properties.

Another example of stereoisomerism is found when the double bond between carbon atoms is taken as a dividing line between atoms that form bonds with carbon. An example is found in dichloroethene.

H   Cl   Cl   Cl

— — C = — C — — — Imaginary dividing line — — — — C = — C — —

Cl   H   H   H

Trans-dichloroethene   **Cis-dichloroethene**

The prefix **trans-** means that the two attached chlorine atoms are linked to the carbon chain "across" the dividing line while the prefix **cis-** indicates the attachment on the "same" side.

A type of isomerism that is very important in biochemistry is known as **optical isomerism**. Optical isomers may have very similar physical properties, but their effects on living organisms may vary from necessary to poisonous. Optical isomers are called mirror images of each other. They differ in the way they cause plane polarized light to rotate. Plane polarized light is produced when ordinary light passes through a film or crystal that only permits passage of light that is vibrating in one direction (Fig. 11-5).

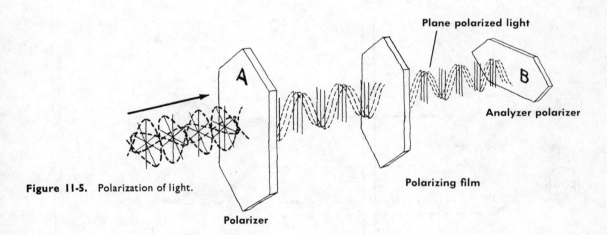

Figure 11-5. Polarization of light.

If a solution containing an optical isomer is placed in a **polarimeter**, the plane polarized light is observed to be rotated to the right or left. If the rotation is to the right (clockwise), the isomer is called **dextrorotatory**. This is symbolized by d or $+$. If the rotation is to the left (counterclockwise), the isomer is **levorotatory**, symbolized by l or $-$. The d- and l-isomers are called **enantiomers**. A simplified diagram of a polarimeter is illustrated in Figure 11-6.

An example of d- and l-isomers is lactic acid (see Fig. 11-7).

As an illustration of the importance of optical isomers, it is interesting to note that the human body needs and uses only d- forms of sugars and l- forms of amino acids. The mirror images of these forms cannot be metabolized. Whenever a 50–50 mixture of d- and l- forms occurs, it is called a **racemic** mixture (belonging to the same "race"), and a separation may be necessary.

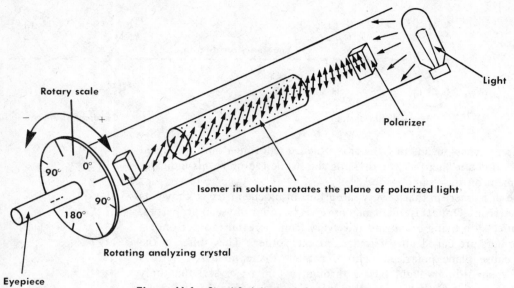

Figure 11-6. Simplified diagram of a polarimeter.

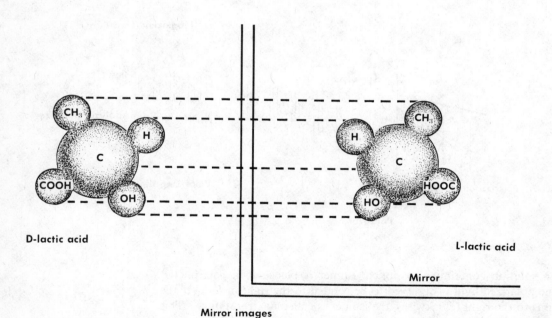

D-lactic acid

L-lactic acid

Mirror

Mirror images

Figure 11-7.

SATURATED HYDROCARBONS

The simplest organic compounds are called saturated hydrocarbons because there are only single covalent bonds present. The other names applied to this group are **alkanes** and **paraffins**. The alkane series provides the basic system of nomenclature in organic chemistry. The simplest alkane is methane gas, CH_4.

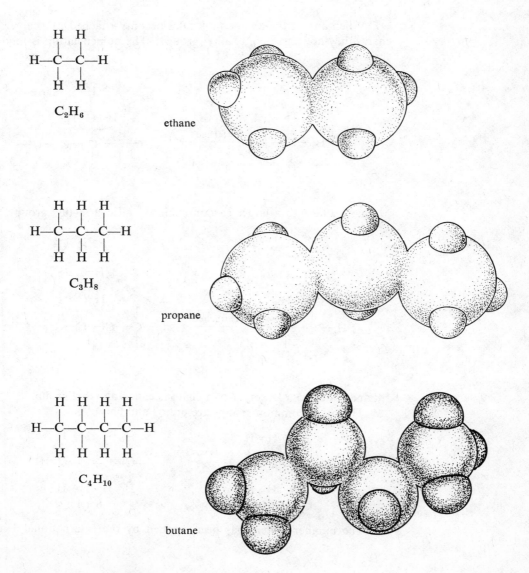

$$H—C—H$$

$$CH_4$$

methane

The suffix, **-ane** identifies the compound as a member of the alkane series.

Adding carbon atoms in the chain develops the next members of the series.

$$C_2H_6$$

ethane

$$C_3H_8$$

propane

$$C_4H_{10}$$

butane

A few more examples of this related, or **homologous**, series:

$$C_5H_{12}$$

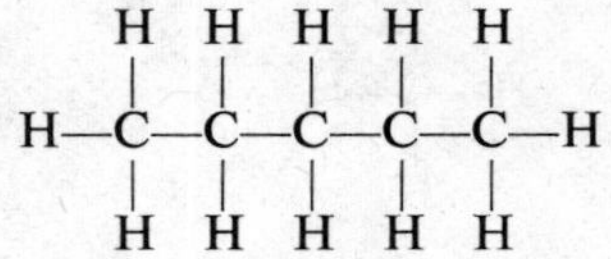

H—C—C—C—C—C—H

pentane

$$C_6H_{14}$$

H—C—C—C—C—C—C—H

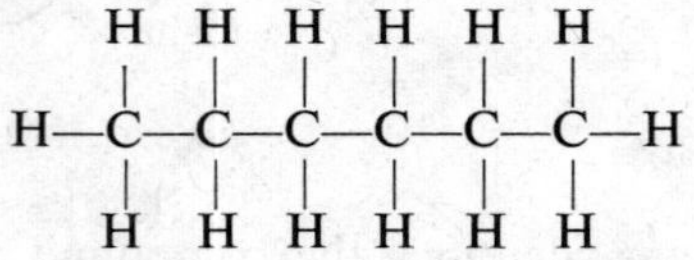

hexane

C_7H_{16} is heptane;
C_8H_{18} is octane;
C_9H_{20} is nonane;
$C_{10}H_{22}$ is decane.

When any of the alkane molecules become attached to a larger carbon chain, they are regarded as **alkyl** groups. The nomenclature is modified as shown.

methane

Methane added to a larger carbon chain is called a **methyl** group.

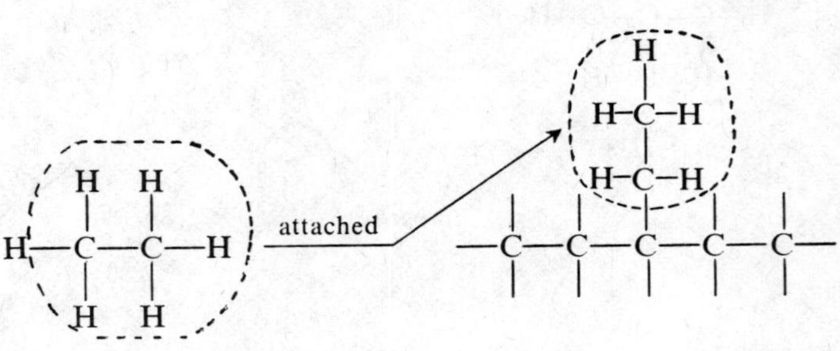

ethane

Ethane added to a larger carbon chain is called an **ethyl** group.
The other common groups are **propyl** and **butyl**.

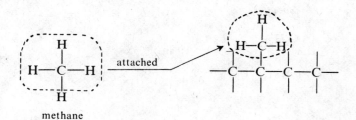

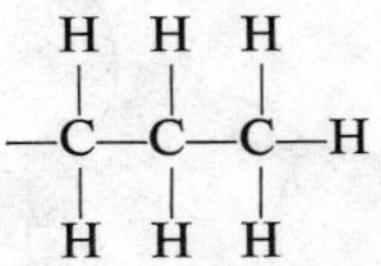

propyl 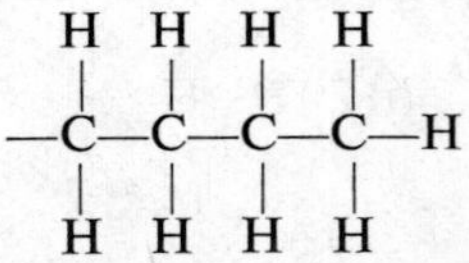   butyl

The number system has been applied by the International Union for

Pure and Applied Chemistry (IUPAC) to designate the position of an attached group on a carbon chain. The rules are:

1. The attached groups are listed *alphabetically*. If the same group appears more than once, the prefixes *di* (2), *tri* (3), *tetra* (4), *penta* (5) and so on are used.
2. The carbon atoms are numbered from the end of the longest chain (trunk) of the structural formula.

Example 1

$$\begin{array}{ccccccc} & & & & CH_3 & & \\ & & & & | & & \\ CH_3 & - & CH_2 & - & CH & - & CH_3 \\ \uparrow & & \uparrow & & \uparrow & & \uparrow \\ 4 & & 3 & & 2 & & 1 \end{array}$$

2-methylbutane

1. The suffix -ane indicates an alkane.
2. The stem but- indicates a four carbon chain as the longest one.
3. The methyl group is attached to the second carbon atom, hence the prefix 2-methyl.

Example 2

$$\begin{array}{ccccccccc} & & & & CH_3 & & CH_3 & & \\ & & & & | & & | & & \\ CH_3 & - & CH_2 & - & CH & - & CH & - & CH_3 \\ \uparrow & & \uparrow & & \uparrow & & \uparrow & & \uparrow \\ 5 & & 4 & & 3 & & 2 & & 1 \end{array}$$

2,3-dimethylpentane

1. Pentane indicates the longest continuous chain as a five carbon alkane.
2. 2,3-dimethyl illustrates the attachment of two methyl groups—one on the number 2 carbon atom and the other on the number 3 atom.

Example 3

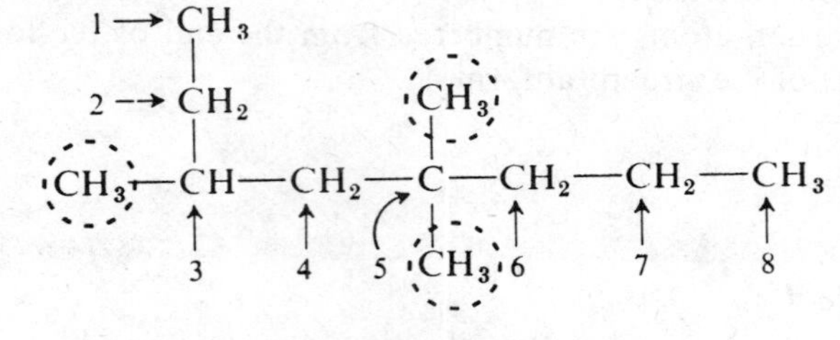

$$1 \rightarrow CH_3$$
$$2 \rightarrow CH_2$$

$$CH_3 - CH - CH_2 - C - CH_2 - CH_2 - CH_3$$

$$3 \quad 4 \quad 5 \quad CH_3 \; 6 \quad 7 \quad 8$$

(with CH_3 groups circled at positions 3, 5, and 5)

3,5,5-trimethyloctane

When halogens substitute for hydrogen atoms in a process called **halogenation**, the IUPAC rules still hold with regard to an alphabetical order for naming the attached groups.

Example 4

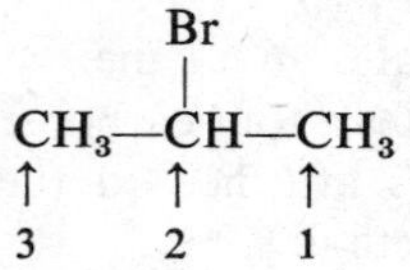

$$\begin{array}{c} Br \\ | \\ CH_3 - CH - CH_3 \end{array}$$

$$3 \quad 2 \quad 1$$

2-bromopropane

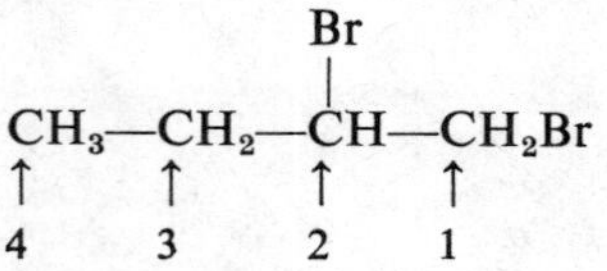

$$\begin{array}{c} Br \\ | \\ CH_3 - CH_2 - CH - CH_2Br \end{array}$$

$$4 \quad 3 \quad 2 \quad 1$$

1,2-dibromobutane

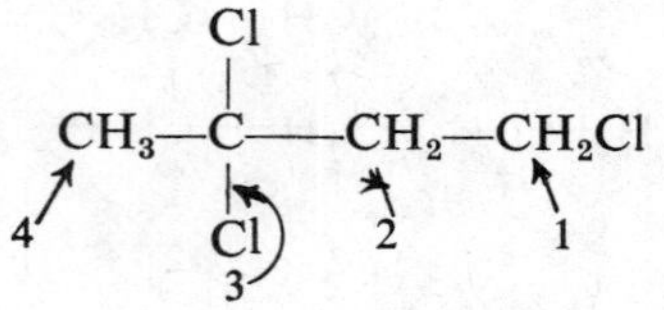

$$\begin{array}{c} Cl \\ | \\ CH_3 - C - CH_2 - CH_2Cl \\ | \\ Cl \end{array}$$

$$4 \quad\quad 3 \quad\quad 2 \quad\quad 1$$

1,3,3-trichlorobutane

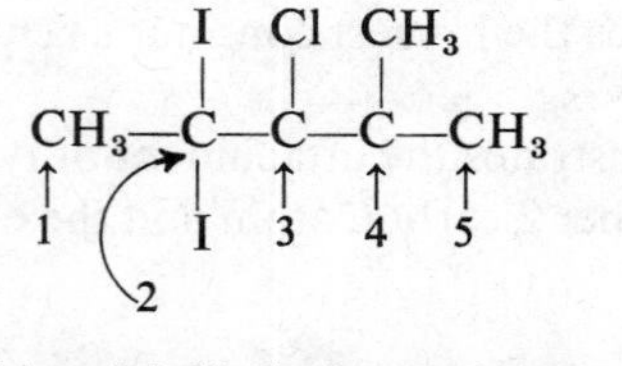

$$\begin{array}{ccc} I & Cl & CH_3 \\ | & | & | \\ CH_3 - C - C - C - CH_3 \\ | \\ I \end{array}$$

$$1 \quad 2 \quad 3 \quad 4 \quad 5$$

3-chloro-2,2-diiodo-4-methylpentane

**TABLE 11-1. PHYSICAL PROPERTIES OF SOME NORMAL
SATURATED HYDROCARBONS**

Name	Formula	M.P., °C	B.P., °C	Specific Gravity	Normal State
Methane	CH_4	-182.6	-161.4	—	gas
Ethane	C_2H_6	-172.0	-88.3	—	gas
Propane	C_3H_8	-187.1	-44.5	—	gas
n-Butane	C_4H_{10}	-135.0	-0.5	—	gas
n-Pentane	C_5H_{12}	-129.7	36.2	0.6264	liquid
n-Hexane	C_6H_{14}	-94.0	69	0.6594	liquid
n-Heptane	C_7H_{16}	-90.5	98.4	0.6837	liquid
n-Octane	C_8H_{18}	-56.8	124.6	0.7028	liquid
n-Decane	$C_{10}H_{22}$	-32	175	0.730	liquid
n-Pentadecane	$C_{15}H_{32}$	10	271	0.772	liquid
n-Octadecane	$C_{18}H_{38}$	28	308	0.77	solid

A stepwise substitution of chlorine on the methane molecule yields several important compounds

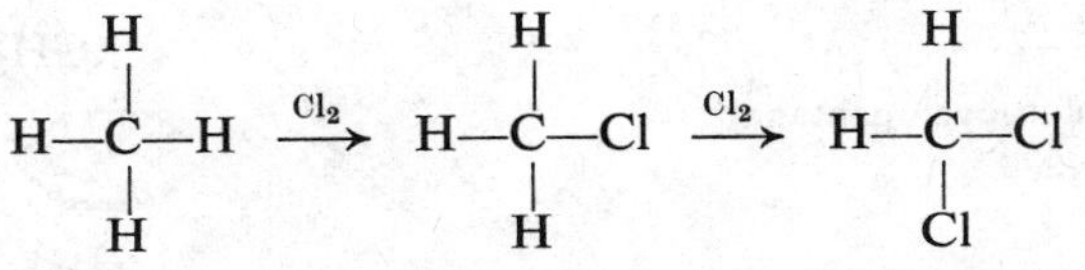

methane   chloromethane   dichloromethane
                 (methyl chloride)

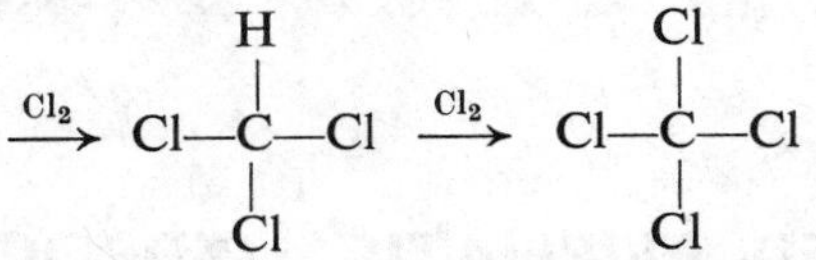

trichloromethane   tetrachloromethane
(chloroform)   (carbon tetrachloride)

Substitutions other than halogens occur commonly. These functional groups, as they are called, will be discussed later in this chapter.

Cycloalkanes

Although most alkanes are open chain (continuous or branched), there are a few common examples of cyclic alkanes. The alkanes having a cyclic structure exhibit nearly the same physical and chemical properties as open chain structures. The fact that they are compounds containing only single bonds (saturated) marks the important distinction between the cycloalkanes and a group yet to be discussed, called the aromatics. A few examples of cycloalkanes are illustrated in Figure 11-8.

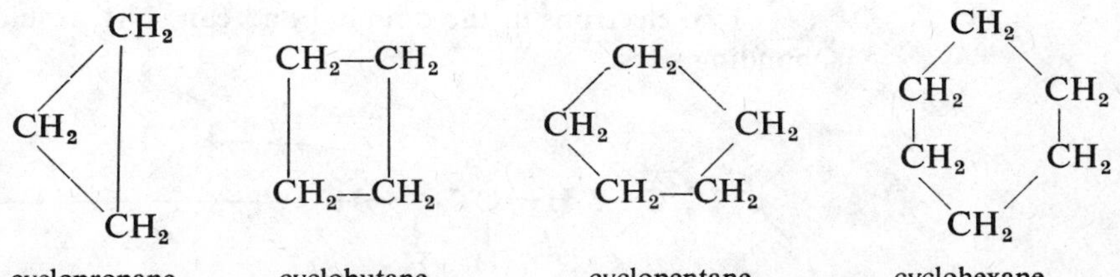

Figure 11-8.

cyclopropane   cyclobutane   cyclopentane   cyclohexane

cyclopropane   cyclobutane   cyclopentane   cyclohexane   **Figure 11-9.**

Ring systems are commonly simplified by organic chemists by the use of geometrical figures (Fig. 11-9).

Example 5

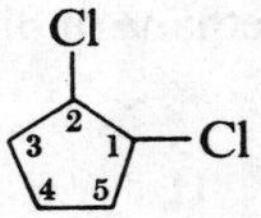

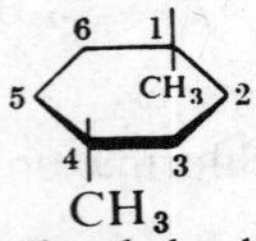

cis-1,4-dimethylcyclohexane

1,2-dichlorocyclopentane

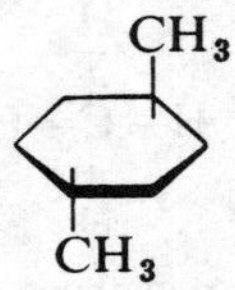

trans-1,4-dimethylcyclohexane

UNSATURATED ALIPHATIC HYDROCARBONS

The Alkenes (Olefins)

Whenever the carbon chain contains one or more double bonds between carbon atoms, the distinguishing suffix **-ene** is applied. If two double bonds are in the carbon chain, the suffix **-diene** is indicated. A **-triene** applies to three double bonds.

The double bond is the basis for the term unsaturated. It indicates an extra pair of electrons available for the addition of hydrogen atoms, halogen atoms, or other groups. For example, the structure that follows illustrates the addition of two hydrogen atoms to an unsaturated hydrocarbon.

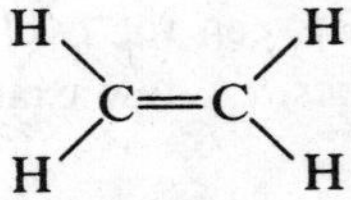

is equivalent to

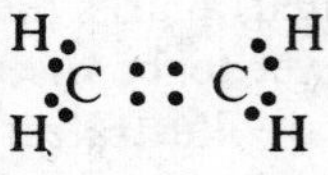

Two electrons in the double bond can shift, resulting in two additional bonding sites.

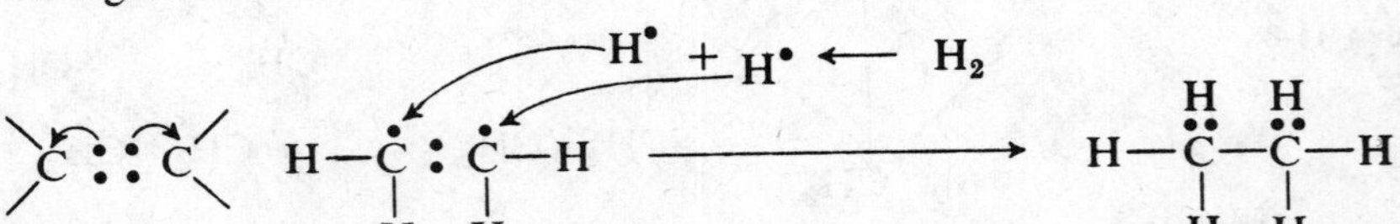

The popular dietary term, "polyunsaturated fats or oils," means long chain hydrocarbons which have numerous double bonds.

There is a common, or trivial, nomenclature applied to the alkenes which give them the suffix, **-ylene**. The position of the double bond in the carbon chain is designated by the Greek letters α (alpha), β (beta), γ (gamma) and so on through the alphabet. The prefix is the same as the alkane series.

Example 6

$$CH_3—CH_2—CH_2{=}CH_2$$
α-butylene

1. The α means that the double bond is attached to the first carbon atom in the chain.
2. The prefix but- means a four carbon chain (from butane).
3. The suffix -ylene means the compound is an alkene.

Example 7

cis-β-butylene

trans-β-butylene

The IUPAC rules for naming the alkenes are:
1. The identifying suffix is -ene.
2. The carbon atoms are counted from the end of the chain nearest the double bond.

Polymerization   The process of making a long chain out of simple units is called polymerization. The single link of the chain is the monomer (meaning one part). The making of a polymer (meaning many parts) is a common procedure in modern organic chemistry.

The polymerization of ethylene results in long chain molecules of considerable molecular weight. The properties of the well-known plastic, polyethylene, are strikingly different from the gas ethylene.

$$CH_2{=}CH_2 \xrightarrow{\text{polymerization}} \overset{\text{repeated many times}}{—(CH_2—CH)—}$$
ethylene   polyethylene

TABLE II-2. EXAMPLES OF ALKENES

Structural Formula	Common Name	IUPAC Name
	ethylene	ethene
	propylene	propene
	β-butylene	2-butene
	isobutylene	2-methylpropene
	isoprene	2-methyl-1,3-butadiene
	none	2,4-dimethyl-2-pentene
	none	3-bromocyclopentene

Another common plastic is polypropylene.

$$\underset{\text{propylene}}{\overset{\overset{\textstyle CH_3}{\textstyle |}}{CH{=}CH_2}} \quad \xrightarrow{\text{polymerization}} \quad \underset{\text{polypropylene}}{\overset{\overset{\textstyle CH_3}{\textstyle |}}{-(CH{-}CH_2)-}} \quad \substack{\text{repeated} \\ \text{many times}}$$

A few more examples:

$$\underset{\text{tetrafluoroethene}}{CF_2{=}CF_2} \quad \xrightarrow{\text{polymerization}} \quad \underset{\text{teflon}}{-(CF_2{-}CF_2)-} \quad \substack{\text{repeated} \\ \text{many times}}$$

$$CH_2{=}C\overset{\displaystyle H}{\underset{\displaystyle Cl}{\Big\langle}}$$ is known as vinyl chloride as well as by the IUPAC name chloroethene.

$$CH_3\!=\!\underset{\underset{Cl}{|}}{\overset{\overset{H}{|}}{C}} \quad \xrightarrow{polymerization} \quad -(CH_2\!-\!\underset{\underset{Cl}{|}}{\overset{\overset{H}{|}}{C}}\ )- \quad \begin{array}{l}\text{repeated}\\\text{many times}\end{array}$$

polyvinyl chloride

Polyvinyl chloride (PVC) is modified industrially to resemble products maed of leather, tile, and rubber.

$$CH_2\!=\!\underset{\underset{Cl}{|}}{\overset{\overset{Cl}{|}}{C}}$$
vinylidene chloride
(1,1-dichloroethene)

$$CH_2\!=\!\underset{\underset{Cl}{|}}{\overset{\overset{Cl}{|}}{C}} \quad \xrightarrow{polymerization} \quad -(CH_2\!-\!\underset{\underset{Cl}{|}}{\overset{\overset{Cl}{|}}{C}}\ )- \quad \begin{array}{l}\text{repeated}\\\text{many times}\end{array}$$

vinylidene chloride   saran wrap

A cyanide substituted ethylene molecule

$$CH_2\!=\!\underset{\underset{CN}{|}}{\overset{\overset{H}{|}}{C}}$$

is commonly called acrylonitrile.

$$CH_2\!=\!\underset{\underset{CN}{|}}{\overset{\overset{H}{|}}{C}} \quad \xrightarrow{polymerization} \quad -(CH_2\!-\!\underset{\underset{CN}{|}}{\overset{\overset{H}{|}}{C}}\ )- \quad \begin{array}{l}\text{repeated}\\\text{many times}\end{array}$$

acrylonitrile   orlon

TABLE 11-3. PHYSICAL CONSTANTS OF THE ALKENES

Name	Formula	M.P., °C	B.P., °C	Density as Liquid
Ethene	$H_2C\!=\!CH_2$	−169.4	−103.8	0.566
Propene	$CH_3CH\!=\!CH_2$	−185.2	−47.7	0.609
1-Butene	$CH_3CH_2CH\!=\!CH_2$	−130	−6.5	0.625
2-Butene	$CH_3CH\!=\!CHCH_3$	−127	1.4	0.630
2-Methylpropene (Isobutylene)	$CH_3\!-\!\underset{\underset{CH_3}{|}}{C}\!=\!CH_2$	−140.7	−6.9	0.594
1,3-Butadiene	$CH_2\!=\!CH\!-\!CH\!=\!CH_2$	−113	−4.5	0.650
2-Methyl-1,3-butadiene (Isoprene)	$CH_2\!=\!\underset{\underset{CH_3}{|}}{C}\!-\!CH\!=\!CH_2$	−120	35	0.681
1-Pentene	$CH_2\!=\!CH(CH_2)_2CH_3$	−138	30.1	0.641
1-Hexene	$CH_2\!=\!CH(CH_2)_3CH_3$	−141	64.1	0.673

The Alkynes (Acetylenes)

The maximum degree of unsaturation that can exist between two carbon atoms is a triple bond. This means that there are two pairs of electrons available for the attachment of other groups. Compounds containing two and three triple bonds are known as **-diynes** and **-triynes**.

Some examples of alkynes are listed in Table 11-4.

TABLE 11-4. EXAMPLES OF ALKYNES

Structural Formula	Common Name	IUPAC Name
H—C≡C—H	acetylene	ethyne
H—C≡C—C—H (with H above and below center C)	none	propyne
H—C—C—C≡C—H (with H H above positions 4,3 and H H below)	none	1-butyne
H—C≡C—C—C≡C (with Cl CH₃ N above and H H below)	none	3-chloro-2-methyl-1-pentene-4-yne

AROMATIC COMPOUNDS

Historically speaking, some of the first compounds extracted from plants which were sweet smelling oils gave the general name, aromatic, to a large group of related substances. It was principally the work of the German chemist, August Kekulé, in his theoretical explanation of the ring structure of benzene, that provided the structural basis of this group.

The benzene structure is a ring of six carbon atoms sharing a number of electrons that is equivalent to 3 single bonds and 3 double bonds. The formulas suggested by Kekulé are somewhat misleading.

and

The Kekulé formulas suggest shifting double bonds which is contrary to experimental evidence. Many chemists agree that a circle within the hexagon represents a more accurate distribution of the electrons among the carbon atoms.

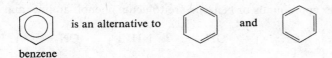

benzene is an alternative to and

When more than one group is attached to a benzene ring, the nomenclature becomes more complicated. A common system of nomenclature describes the positions of the attached groups by the terms **ortho** (o), **meta** (m), and **para** (p). Here are three examples:

o-dichlorobenzene   m-dichlorobenzene   p-dichlorobenzene

The IUPAC system of nomenclature numbers the attached groups in an order that uses the smallest numbers possible.

1,2-dichlorobenzene   1,3-dichlorobenzene   1,4-dichlorobenzene

The most commonly used names applied to aromatic hydrocarbon radicals (called **aryls**) are phenyl and benzyl.

phenyl   benzyl

Combinations of the aryls will be investigated later in this chapter.

TABLE 11-5. SIMPLE BENZENE COMPOUNDS

Structural Formula	Name
Cl	chlorobenzene
NO_2	nitrobenzene
Br	bromobenzene

Other compounds of benzene are toluene, phenol, aniline, and o-xylene.

toluene   phenol   aniline   o-xylene

When two groups are attached to the benzene ring, the position of the second group is described by the ortho, meta, or para prefix. If more than two groups are attached, the carbon atoms are numbered according to the IUPAC convention. Some examples are listed in Table 11-6.

TABLE 11-6. EXAMPLES OF BENZENE COMPOUNDS WITH TWO OR MORE GROUPS ATTACHED TO THE BENZENE RING

Structural Formula	Common Name	IUPAC Name
	m-nitrotoluene	1-methyl-4-nitrobenzene
	o-chlorophenol	2-chloro-1-hydroxybenzene
	p-bromo aniline	4-bromo-1-aminobenzene
	none	1,3,5-trinitrobenzene
	2-4-6-trinitrotoluene (TNT)	1-methyl-2,4,6-trinitrobenzene
	3-chloro-4-nitrophenol	3-chloro-1-hydroxy-4-nitrobenzene
	p-xylene	1,4-dimethylbenzene

THE ALCOHOLS

An organic compound containing a **hydroxyl group** attached to a saturated carbon atom is called an **alcohol**. The hydroxyl group is not the same as a hydroxide ion. While the $OH^-(aq)$ is a charged particle, the hydroxyl group is neutral. The molecules are called "enols," and the alcohol is identified by the suffix -ol. The compound

OH

is called phenol, being made of phenyl radical and a hydroxyl group. Phenyl -ol contracts to phenol.

The hydroxyl group is the first of a series of **functional groups** to be investigated in this chapter. A functional group is the most reactive part of an organic molecule. Functional groups are used in classification to characterize organic species. A table of functional groups is found at the end of this chapter in Table 11-16. The hydroxyl radical, a functional group, characterizes the alcohols.

The IUPAC Rules

1. Adopt the corresponding name of the alkane related to the alcohol by the number of carbon atoms in the chain. Add -ol to the end of the name, while dropping the final -e.
2. Number the location of any attached groups.
3. The common name, **glycol**, is applied to compounds containing two hydroxyl groups.

The terms **primary** (n), **secondary** (sec), and **tertiary** (tert), are applied to alcohols to indicate the number of alkyl groups that are attached to the carbon having the hydroxyl group.

The following are some examples of this nomenclature;

$$CH_3-CH_2-CH_2-CH_2OH$$
n-butyl alcohol

$$CH_3-CH_2-\overset{\overset{\textstyle CH_3}{|}}{C}HOH$$
sec-butyl alcohol

$$CH_3-\overset{\overset{\textstyle CH_3}{|}}{\underset{\underset{\textstyle CH_3}{|}}{C}}OH$$
tert-butyl alcohol

TABLE 11-7. EXAMPLES OF ALCOHOLS

Structural Formula (condensed)	IUPAC Name (common name in parentheses)
CH_3OH	methanol (methyl alcohol)
CH_3-CH_2OH	ethanol (ethyl alcohol)
$CH_3-CH_2-CH_2OH$	1-propanol (n-propyl alcohol)
$CH_3-\overset{\displaystyle OH}{\overset{\displaystyle |}{CH}}-CH_3$	2-propanol (isopropyl alcohol)
$\overset{\displaystyle OH}{\overset{\displaystyle |}{CH_2}}-\overset{\displaystyle OH}{\overset{\displaystyle |}{CH_2}}$	1,2-ethanediol (ethylene glycol)
$\underset{\underset{OH}{|}}{CH_2}-\underset{\underset{OH}{|}}{CH}-\underset{\underset{OH}{|}}{CH_2}$	1,2,3-propanetriol (glycerol)
$CH_3-CH_2-CH_2-CH_2-CH_2OH$	1-pentanol (amyl alcohol)
$CH_3-CH_2-CH_2-CH_2-CH_2-CH_2OH$	1-hexanol
cyclopentanol (ring structure with OH)	cyclopentanol
$CH_2=CH-CH_2OH$	1-propene-3-ol (allyl alcohol)
benzene ring $-CH_2OH$	1-phenylmethanol (benzyl alcohol)

This distinction is very useful because of the different chemical properties shown by these compounds. The formation of ketones or aldehydes depends on which form of the alcohol is used as a reactant.

$$\text{primary alcohol} \xrightarrow{\text{oxidation}} \text{aldehyde}$$

$$\text{secondary alcohol} \xrightarrow{\text{oxidation}} \text{ketone}$$

The Phenols and Ethers

There are several interesting phenols. Some members of this group are best known for their activity as insecticides and antiseptics. The attachment of a hydroxyl group to an aromatic ring characterizes the compounds. Some common phenols are illustrated in Table 11-8.

The only structural difference between an ether and an alcohol is the

TABLE 11-8. SOME COMMON EXAMPLES OF PHENOLS

Structural Formula	Common Name	IUPAC Name
⬡—OH	phenol	hydroxybenzene
⬡—CH₃ (OH)	o-cresol	2-hydroxytoluene
OH ⬡ OH	resorcinol	1,3-dihydroxybenzene
HO—⬡—OH	hydroquinone	1,4-dihydroxybenzene
⬡⬡—OH	β-naphthol	2-hydroxynaphthalene

position of the oxygen atom and whatever may be attached to the oxygen atom. The IUPAC names are not usually used for the ethers. The common name is taken from the two groups on either side of the oxygen. An ether may be thought of as a dehydrated alcohol. The particular compound, diethyl ether, is famous as an anesthetic.

A few examples of ethers are:

$$CH_3—O—CH_3 \qquad CH_3—O—CH_2—CH_3$$
dimethyl ether   methyl ethyl ether

⬡—O—CH₃   ⬡—O—⬡

methyl phenyl ether   diphenyl ether

THE ALDEHYDES AND KETONES

Aldehydes and ketones are important because of their routine use in the laboratory for cleaning, drying, and preserving. Some are common solvents, and some are employed in the production of hard plastics.

In the same manner in which the hydroxyl (—OH) group characterized the alcohols, the aldehydes and ketones have their own identifying functional groups. The carbon chain, which is of secondary importance and which may be short or long, is often symbolized by the letter R. If there are two carbon chains on either side of the functional group, the second chain will be

TABLE 11-9. COMPARISONS OF COMMON AND IUPAC NAMING METHODS

	Prefixes	
Number of Carbon Atoms	**IUPAC**	**Common**
—C—	meth-	form-
—C—C—	meth-	acet-
—C—C—C—	prop-	propion-
—C—C—C—C—	but-	butyr-

TABLE 11-10. EXAMPLES OF ALDEHYDES

Condensed Structural Formula	Common Name	IUPAC Name
H—C(=O)H	formaldehyde	methanal
CH₃—C(=O)H	acetaldehyde	ethanal
CH₃—CH₂—C(=O)H	propionaldehyde	propanal
CH₃—CH₂—CH₂—C(=O)H	n-butyraldehyde	butanal
CH₃—CH(CH₃)—C(=O)H	isobutyraldehyde	2-methylpropanal
CH₃—CH(OH)—CH₂—C(=O)H (γ β α)	β-hydroxybutyraldehyde	3-hydroxybutanal
C₆H₅—C(=O)H	benzaldehyde	phenylmethanal

distinguished from the first by labeling it, R′ (R prime). The functional group of the aldehydes is

$$-\overset{\displaystyle O}{\underset{\displaystyle H}{C}}$$

and it is called a **carbonyl**. More precisely, the carbonyl is

$$\diagdown C = O$$

In the specific case of the aldehydes, a hydrogen is one of the atoms bonded to the carbonyl group. The simple carbonyl is sometimes called a **keto** group because it identifies the ketones specifically. The general formula for the ketones is

$$R - \overset{\displaystyle O}{\overset{\|}{C}} - R'$$

In the common naming of carbonyl compounds, the positions of groups attached to the main carbon chain are described by the Greek letters α, β, γ, δ. The alpha designation begins with the first carbon atom that is not part of the carbonyl group.

$$-\overset{\displaystyle O}{\overset{\|}{C}} - \underset{\alpha}{C} - \underset{\beta}{C} - \underset{\gamma}{C}$$
carbonyl

TABLE 11-11. EXAMPLES OF KETONES

Condensed Structural Formula	Common Name	IUPAC Name
$CH_3-\overset{O}{\overset{\|}{C}}-CH_3$	acetone (dimethyl ketone)	propanone
$CH_3-\overset{O}{\overset{\|}{C}}-CH_2-CH_3$	methyl ethyl ketone	butanone
$CH_3-CH_2-\overset{O}{\overset{\|}{C}}-CH_2-CH_3$	diethyl ketone	3-pentanone
(cyclohexanone ring)=O	none	cyclohexanone
(phenyl)$-\overset{O}{\overset{\|}{C}}-CH_3$	acetophenone	phenylethanone
(phenyl)$-\overset{O}{\overset{\|}{C}}-$(phenyl)	benzophenone	diphenylmethanone

Furthermore, common names use a different set of prefixes to describe the number of carbon atoms in the chain. Comparing the common and IUPAC methods in a chart will illustrate the differences.

The IUPAC system uses a suffix of -al to indicate the aldehydes and the suffix -one describes the ketones. The suffix -one is applied to the longer chain of carbon atoms since there are two carbon atom branches from the carbonyl.

Some examples of aldehydes and ketones are listed in Tables 11-10 and 11-11.

ORGANIC ACIDS AND ESTERS

Organic acids are so named because of their proton donating action which is a typical property of all acids. The variety of organic acids and their related compounds is vast compared to inorganic acids.

The great significance of organic acids is immediately recognized when amino acids are considered. Some descriptive chemistry of organic acids is helpful in more fully appreciating the role of amino acids as the building blocks of protein.

TABLE 11-12. SOME EXAMPLES OF ORGANIC ACIDS

Condensed Structural Formula	Common Name	IUPAC Name
$HC\begin{smallmatrix}O\\\\OH\end{smallmatrix}$	formic acid	methanoic
$CH_3-C\begin{smallmatrix}O\\\\OH\end{smallmatrix}$	acetic acid	ethanoic
$CH_3-CH_2-C\begin{smallmatrix}O\\\\OH\end{smallmatrix}$	propionic acid	propanoic
$CH_3-CH_2-CH_2-C\begin{smallmatrix}O\\\\OH\end{smallmatrix}$	butyric acid	butanoic
$CH_3-CH_2-CH_2-CH_2-C\begin{smallmatrix}O\\\\OH\end{smallmatrix}$	valeric acid	pentanoic
$CH_3-(CH_2)_4-C\begin{smallmatrix}O\\\\OH\end{smallmatrix}$	caproic acid	hexanoic
$CH_3-(CH_2)_{14}=C\begin{smallmatrix}O\\\\OH\end{smallmatrix}$	palmitic acid (from palm oil)	hexadecanoic

The functional group that identifies organic acids is basically a carbonyl group plus a hydroxyl group:

carbonyl   hydroxyl

The specific name assigned to this functional group is **carboxyl**.

Some organic acids have been known for centuries. Formic acid is responsible for the sharp smell of squashed ants. Acetic acid is essence of vinegar. Valeric acid means strong, and this aptly describes the odor. Butyric acid is the smell of rancid butter, while the name caproic acid comes from the Latin word for goat. If a sweaty athlete smells like an "old goat," it is because of the caproic acid in his perspiration.

A representative table of organic acids is presented in Table 11-12.

TABLE 11-12. (*Continued*)

Condensed Structural Formula	Common Name	IUPAC Name
$CH_3—(CH_2)_{16}—C$ (with $=O$ and OH)	stearic acid (from animal fat)	octadecanoic
benzene ring—C (with $=O$ and OH)	benzoic acid	benzenecarboxylic
benzene ring with two $C(=O)OH$ groups ortho	o-pthalic acid	benzene-1,2-dicarboxylic
$CH_2=CH—C$ (with $=O$ and OH)	acrylic acid	propenoic
benzene ring with OH and $C(=O)OH$	salicylic acid	1-hydroxybenzenecarboxylic
$HO—C(=O)—C(=O)—OH$	oxalic acid	ethandioic

Many of the common organic acids are derived from fats and oils. This is the reason for the biological term **fatty acid**.

The reactions between organic acids and alcohols gives rise to a family of organic compounds noted for its odors, many of which are quite pleasant. The group is known as the **esters**.

Esters

Esters result when acids combine with alcohols with the resulting loss of water. Bring the two formulas together after removing the ⸺OH and

$$CH_3 - C \overset{\displaystyle O}{\underset{\displaystyle OH}{<}} + HO - \overset{\displaystyle H}{\underset{\displaystyle H}{\overset{|}{\underset{|}{C}}}} - H$$

remove

$$\text{acetic acid} \qquad\qquad \text{methanol}$$

H⸺ as water ⸺HOH. The first part of the new formula name is derived

$$CH_3 - \overset{\displaystyle O}{\overset{\displaystyle \|}{C}} - O - CH_3 + HOH$$

acid part   alcohol part

from the alcohol while the suffix, -ate is attached to the acid name in order to identify the compound as an ester. The common name becomes methyl

$$CH_3 - \overset{\displaystyle O}{\overset{\displaystyle \|}{C}} - O - CH_3$$

acetate   methyl

acetate, or methyl ethanoate in the IUPAC system.

The general formula for esters is

$$R - \overset{\displaystyle O}{\overset{\displaystyle \|}{C}} - O - R'$$

The R represents the part of the formula derived from the acid and R' comes from the alcohol. The pleasant odors of many esters is a sharp contrast to the obnoxious odors of their related acids. Examples of some common esters are organized in Table 11-13.

TABLE 11-13. EXAMPLES OF COMMON ESTERS

Condensed Structural Formula	Common Name (characteristic odor)	IUPAC Name
$\overset{\text{O}}{\overset{\|}{\text{HC}}}$—O—$CH_2CH_3$	ethyl formate (rum)	ethyl methanoate
$\overset{\text{O}}{\overset{\|}{\text{HC}}}$—O—$CH_2$—$\overset{\text{CH}_3}{\overset{\|}{\text{CH}}}$—$CH_3$	isobutyl formate (raspberries)	2-methyl propyl methanoate
CH_3—$\overset{\text{O}}{\overset{\|}{\text{C}}}$—O—$(CH_2)_4$—$CH_3$	n-amyl acetate (bananas)	pentyl ethanoate
CH_3—$\overset{\text{O}}{\overset{\|}{\text{C}}}$—O—$(CH_2)_7$—$CH_3$	n-octyl acetate (oranges)	octyl methanoate
CH_3—CH_2—CH_2—$\overset{\text{O}}{\overset{\|}{\text{C}}}$—O—$CH_2$—$CH_3$	ethyl butyrate (pineapples)	ethyl butanoate
phenyl ring with OH, $\overset{\text{O}}{\overset{\|}{\text{C}}}$—O—$CH_3$	methyl salicylate (wintergreen)	methyl-1-hydroxy benzoate

MERCAPTANS AND AMINES

Mercaptans are sulfur analogs of alcohols. **Analogs** are compounds that are similar in structure but show differences in the atoms composing the structure. In the case of mercaptans, the molecular structure is the same as alcohol, except for the fact that a sulfur atom replaces the oxygen in the hydroxyl group. Mercaptans are called **thioalcohols** in the IUPAC nomenclature systems.

The -SH attachment, which characterizes the mercaptans, is called the **sulfhydryl** group. This -SH group is found in many proteins and has a certain infamy because of the foul odor associated with it. The odor of skunk is n-butyl mercaptan. A few examples of mercaptans are found in Table 11-14.

Amines are compounds related to ammonia. When a single hydrogen of the NH_3 molecule is substituted by a carbon chain, it is called a primary amine. A two hydrogen substitution is a secondary amine (sec-), and a three hydrogen substitution is a tertiary amine (tert-).

TABLE 11-14. EXAMPLES OF MERCAPTANS

Condensed Structural Formula	Common Name	IUPAC Name
CH_3SH	methyl mercaptan	methanethiol
CH_3—CH_2SH	ethyl mercaptan	ethanethiol
CH_3—CH_2—CH_2—CH_2SH	n-butyl mercaptan	1-butanethiol

TABLE 11-15. EXAMPLES OF AMINES

Condensed Structural Formula	Common Name	IUPAC Name
CH_3NH_2	methylamine (primary)	aminomethane
(structure with NH₂ on benzene ring)	aniline	aminobenzene
(structure: CH_3, CH_2, $CH_3—N—H$)	ethylmethylamine (secondary)	1-(N-methylamino)ethane*
(structure: triphenylamine)	triphenylamine (tertiary)	aminotribenzene

* The capital letter N means that the next section of carbon chain is attached to the nitrogen.

$$ \underset{\text{primary amine}}{R—\overset{\overset{H}{|}}{N}—H} \qquad \underset{\text{secondary amine}}{R—\overset{\overset{R'}{|}}{N}—H} \qquad \underset{\text{tertiary amine}}{R—\overset{\overset{R'}{|}}{N}—R''} $$

Amines, especially in the form of amino acids, have extensive application to biological chemistry. The proteins will be investigated in the next chapter.

The common names for amines are formed by adding the suffix -amine to the regular nomenclature that describes the number of carbon atoms in the chain.

HETEROCYCLIC COMPOUNDS

Heterocyclic compounds are fundamentally carbon ring structures where one or more carbon atoms are replaced by another type of atom. Examples in Figure 11-10 are taken from an enormous group of compounds. Many heterocyclic compounds are of great importance in biological chemistry because of their involvement in body chemistry and medicine.

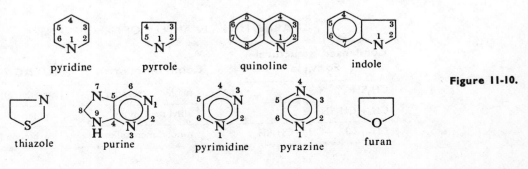

pyridine   pyrrole   quinoline   indole

thiazole   purine   pyrimidine   pyrazine   furan

Figure 11-10.

TABLE II-I6. SUMMARY OF FUNCTIONAL GROUPS

Functional Group	Type of Compound	IUPAC Suffix	Example
—OH hydroxyl	alcohol	-ol	methanol (methyl alcohol)
carbonyl	aldehyde	-al	methanal (formaldehyde)
C=O carbonyl	ketone	-one	propanone (acetone)
carboxyl	acid	-oic acid	methanoic acid (formic acid)
oxy	ether	depends on alkyl or aryl chain	methoxymethane (dimethyl ether)
—C—O— oxyketone	ester	-ate	methylmethanoate (methyl formate)
amide	amine	depends on the alkyl or aryl group	aminomethane (methylamine)

Summary of the IUPAC Rules

1. Use the longest continuous carbon chain as the main name. Example:

$$-\overset{|}{\underset{|}{C}}-\overset{|}{\underset{|}{C}}-\overset{|}{\underset{|}{C}}-\overset{|}{\underset{|}{C}}-$$

four carbon atoms = but-

2. Number the carbon atoms from the end containing the functional group. Use the suffix that identifies the functional group.

$$-\underset{4}{\overset{|}{\underset{|}{C}}}-\underset{3}{\overset{|}{\underset{|}{C}}}-\underset{2}{\overset{|}{\underset{|}{C}}}-\underset{1}{\overset{|}{\underset{|}{C}}}-OH$$

1-butanOL

3. Adjust the suffix to identify a double or triple bond:

$$-\underset{4}{\overset{|}{\underset{|}{C}}}-\underset{3}{\overset{|}{\underset{|}{C}}}-\underset{2}{\overset{|}{\underset{|}{C}}}=\underset{1}{\overset{|}{C}}-OH$$

1-butene-1-ol

4. Add any attached group as a prefix which is numbered accordingly.

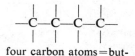

$$-\underset{4}{\overset{\overset{\displaystyle CH_3}{|}}{\underset{|}{C}}}-\underset{3}{\overset{|}{\underset{|}{C}}}-\underset{2}{\overset{|}{C}}=\underset{1}{\overset{|}{C}}-OH$$

4-methyl-1-butene-1-o-1

PROBLEMS

Set A

1. Define or explain the following terms:

a. alkyl group   e. olefin
b. diene   f. racemic
c. hydrocarbon   g. dextrorotatory
d. isomer   h. halogenation

2. What is a functional group? Name four functional groups.

3. Write the names and structural formulas of the first four members of the alkane, alkene, and alkyne series.

4. Write structural formulas for the cis- and trans- isomers of dichloro-ethene.

5. What is meant by the term "polyunsaturated hydrocarbon"?

6. Draw structural formulas for the following:

a. 1-butene

b. 1,3-pentadiene

c. 2,2,4-trimethylpentane

7. Name the following compounds

a. 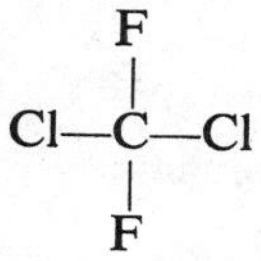 $Cl-\underset{\underset{F}{|}}{\overset{\overset{F}{|}}{C}}-Cl$ b. CH_3-CH_2OH c. 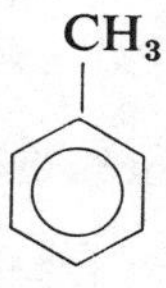 (structure of toluene with CH_3)

8. Identify the family to which each of the following compounds belong.

a. 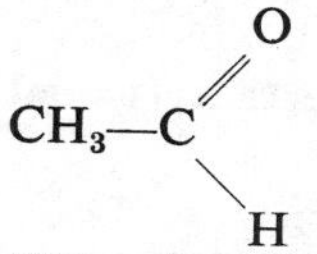 $CH_3-\overset{\overset{O}{\parallel}}{C}\diagdown H$

b. CH_3-O-CH_3

c. 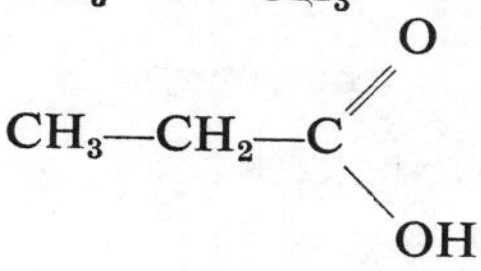 $CH_3-CH_2-\overset{\overset{O}{\parallel}}{C}\diagdown OH$

d. 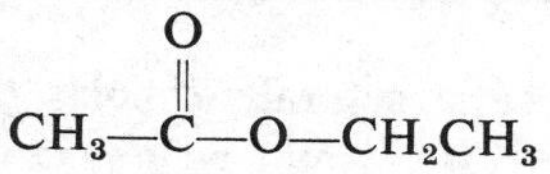 $CH_3-\overset{\overset{O}{\parallel}}{C}-O-CH_2CH_3$

e. CH_3SH

f. $CH_3-CH_2-NH_2$

9. Write structural formulas for the following compounds:

a. formaldehyde

b. 3-methylbutanol

c. propionic acid

d. butanoic acid

e. acetyl chloride

f. aniline

g. toluene

h. p-xylene

10. Write all the possible structural formulas for C_4H_9Cl.

11. Write condensed structural formulas for the following compounds:

a. cyclopentane

b. 2-bromo-1-butene

c. 2-pentyne

d. metadichlorobenzene

e. 1,4-dinitrobenzene

12. Write the names of the following structures:

a. 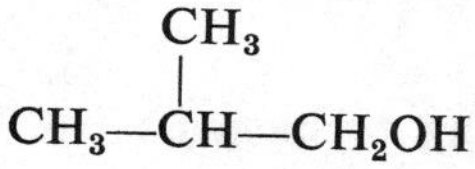 $CH_3-\underset{\underset{CH_3}{|}}{CH}-CH_2OH$

b. 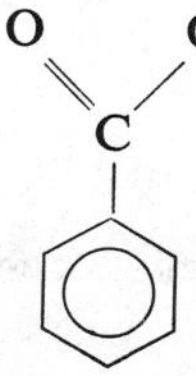 (benzene ring with $\overset{\overset{O}{\diagup}}{C}\diagdown OH$ group)

c. (benzene ring with NO_2 and $-OH$)

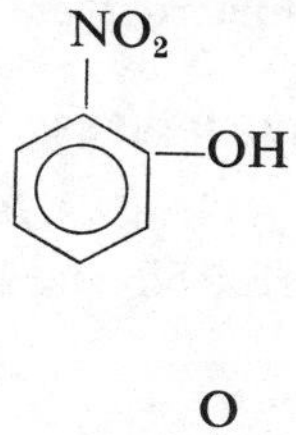

d. 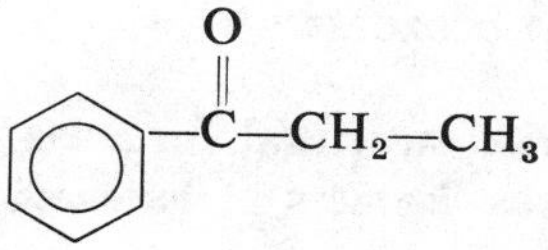 (benzene ring with $\overset{\overset{O}{\parallel}}{C}-CH_2-CH_3$)

Set B

13. Define or explain the following terms:

a. polymer
b. paraffin
c. aryl
d. analog

e. ortho-
f. enantiomer
g. optical isomer
h. fatty acid

14. Write the structures of the following functional groups:

a. carbonyl
b. carboxyl
c. amine

d. sulfhydryl
e. mercaptan
f. ketone

15. Write structural formulas for the following compounds:

a. β-butylene
b. isopropyl alcohol
c. 2-bromo-1,3-butadiene

d. caproic acid
e. dimethyl ether
f. propanone

16. Use structural formulas to differentiate between the cis- and trans-isomers of 1,4-dibromocyclohexane.

17. What results from the polymerization of tetrafluoroethene?

18. Write structural formulas for the following compounds:

a. 1,2,3-propantriol
b. diphenyl ketone
c. sec-butyl alcohol

d. tert-butyl alcohol
e. propanal
f. ethandioic acid

19. Identify the following compounds by their common and IUPAC names.

a. $HC \overset{O}{\underset{}{\parallel}}\!\!-O-CH_2-CH_3-CH_2-CH_3$
b. $CH_3-CH_2-CH_2-CH_2-CH_2SH$
c. CH_3NH_2

d.

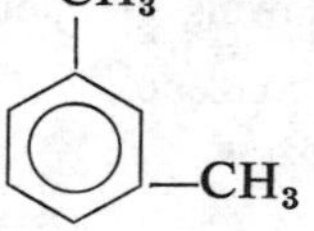

e.

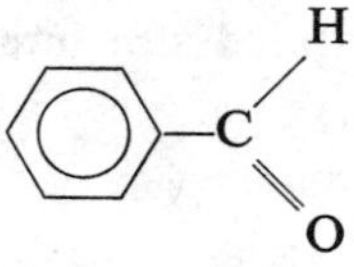

20. Use structural formulas to explain why 1,2-difluoroethene can have two isomers while 1,1-difluoroethene cannot.

21. Write structural formulas for the following compounds:

a. propandioic acid
b. benzene-1,2,4-tricarboxylic acid
c. diphenylamine
d. ethyl benzoate

22. Write structural formulas for maleic and fumaric acid. They are cis- and trans- isomers of butenedioic acid respectively.

23. Write the names of the following structures:

a. $CH_3-\overset{\overset{\displaystyle O}{\|}}{C}-O-CH_2-CH_2-CH_3$

b. 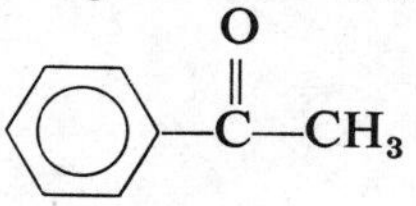 ⟨benzene ring⟩$-\overset{\overset{\displaystyle O}{\|}}{C}-CH_3$

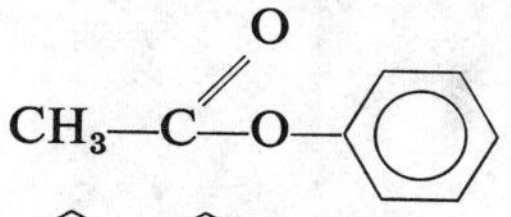

c. ⟨benzene ring⟩$-O-$⟨benzene ring⟩

d. $CH_3-\overset{\overset{\displaystyle O}{\|}}{C}-O-$⟨benzene ring⟩

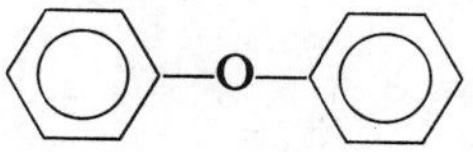

e. 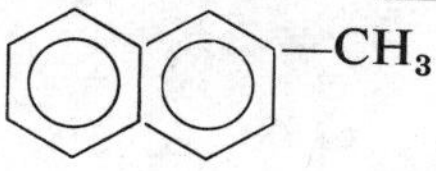 ⟨naphthalene ring⟩$-CH_3$

SUPPLEMENTARY READING

Schellman, J. A., and Schellman, C.: Optical Rotation and the Shape of Molecules. *Chemistry*, May, 1967.

A short paper that describes the theory and application of polarimetry.

Jones, P. R.: Infrared Spectroscopy and Molecular Architecture. *Chemistry*, February, 1965.

A fine introduction to the subject of IR spectroscopy.

Van Orden, H. O., and Lee, G. L.: *Elementary Organic Chemistry: A Brief Course.* W. B. Saunders Company, Philadelphia, 1969.

A clearly written and well illustrated introductory text.

Hart, H., and Schuetz, R. D.: *Organic Chemistry*, 3rd ed. Houghton Mifflin Co., Boston, 1966.

A well organized and very readable text that emphasizes practical applications of organic chemistry.

12 An Introduction to Biochemistry

Biochemistry is the chemistry of life. A great deal of the laboratory investigations being carried out in universities, hospitals, pharmaceutical companies, and food control centers is largely biochemistry. Biochemistry is also concerned with much of the nonbiological organic chemistry of plastics, dyes, fabrics, and fuels in both manufacture and research because the question of industrial pollution is critical to life. The biochemistry laboratory is facing the problems of carbon monoxide and sulfur dioxide as products of organic compound combustion. What can be done with detergent foam in rivers and accumulating plastic containers that can't be safely burned? This decade of the seventies is witnessing the concern and activity of biochemists in an effort to save life on this planet.

The pharmaceutical companies are attacking the problems of disease, pain, malnutrition, and population control. This is the chemistry of life. They are deeply involved in the abused chemistry of drugs, narcotics, and pest control so that biochemistry may not become the organic spectre of deformity, insanity, and death.

The laboratories of our colleges and universities are exploring the chemistry of life from living organisms down to the submolecular activities of virus composition and gene structure—matter which lies on the ill-defined boundary of what is described as life or nonlife. The incredibly complex series of chemical events leading from the production of basic organic compounds by green plants in the process of photosynthesis to the formation of living tissue in animals is the object of biochemical research. The intensive investigations of small areas of biochemistry have produced thousands of books and many thousands of articles in scientific journals, and we are still at the threshold of discovering what life is all about. Disease, hunger, birth defects, and mental illness remain unconquered. When Isaac Newton said that he felt like a small boy playing with pebbles on a beach while the whole ocean before him remained unexplored, he was speaking for our time, too.

This chapter attempts no more than a brief survey of biochemistry. To attempt more would be unrealistic. The emphasis will be on descriptive biochemistry rather than reaction mechanisms. It is important to start with some familiarity with representative types of compounds classified as carbohydrates, lipids, proteins, nucleic acids, enzymes, vitamins, hormones, and pharmaceuticals.

CARBOHYDRATES

Carbohydrates are compounds composed of carbon, hydrogen, and oxygen. The hydrogens and oxygens branch from a carbon chain or a carbon ring as hydroxyl, aldehyde, or ketone functional groups. The carbohydrates are generally known as sugars and starches, all of which owe their existence to the photosynthetic process of green plants where the reaction between CO_2 and water is catalyzed by chlorophyll (the green pigment in leaves) to form simple sugars. It is an energy-absorbing reaction in which solar energy is the source. The carbohydrates are usually classified as the following:

1. **Monosaccharides** are divided into **aldoses** and **ketoses**, depending on what kind of carbonyl group is present. The most common monosaccharides have five or six atoms composing the structure.
2. **Disaccharides** are double structures (two connected monosaccharides). They can be broken down into monosaccharides.
3. **Polysaccharides**, which may also be broken down into monosaccharides, are composed of many monosaccharide units.

Monosaccharides

The most important monosaccharide for human energy production is **glucose** ($C_6H_{12}O_6$). It is also known as **dextrose**, which suggests the dextrorotatory optical rotation it exhibits. The $l(-)$ enantiomer has no nutritional value. There are two common forms of D-glucose because of the position of the hydroxyl group that forms when the polyhydroxy aldehyde structure is in water solution.

A two dimensional (plane projection) formula of D-(+) glucose is illustrated in Figure 12-1. Six carbon atoms indicate the hexose designation.

Figure 12-1.

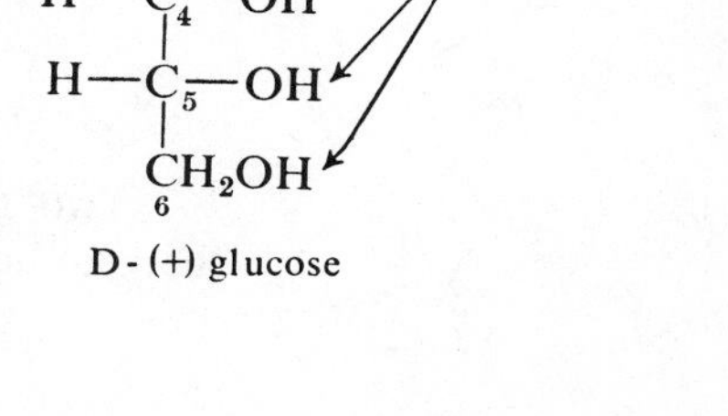

D- (+) glucose

In water, the hydrogen atom of the hydroxyl group shifts to the carbonyl. The resulting formation of a bond between the remaining oxygen atom and the carbon of the carbonyl group causes a ring structure with two possible isomers (Fig. 12-2).

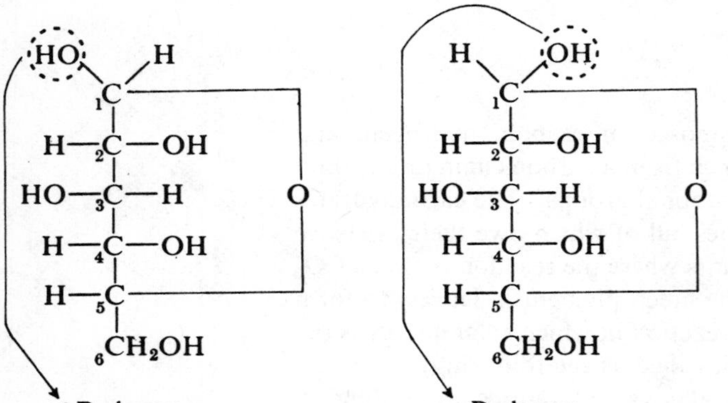

β-D-glucose   α-D-glucose

Figure 12-2.

When a carbon atom is bonded to different numbers of atoms, it is described as **asymmetrical** (unbalanced). For example, $H—C—OH$ is asymmetrical while $H—C—H$ is symmetrical. Asymmetry is a direct cause of geometrical isomerism and optical isomerism.

Chemists tend to favor a structural formula that more closely simulates the three dimensional ring structures of the saccharides in solutions (Fig. 12-3).

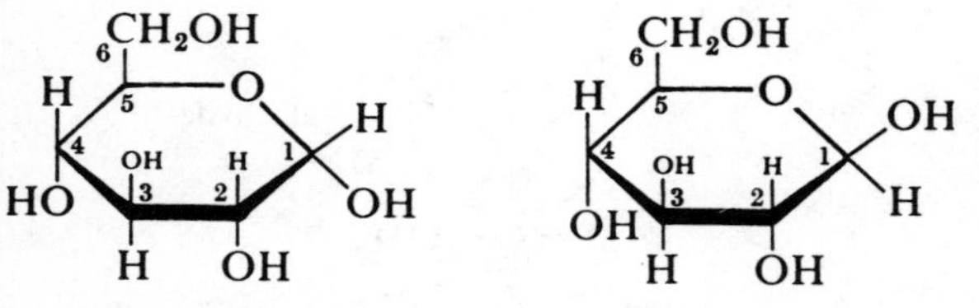

α-D-glucose   β-D-glucose

Figure 12-3.

Two more examples of common monosaccharides are **galactose** and the keto hexose, **fructose** (Fig. 12-4).

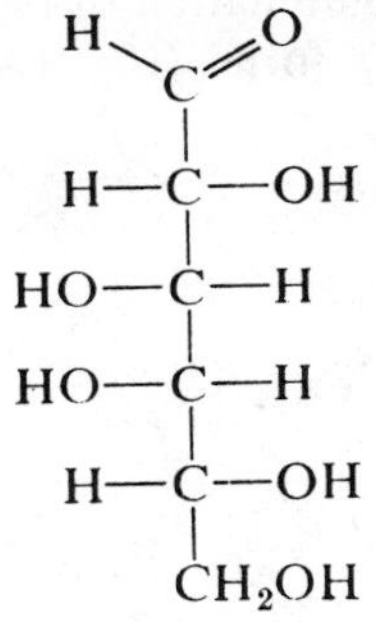

D-galactose (open chain)

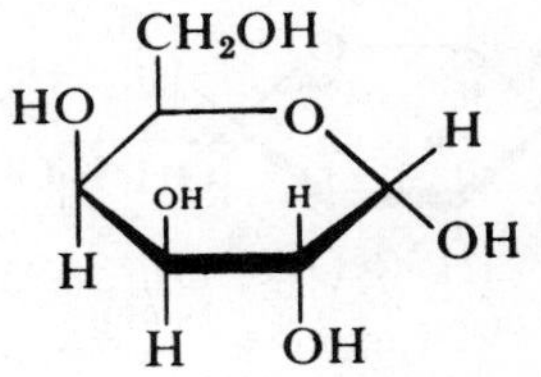

α-D-galactose (cyclic)

Figure 12-4.

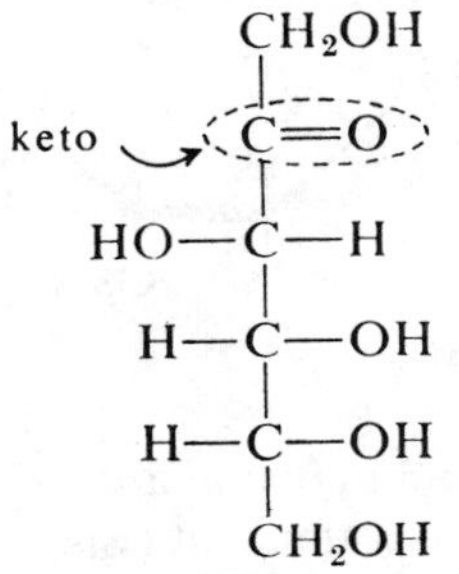

D-fructose (open chain)

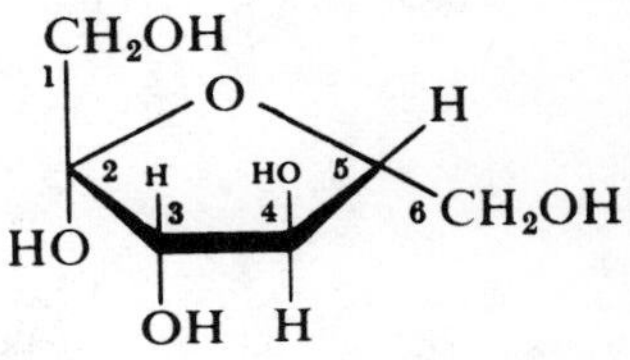

β-D-fructose (cyclic)

Disaccharides

Disaccharides are carbohydrates having two monosaccharide units per molecule. Disaccharides are the most common examples of a group known as **oligosaccharides**. The prefix oligo- means few. The three most common disaccharides are **sucrose** (table sugar from sugar cane and sugar beets), **maltose** (from the breakdown of starch), and **lactose** (from milk).

Sucrose is a combination of D-glucose and D-fructose minus a water molecule (Fig. 12-5).

D-glucose

D-fructose

sucrose $-C_{12}H_{22}O_{11}$

Figure 12-5.

H_2O

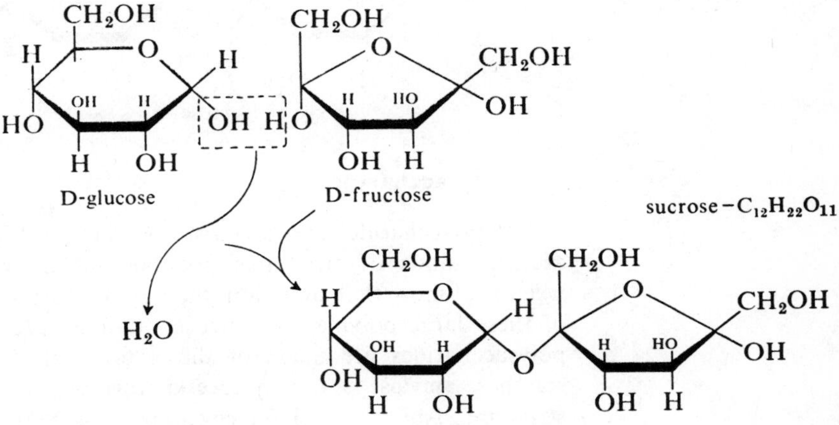

The addition of water allows the disaccharide to be converted to its monosaccharide components in the presence of the enzyme **sucrase**. The splitting of a water molecule in such a reaction is called **hydrolysis**.

Maltose is a combination of two units of D-glucose (minus H_2O) which may be either α or β forms (Fig. 12-6).

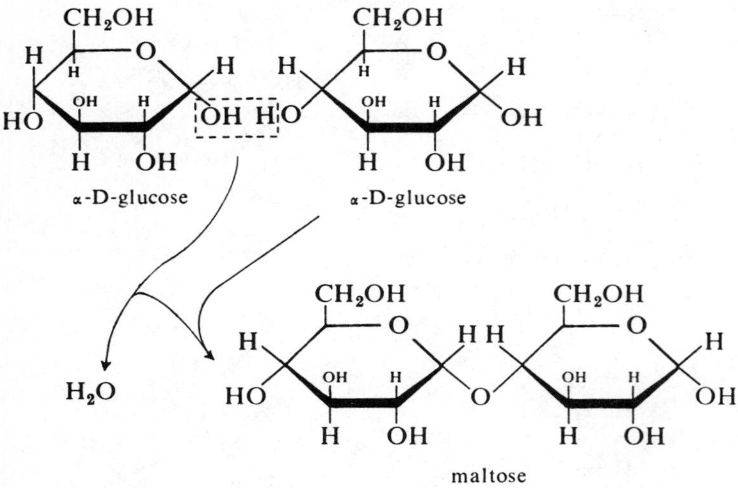

Figure 12-6.

Lactose is found in mammalian milk. It is a combination of glucose and β-galactose minus a water unit (Fig. 12-7).

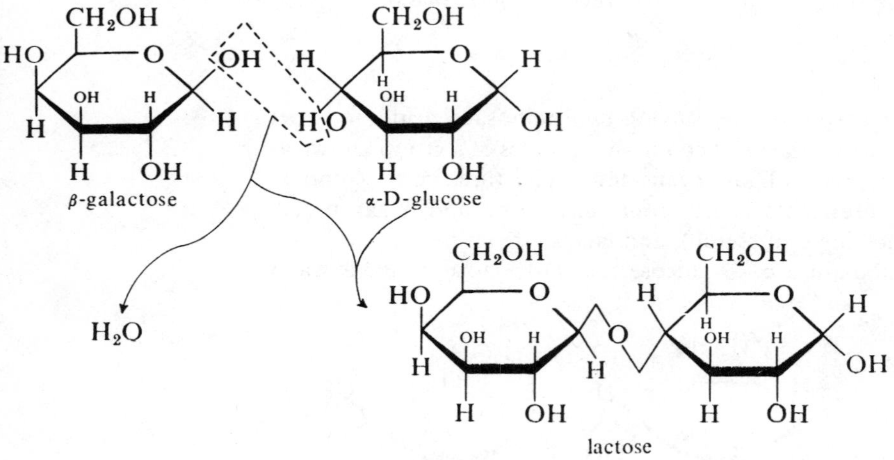

Figure 12-7.

Polysaccharides

Polysaccharides are the most abundant carbohydrates. Although there are thousands of structures possible, including some having molecular weights of more than a million, the most important unit structure (monomer) of these large polymers is glucose. Biologically speaking, the most useful polysaccharides are made of the α form of the monosaccharide. Plant starches, **amylose** and **amylopectin**, form up to 90 per cent of a plant's structure. Animal starch (**glycogen**) is found in the liver and muscles of most animals. Green plants effectively polymerize the glucose resulting from photosynthesis. The **cellulose**, which composes the cell walls of plants, is the result of the polymerization of the β form of glucose. This is the reason for cellulose not being readily digested by most animals. Cellulose is made of

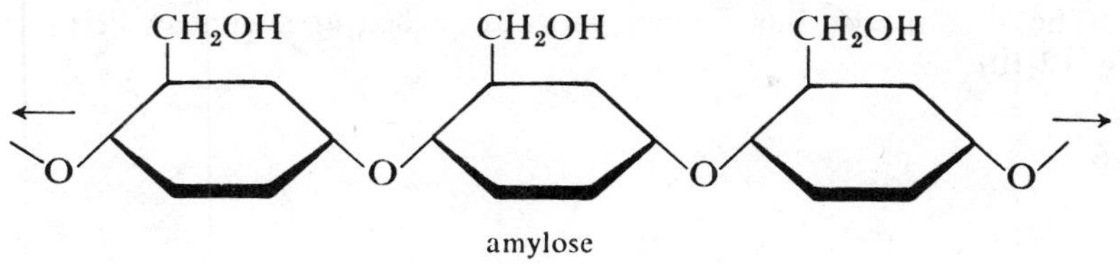

$$CH_2OH \qquad CH_2OH \qquad CH_2OH$$

amylose

Figure 12-8.

several thousand β-glucose units in long chain polymers. It is a rigid structure and it is the most abundant organic compound on earth.

Amylose is a coil arrangement of the α-glucose polymer. The long chain of α-glucose units (Fig. 12-8), may be visualized as a coil (Fig. 12-9).

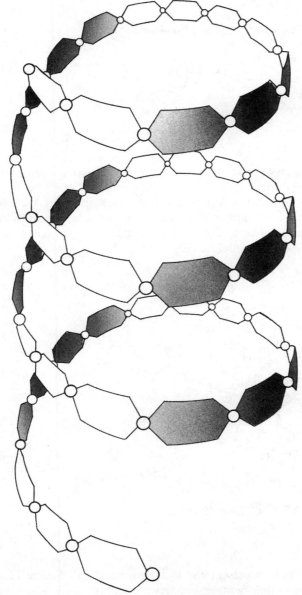

Figure 12-9.   **Amylose coil**

The amylopectin has a much more branched arrangement of the coils (Fig. 12-10).

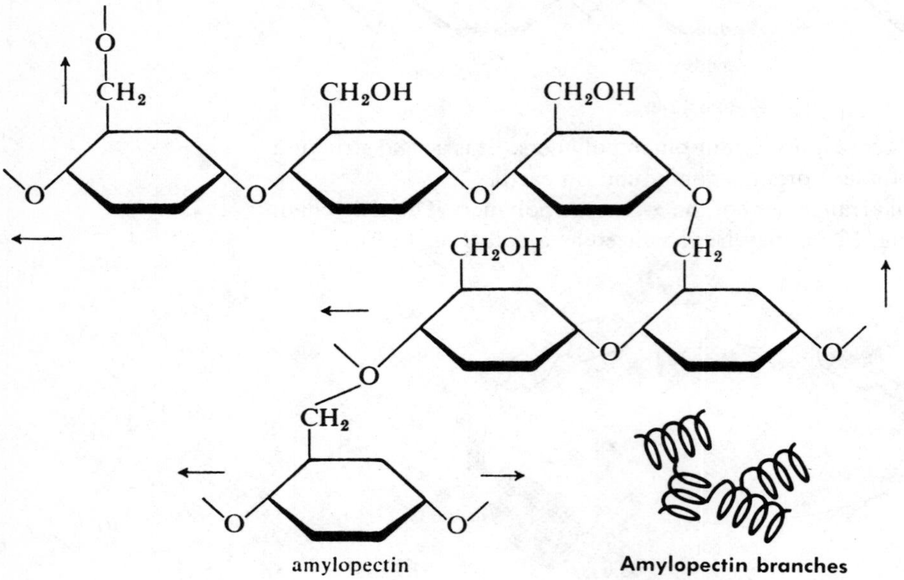

amylopectin   Amylopectin branches

Figure 12-10.

Glycogen is very similar to amylopectin except that it has an even more branched arrangement.

Cellulose results from a twisting and overlapping of the chains of β-glucose polymers. The strength and rigidity is due in large part to hydrogen bonding between the chains. Cotton, paper, and some synthetic products are composed principally of cellulose (Fig. 12-11).

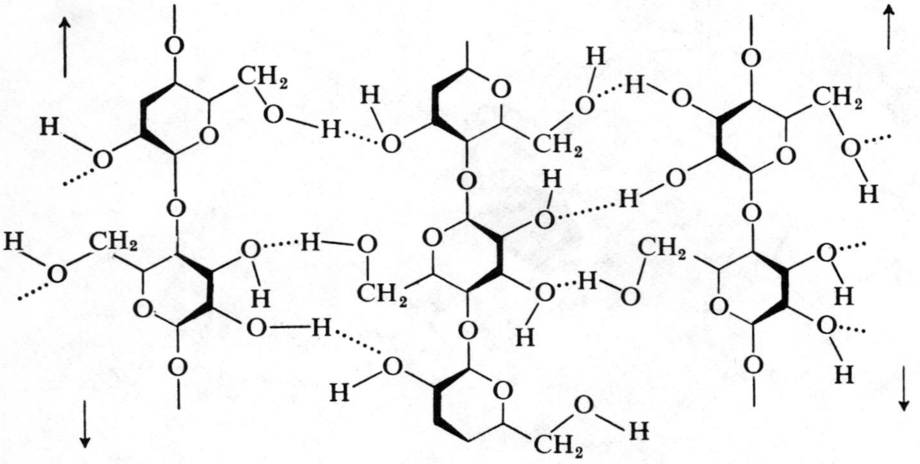

Figure 12-11.
Cellulose.

The dotted lines represent the hydrogen bonding. Figure 12-12 illustrates the hydrogen link between two oxygen atoms.

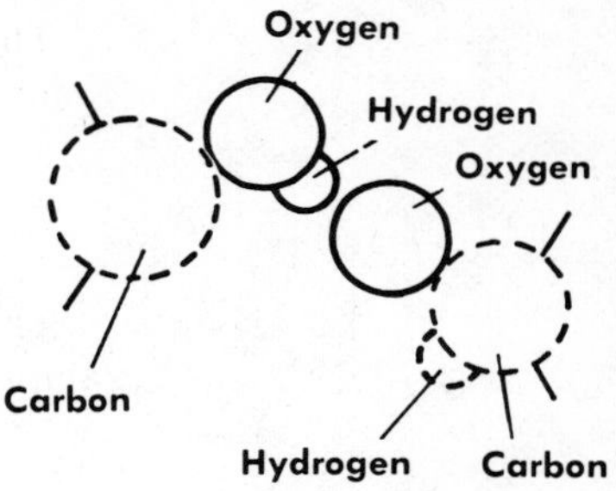

Figure 12-12. Hydrogen bonding.

Cellulose is indirectly available to man's food requirements because of the ability of cattle to decompose the polymers by the action of digestive acid in the warm environment of a peculiar stomach over an extended period of time.

LIPIDS

Lipids may be thought of as **fats** and **oils**. The structure of lipids is not especially easy to describe because of the variety of combinations of long chain hydrocarbons containing **phosphorus** (phospholipids), **alcohols**, **waxes**, **steroids** (polycyclic carbon structures), and **terpenes** (isoprene rings). A striking physical property of lipids is their tendency to dissolve readily in common organic solvents such as carbon tetrachloride and benzene.

Many lipids are esters of long chain organic acids and alcohols. The most common lipids are **triglycerides**, which are relatively simple structures composed of the ester of glycerol with long chain fatty acids. Some examples are lard, butterfat, linseed oil, and peanut oil. The most common fatty acids are:

Palmitic acid (saturated) $CH_3(CH_2)_{14}C\begin{matrix} \diagup\!\!\!\diagup O \\ \diagdown OH \end{matrix}$

Oleic acid (unsaturated) $CH_3(CH_2)_7CH{=}CH(CH_2)_7C\begin{matrix} \diagup\!\!\!\diagup O \\ \diagdown OH \end{matrix}$

Stearic acid (saturated) $CH_3(CH_2)_{15}C\begin{matrix} \diagup\!\!\!\diagup O \\ \diagdown OH \end{matrix}$

Linoleic acid (unsaturated)

$$CH_3(CH_2)_4CH{=}CHCH_2CH{=}CH(CH_2)_7C\begin{matrix} \diagup\!\!\!\diagup O \\ \diagdown OH \end{matrix}$$

The structure of glycerol is

$$CH_2—CH—CH_2$$
$$\quad|\quad\quad|\quad\quad|$$
$$OH\quad OH\quad OH$$

The formation of a triglyceride involves the formation of the ester, an example of which is illustrated:

$$
\text{from stearic acid}\quad —\overset{\displaystyle O}{\underset{}{C}}—O—CH_2
$$

$$
\text{from oleic acid}\quad —\overset{\displaystyle O}{\underset{}{C}}—O—CH
$$

$$
\text{from palmitic acid}\quad —\overset{\displaystyle O}{\underset{}{C}}—O—CH_2
$$

$$\}\ \text{from glycerol}$$

A **wax** results from the ester formation between a simple alcohol containing one hydroxyl group and a fatty acid. For example,

$$
CH_3(CH_2)_{14}CHOH + CH_3(CH_2)_{14}C\overset{\displaystyle O}{\underset{OH}{}}
$$

cetyl alcohol   palmitic acid
(1-decahexanol)

$$
CH_3(CH_2)_{14}—\overset{\displaystyle O}{\underset{}{C}}—O—(CH_2)_{15}—CH_3
$$

cetyl palmitate

The formation of room temperature solids such as peanut butter and commercial products of the Spry, Crisco, and margarine are accomplished by **hydrogenation**. Hydrogenation is the addition of hydrogen atoms at the unsaturated double bond sites in the long chain molecules. Products which maintain a number of double bonds are popularly referred to as **polyunsaturated** fats and oils. When fats and oils are exposed to oxygen- and enzyme-bearing bacteria at room temperature, the lipids become oxidized. This leads to the decomposition into fatty acids and aldehydes which have the unpleasant odor associated with **rancid** fats and oils.

Soaps and **detergents** work by loosening the dirt that is bound to fabrics by oils. The cleaning agents have long hydrocarbon tails which mix with the dirt-binding oils in the process of setting them free. The head of the hydrocarbon tail is a metal ion, such as sodium, which mixes well with polar water molecules. The force of repulsion among the like charged metal ions prevents clotting and allows the suspension of dirt, oil, and cleaning agent to be washed away. The general name—**syndet**—means synthetic detergent. The reaction between a strong base, such as NaOH and an organic acid to form a soap is called **saponification**. Detergents which can be decomposed by bacterial action are described as being **biodegradable**, and such detergents are an aid in preventing "suds" pollution of urban waterways.

The diagram in Figure 12-13 idealizes the action of soap and detergent molecules in cleaning.

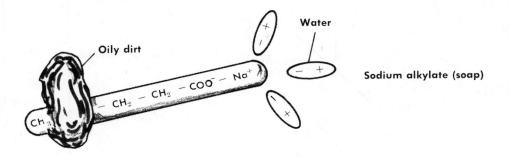

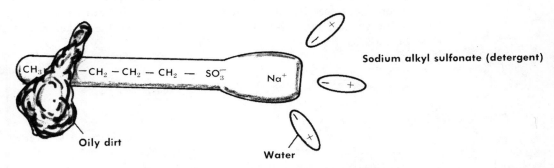

Figure 12-13. Cleaning action of soap and detergent molecules.

Phospholipids are naturally-occurring detergents or **emulsifiers**. An emulsifier is an agent that allows the break-up of insoluble fats and oils into tiny droplets that are evenly suspended throughout the solvent. It is the polar phosphate group that gives the phospholipid its name as well as its emulsifying property. **Lecithins**, found in living cells and tissues, are phospholipids which are thought to have some action in fat digestion. Lecithins are also used in food products to prevent the separation of oils from other ingredients.

A biologically important group of lipids is the **steroids**. The steroids have a polycyclic structure as shown in Figure 12-14. Steroids commonly

Figure 12-14.

have double bonds and keto groups, and the presence of a hydroxyl group gives the special name **sterol**. The best known sterol, because of its popular association with hardening of the arteries, is **cholesterol**. The structure of cholesterol is illustrated in Figure 12-15.

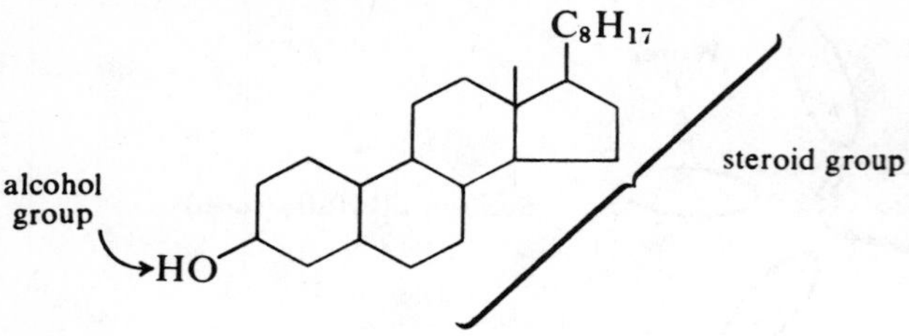

Figure 12-15. Cholesterol.

Steroids have been a subject of intensive investigation in recent years because of the connection between sex **hormones** and the search for oral contraceptives. The male hormone, **testosterone**, and the female hormone, **estradiol**, are primarily responsible for the development of the male and female sex characteristics. Synthetic compounds which have been developed as oral contraceptives are analogs of estradiol and another hormone related to pregnancy, **progesterone**. Figure 12-16 shows the structures of these hormones.

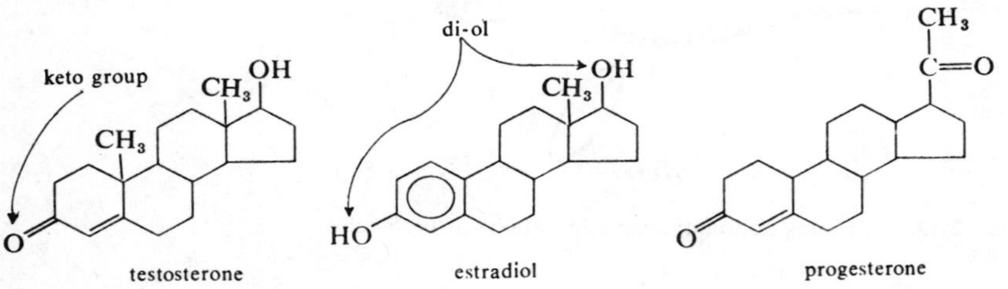

Figure 12-16.

One other example worth special mention is the steroid **cortisone** because of its ability to give relief from the pain and symptoms of joint inflammation and arthritis. The structure of cortisone is shown in Figure 12-17.

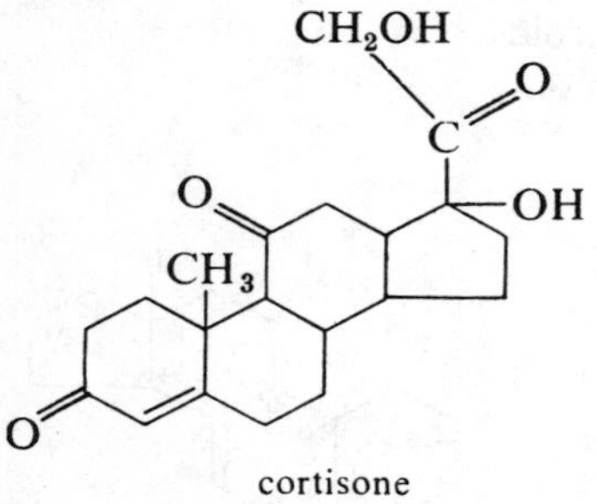

Figure 12-17.

Terpenes are a group of oils characterized by pleasant odors. These low molecular weight, volatile oils are commonly distilled from evergreens and other plants. The odors of pine, turpentine, menthol, lemon, and camphor are examples. The structure of menthol is shown as an example.

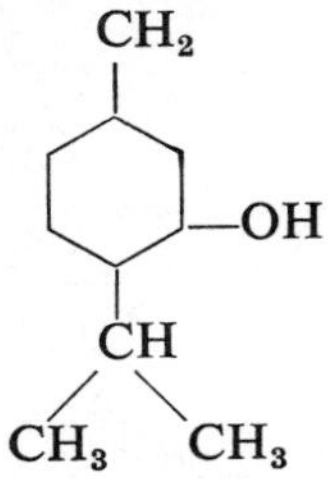

$$CH_2$$

menthol

Vitamin A, important to vision, is a biologically significant terpene.

PROTEINS

Proteins are the most abundant and active group of compounds in the life processes (e.g., digestion, circulation, sensitivity) of living organisms. Proteins make up the skin, muscles, and tendons. Hemoglobin, antibodies, enzymes, and nerve tissue are proteins. Proteins are responsible for growth, motion, defense, and metabolic control.

Proteins are polymers of amino acids. The amino acid monomer is composed of a carbon chain to which are attached amino and carboxyl groups. The general formula is

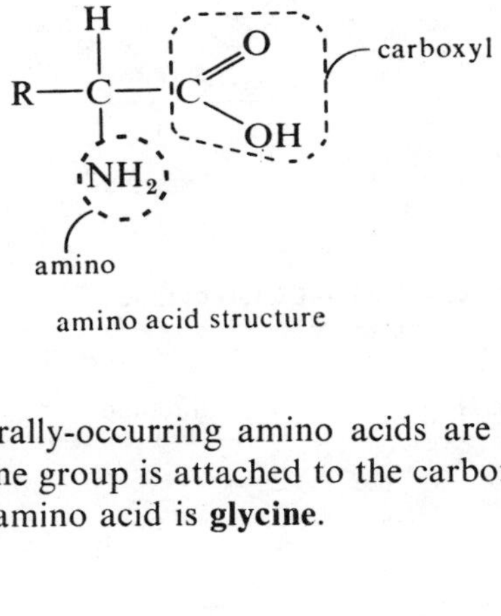

amino acid structure

The twenty naturally-occurring amino acids are classified as α-amino acids because the amine group is attached to the carbon next to the carboxyl group. The simplest amino acid is **glycine**.

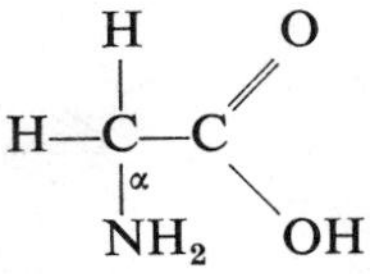

glycine

Living organisms synthesize protein by the linkage of long chains of amino acids (polymers) into **peptides**. A peptide linkage is the term used to describe the bonding of an amine (basic) group of one amino acid to the carboxyl (acidic) group of another amino acid. For example, the **dipeptide** (two amino acid units) **glycylalanine** formation is illustrated in Figure 12-18.

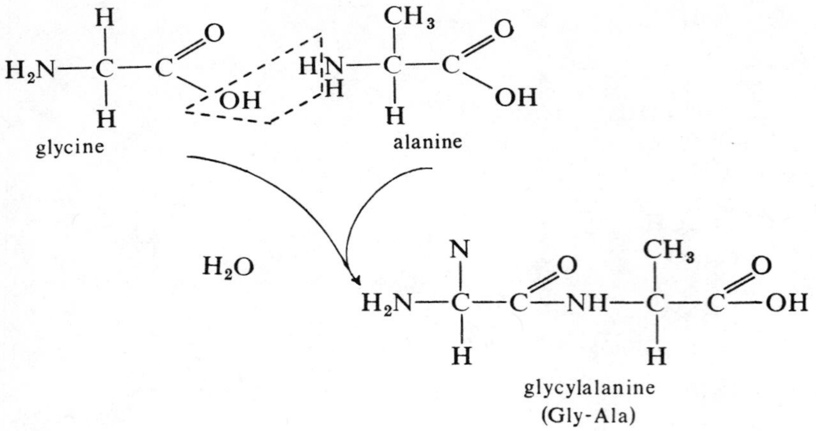

Figure 12-18.

TABLE 12-1. AMINO ACIDS

Formula	Name	Abbreviation
NH_2—CH_2—COOH	glycine	Gly
CH_3—$\overset{\overset{\displaystyle NH_2}{\mid}}{CH}$—COOH	alanine	Ala
CH_2OH—$\overset{\overset{\displaystyle NH_2}{\mid}}{CH}$—COOH	serine	Ser
CH_2SH—$\overset{\overset{\displaystyle NH_2}{\mid}}{CH}$—COOH	cysteine	Cys
$HOOC$—$\overset{\overset{\displaystyle NH_2}{\mid}}{CH}$—$CH_2$—S—S—$CH_2$—$\overset{\overset{\displaystyle NH_2}{\mid}}{CH}$—COOH	cystine	Cys
$CHOH$—CH_3—$\overset{\overset{\displaystyle NH_2}{\mid}}{CH}$—COOH	threonine	Thr
$\overset{\displaystyle CH_3}{\underset{\displaystyle CH_3}{\diagdown CH}}$—$\overset{\overset{\displaystyle NH_2}{\mid}}{CH}$—COOH	valine	Val
CH_3—$\overset{\overset{\displaystyle CH_3}{\mid}}{CH}$—$CH_2$—$\overset{\overset{\displaystyle NH_2}{\mid}}{CH}$—COOH	leucine	Leu
$\overset{\displaystyle CH_3-CH_2}{\underset{\displaystyle CH_3}{\diagdown CH}}$—$\overset{\overset{\displaystyle NH_2}{\mid}}{CH}$—COOH	isoleucine	Ile
CH_3—S—CH_2—CH_2—$\overset{\overset{\displaystyle NH_2}{\mid}}{CH}$—COOH	methionine	Met
$COOH$—CH_2—$\overset{\overset{\displaystyle NH_2}{\mid}}{CH}$—COOH	aspartic acid	Asp

The symbol **Gly-Ala** is often used in place of the larger name for convenience. Table 12-1 lists the twenty common amino acids with their abbreviations.

When an amino acid polymer exceeds 100 amino acid units, the name **protein** is more appropriate than polypeptides. Experimental evidence, especially the famous investigations of Linus Pauling, indicate the spiral nature of the long chain amino acid polymers. While peptide bonding is most significant in polymerization, the spiral nature of the protein structure is related to hydrogen bonding between the oxygens of the keto groups and the hydrogens of the amine groups, $=O \ldots HN-$. The process of rupturing the hydrogen bonds by heating (causing increased molecular vibrations) or chemical agents, leads to a breakdown of the spiral structure without the destruction of the peptide bonds. This action is called protein **denaturation** (Fig. 12-19).

TABLE 12-1. *(Continued)*

Formula	Name	Abbreviation
COOH—CH₂—CH₂—CH—COOH (NH₂)	glutamic acid	Glu
NH₂—(CH₂)₃—CH—COOH (NH₂)	lysine	Lys
NHCNH₂—(CH₂)₃—CH—COOH (NH₂)	argenine	Arg
⬡—CH₂—CH—COOH (NH₂)	phenylalanine	Phe
HO—⬡—CH₂—CH—COOH (NH₂)	tyrosine	Tyr
(ring structure)—CH₂—CH—COOH (NH₂)	tryptophan	Trp
—CH₂—CH—COOH (NH₂) (H—N⎯N)	histidine	His
(HO—C ring, N—H, C=O, OH)	hydroxyproline	Hyp
(ring, N—H, C=O, OH)	proline	Pro

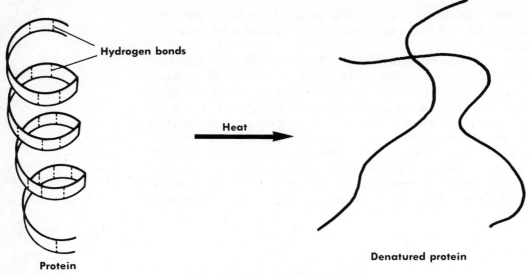

Figure 12-19.   Denaturation of protein.

Types of Proteins

1. *Simple proteins* are those composed of α-amino acids.
2. *Conjugated proteins* are made of α-amino acids in addition to another substance. The nonprotein part of a conjugated protein is called the **prosthetic group**.
   a. *Glycoproteins* have carbohydrates as prosthetic groups
   b. *Lipoproteins* are obviously attached to lipids
   c. *Nucleoproteins* are conjugated with nucleic acids

A tendency of a conjugated protein to divide into two equal parts is called protein **moiety**. A well-known example of a conjugated protein is hemoglobin which has the protein globin and the heme prosthetic group.

Isoelectric Points and Electrophoresis

In Chapter 8 it was pointed out that electrophoresis is a method of separating the components of a complex mixture by the polarization of electric charge. Proteins may be separated by this technique. The amino acids tend to behave as acids or bases depending on the pH of the solution. In basic medium, the amino acid may act as a proton donor. In acid medium, the amino acid is likely to act as a proton acceptor. The negatively charged amino acid (in high pH) will migrate toward the positive electrode while the positively charged amino acid (in low pH) will migrate toward the negative electrode. The pH producing no migration because of a net charge of zero, is called the **isoelectric point**. The amino acid in this condition is referred to as a **zwitterion**. Each amino acid has its own particular isoelectric point. Electrophoresis takes advantage of the variety of migration rates and the directions of migration in order to effect the separation of the amino acids of a protein. This is an excellent analytical device.

Brief mention ought to be made of a few significant processes in protein biochemistry before moving on. When an amine group is transferred from an amino acid to a keto acid by the action of an enzyme called transaminase, the change is known as **transamination**. Transamination may be either oxidative deamination or reductive amination. If a reaction produces a loss of the carboxyl group it is called **decarboxylation**.

NUCLEIC ACIDS

Nucleic acids are few in number but they are of great importance. The molecules are enormous. The molecular weight may go to several million—a structural arrangement which allows an infinite variety of forms. Nucleic acids are fundamental to cellular nuclei, from which the genes, as determiners of hereditary characteristics, are formed.

Nucleic acids are composed of phosphoric acid (H_3PO_4), a simple sugar, either β-D-ribose or β-2-deoxy-D-ribose, and an organic base. The single unit of the nucleic acid polymer is called a **nucleotide**, while the base and sugar structures are called **nucleosides**. A nucleotide of critical importance in supplying energy to living cells is adenosine triphosphate (ATP). Its structure is illustrated in Figure 12-20.

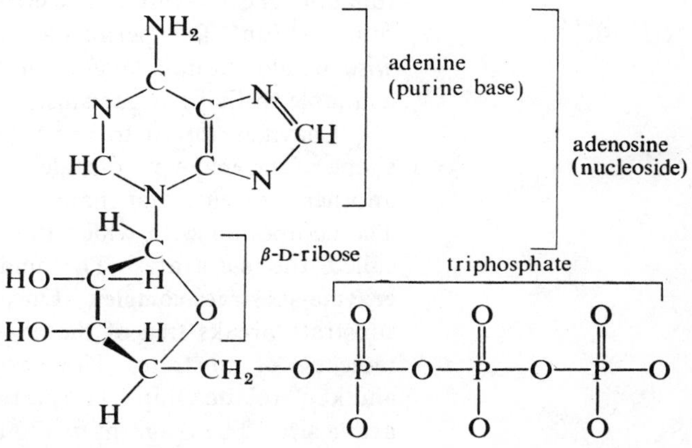

Figure 12-20. ATP.

The two sugars consist of the structures designated in Figure 12-21.

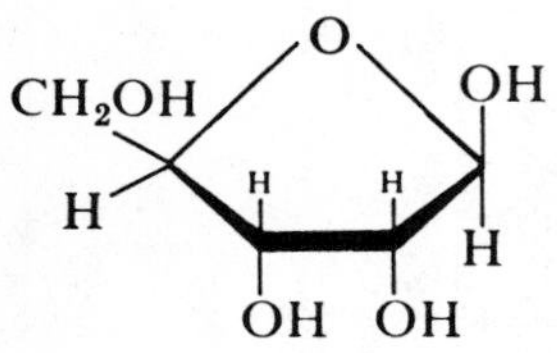

β-D-ribose 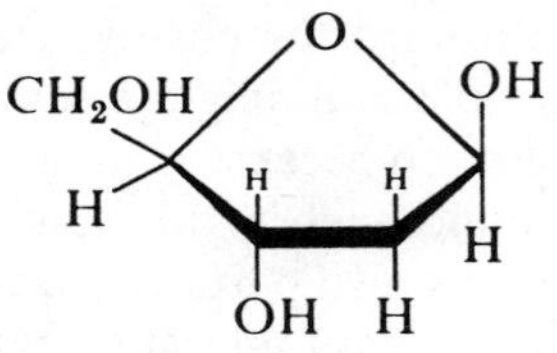   β-2-deoxy-D-ribose

Figure 12-21.

The organic bases are named according to their relationship to **purine** and **pyrimidine** whose structures are shown in Figure 12-22.

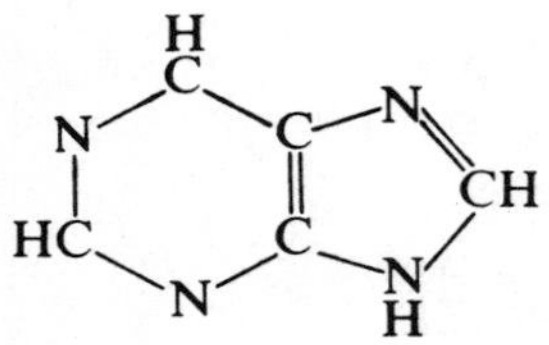

purine 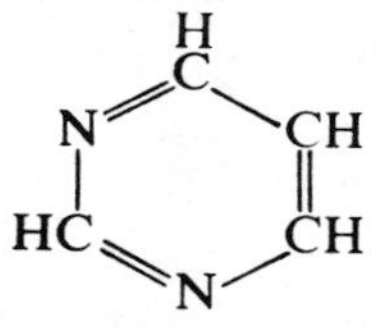   pyrimidine

Figure 12-22.

One of the most famous biochemical achievements in recent years was the determination of the structure of DNA (deoxyribonucleic acid) by Watson and Crick. Their work, and the contributions of others, has led to a greater understanding of the mechanism of hereditary trait transmission by the genes and a better chance for man to develop the knowledge to find preventative measures and cures for viral infections and cancer.

ENZYMES

Catalysts of biochemical activity are known as enzymes. Enzymes are protein or conjugated protein structures. The prosthetic group may be called a **cofactor** or **coenzyme**, and the protein section is known as the **apoenzyme**. Many vitamins necessary to the well-being of living organisms function as coenzymes. The enzyme works by lowering the activation energy of a reaction. This permits a great many vital reactions to occur that otherwise would require lengths of time and amounts of energy than are not available to living organisms.

Enzymes appear to be very specific in their chemical combination. The shape of the enzyme molecule is such that it forms a complex compound with another molecule that matches its physical contour and net electric charge. The compound with which the enzyme reacts in the process of changing is called the **substrate**. The temporary product formed is known as an **enzyme-substrate complex**. Once the chemical change is effected, the altered substrate breaks free of the enzyme, thus allowing it to react with another molecule of substrate. This enzyme specificity is often described as a "lock and key" relationship. The place where the "key" fits the "lock" is called the **active site**. The diagram in Figure 12-23 illustrates enzyme action.

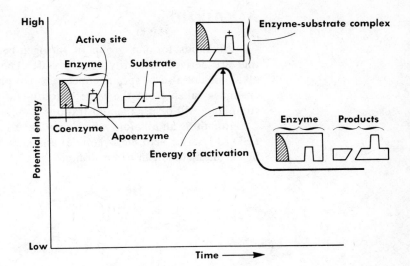

Figure 12-23. Diagrammatic illustration of enzyme action.

Types of Enzymes

Enzymes are characterized by the suffix **-ase.** The prefix is taken from the name of the specific substrate.

1. **Hydrolases** are enzymes that cause the splitting of water molecules. Example: The enzyme sucrase (catalyzes sucrose change) permits the splitting of a water molecule for the purpose of addition in the formation of two glucose molecules from a sucrose molecule.
2. **Oxidoreductases** are enzymes that catalyze oxidation-reduction reactions.
3. **Transferases** catalyze the transfer of groups between compounds.
4. **Lyases** catalyze the loss or gain of parts of molecules.
5. **Isomerases** are involved in changing isomeric forms.
6. **Ligases** catalyze reactions in which bonds are formed between atoms or groups of atoms or both.

When a substance prevents the formation of an enzyme by altering the normal structure of the cofactor, the enzyme is said to be poisoned. This process of poisoning is called **enzyme inhibition.** Cyanide ions prevent the binding of the ion center of cytochrome oxidase with oxygen, and arsenate ions so closely resemble phosphate ions that replacement inhibits the coenzyme function.

VITAMINS

Vitamins are naturally occurring compounds that are essential to normal functioning of living organisms. They work as vital parts of enzyme systems by giving rise to essential cofactors. Vitamins are obviously important to the pharmaceutical industry which is extensively involved in the extraction of vitamins from food substances and the synthesis of new vitamins. Vitamins may be classified as derivatives of carbohydrates, lipids, or proteins, but a more useful classification is based on their solubility characteristics.

ALKALOIDS

Alkaloids form a group of nitrogen-bearing heterocyclic compounds that are derived primarily from plants. The various drugs obtained from alkaloids are also of special concern to the pharmaceutical industry. A few examples taken from the large spectrum of antibiotics, tranquilizers, and stimulants will be used as illustrations.

Serotonin is an alkaloid which may have some relation to brain function. **LSD** (d-lysergic acid diethylamide) and similarly functioning hallucinogens, block the metabolism of serotonin.

serotonin

Related alkaloids are the powerful poison **strychnine** and the tranquilizer **reserpine**.

A famous alkaloid in the treatment of malaria is **quinine**, which is extracted from cinchona bark. The common poppy is a source of the alkaloids **morphine** and **codeine**, which are known for their pain-relieving properties.

The alkaloid **cocaine** is obtained from the cocoa plant, but a synthetic analog, **Novocain**, is preferable as a local anesthetic.

Penicillin is probably the best known of the antibiotics. A variety of analogs of the original mold-extracted penicillin have been synthesized, and some of the synthetic forms are superior in their antibacterial action.

Many pharmaceutical companies are constantly searching the face of the earth for natural products used as remedies for human ills by even the most primitive people. If a herb, root, berry, or leaf seems to hold some promise for an advance against human ills, it is investigated. A search is first made for the **active ingredient**, which is that particular compound among many extracted from the plant which seems to do the job. Then the task is to determine its molecular structure so that analogs may be synthesized. The hope is that they may develop superior related compounds that work more effectively and present a minimal degree of unwanted side effects.

PROBLEMS

Set A

1. Define or explain the following terms:

a. carbohydrate
b. glycogen
c. cellulose
d. amino acid
e. polypeptide

f. denaturation
g. lipid
h. triglyceride
i. terpene

2. What is the difference between an aldose and a ketose?

3. Illustrate the formation of a disaccharide from molecules of β-galactose and D-fructose.

4. What is the essential difference between amylose and amylopectin?

5. Define or explain the following terms:

a. nucleic acid
b. nucleoside
c. purine base
d. coenzyme
e. enzyme

f. inhibitor
g. substrate
h. vitamin
i. moiety

6. What special structural characteristic gives cellulose its strength and rigidity?

7. What types of solvents are most suitable for lipid extractions?

8. What is the difference between a triglyceride and a wax?

9. What effect is observed during the hydrogenation of a polyunsaturated oil?

10. Use a reference text to determine the particular vitamin deficiency that may lead to the following diseases: scurvy, pellagra, and rickets.

Set B

11. Define or explain the following terms:

a. syndet
b. emulsification
c. biodegradable
d. asymmetrical carbon atom
e. ketohexose

f. oligosaccharide
g. polycyclic
h. peptide linkage
i. prosthetic group

12. Use structural formulas to write an equation showing the hydrolysis of sucrose.

13. What is the relationship between triglycerides in perspiration and unpleasant body odor?

14. How would you explain the following cryptic chain of letters to a nonscientist: Gly-Arg-Pro-Cys-Tyr?

15. Define or explain the following terms:

a. isoelectric point
b. conjugated protein
c. zwitterion
d. decarboxylation
e. nucleotide

f. DNA
g. ATP
h. apoenzyme
i. transferase

16. What structural characteristics prevent a particular enzyme from acting as a general biochemical catalyst?

17. What biochemical change occurs to cause potassium cyanide to be a deadly poison?

18. What direction would an amino acid take in an electrophoresis apparatus at a pH of 4.0 when its isoelectric point is at pH 6.8?

19. What is the importance of hydrogen bonding in proteins?

20. Use structural formulas to write an equation for the hydrolysis of the dipeptide, glycylserine.

SUPPLEMENTARY READING

Mazur, A., and Harrow, B.: *Biochemistry: A Brief Course.* W. B. Saunders Company, Philadelphia, 1968.

An excellent introductory text for serious inquiry into biochemistry.

Watson, J. D.: *The Double Helix.* Atheneum, New York, 1968.

A fascinating story of the discovery of the structure of DNA.

Holum, J. R.: *Principles of Physical, Organic, and Biological Chemistry.* J. Wiley & Sons, Inc., New York, 1969.

A very well written reference text, including excellent illustrations.

Baldwin, E.: *Dynamic Aspects of Biochemistry*, 5th ed. Cambridge University Press, 1967.

A classic text in biochemistry. Recommended to the serious student.

Dickerson, R. E., and Geis, I.: *The Structure and Action of Proteins.* (paperback) Harper and Row, New York, 1969.

A clearly written text containing beautiful illustrations.

Radioactivity 13

Radioactivity is a process in which energy is released by an atomic nucleus as it achieves greater stability. Nuclear radiation is most often accomplished by a loss of some nuclear material. This phenomenon called radioactivity, first observed by Henri Becquerel in 1896, raises some fundamental questions. Why do some atoms lose nuclear material as radiation products while others do not? What is the nature of these particles of nuclear decay? What kinds of nuclear reactions are there? And, finally, what use can be made of radioactivity?

Since Becquerel discovered the mysterious exposure effect that a uranium salt had on a covered photographic plate and communicated his observations to the Curies, the topic of radioactivity has fascinated mankind. Man's use and abuse of nuclear energy has generated the fame and infamy of a number of atomic species. On the one hand, uranium 235 conjures up visions of Hiroshima and Nagasaki, and strontium 90 in "fall-out" debris has led to an international test ban treaty because of its threat to life now and life to come. By contrast, cobalt 60 has been effective in the treatment of some types of cancer, iodine 121 attacks diseased thyroid glands, and small samples of other radioactive isotopes have been used as biological tracers as they permit investigators to follow their trails by detectable radiation.

In this chapter the characteristics of radioactive nuclei will be examined. The nature and rates of nuclear decay, nuclear reactions, methods of detection, and practical applications are other topics to be considered.

RADIOACTIVE ISOTOPES

Isotopes, as they have been described previously, are species of a single element that differ in the number of neutrons in the nucleus. The physical and chemical properties are usually quite similar, but the behavior of their nuclei often differ dramatically.

The nature of the "cement" that binds the primary nuclear particles (i.e., protons and neutrons) together is not fully understood. Nevertheless, the mysterious force which binds the nuclear particles, called **nucleons**, is observed to be remarkable. Nuclear forces operate over short distances on the order of a ten thousandth of an ångstrom, but they are the most powerful force known. The nuclear binding force and the nuclear stability dependent on this binding force, seem to be related to the **neutron-to-proton ratios**. It is at this point that isotopes become the focal point. When a particular isotope has a neutron-to-proton ratio that is unstable, the nucleus proceeds to decay and the phenomenon of radioactivity is observed. If the nuclear decay leads to a change in the atomic number, the original element becomes transformed to a new element.

Investigations of large numbers of isotopes have provided evidence in support of the following conclusions:

1. The stability of isotopes is related to its odd or even numbers of protons. If the numbers of protons are even, the isotopes tend to be stable. If the total number of protons is odd, more isotopes tend to be unstable:

    a. Phosphorus 32 $^{32}_{15}P$ 15 protons—unstable
    b. Cobalt 60 $^{60}_{27}Co$ 27 protons—unstable
    c. Iodine 121 $^{121}_{53}I$ 53 protons—unstable

2. Isotopes with an odd number of neutrons are more likely to be unstable.

3. The nuclei of atoms having atomic numbers greater than 83 are generally unstable.

4. When the number of neutrons is approximately equal to the number of protons, the isotopes tend to have stable nuclei.

5. A high neutron to proton ratio ($n°/p^+$) of an isotope, compared to the ratio of a stable arrangement, leads to decay.

    a. $^{14}_6C$ has an $8n°/6p^+$ ratio. This is high compared to $^{12}_6C$ where the ratio is $6n°/6p^+$. Carbon 14 is unstable.
    b. $^{90}_{38}Sr$ has a $52n°/38p^+$ ratio compared to the stable $^{88}_{38}Sr$ with a ratio of $50n°/38p^+$. Strontium 90 is unstable.

6. A comparatively low $n°/p^+$ ratio also tends toward instability in isotopes.

    a. $^{10}_6C$ has a $4n°/6p^+$ ratio compared to the stable $^{12}_6C$ with a $6n°/6p^+$ ratio.
    b. $^{121}_{53}I$ has a $68n°/53p^+$ ratio compared to the stable $^{127}_{53}I$ with a $64n°/53p^+$ ratio.

The next question to be considered is the nature of the radiation that nuclear decay causes.

TYPES OF RADIATION

Alpha radiation is composed of a stream of alpha particles from the nuclei. Alpha particle emission is characteristic of large nuclei. The alpha particle is identical to a helium nucleus, being composed of two protons and two neutrons. A loss of an alpha particle causes a transformation of the elemental atom as well as a loss in mass. The alpha particle is symbolized as α or ^4_2He. Examples of α particle emission are:

$$^{234}_{90}\text{Th} \rightarrow {}^{230}_{88}\text{Ra} + {}^4_2\text{He}$$

$$^{226}_{88}\text{Ra} \rightarrow {}^{222}_{86}\text{Rn} + {}^4_2\text{He}$$

Notice that the sum of the charges and the sum of the masses on both sides of the equation are equal.

Alpha radiation is not as hazardous as the others to be discussed. The massive alpha particle is relatively slow moving and its range is only a few centimeters. However, an alpha-emitting substance that is breathed in or ingested is damaging because it destroys living cells by ionizing atoms of protein molecules which compose the cell. The ionization is produced when the alpha particle knocks an electron out of an atom in its path. This phenomenon is called **ion pair production**. The freed electron is one part of the pair and the remaining charged atom (ion) is the other part (Fig. 13-1).

Beta radiation consists of a stream of electrons (if negative) or a stream of **positrons** (if positive). Beta emission usually occurs when the nucleus of the isotope has a high n°/p^+ ratio.

The beta particle (β) is produced when a neutron is transformed into a proton and an electron. If two beta particles are produced in the process of decay, one particle transforms a proton into a neutron while the other is emitted as a positron.

The symbols for positive and negative beta particles are $_{-1}\beta^\circ$ or $_{-1}^\circ\text{e}$ for the electron and $_{+1}\beta^\circ$ or $_{+1}^\circ\text{e}$ for the positron. The superscript of zero indicates the mass of a beta particle. The electron produced by nuclear decay is identical to the normal electrons outside the nucleus. Some examples of beta emission are:

a. $$^{234}_{90}\text{Th} \rightarrow {}^{234}_{91}\text{Pa} + {}^{\ \circ}_{-1}\text{e}$$

Note once again that the sums of the masses and charges are equal.

b. $$^{14}_{6}\text{C} \rightarrow {}^{14}_{7}\text{N} + {}^{\ \circ}_{-1}\text{e}$$

c. $$^{133}_{53}\text{I} \rightarrow {}^{133}_{54}\text{Xe} + {}^{\ \circ}_{-1}\text{e}$$

Decay is also possible by electron "capture" among some heavy metal isotopes.

$$^{194}_{79}\text{Au} + {}^{\ \circ}_{-1}\text{e} \rightarrow {}^{194}_{78}\text{Pt}$$

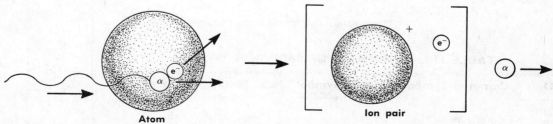

Atom   **Ion pair**

Figure 13-1. Ion pair production.

Examples of positron emission are:

a. $$^{10}_{6}C \rightarrow {}^{10}_{5}B + {}^{0}_{+1}e$$

b. $$^{121}_{53}I \rightarrow {}^{121}_{52}Te + {}^{0}_{+1}e$$

Because of the negligible mass of the β particle, it has only the slightest ionizing power. The range of beta particles depend on their energy, but the average range is limited to a few meters. Beta radiation as a hazard is not great except for the possible exposure to a direct beam, in which case the ionization of protein atoms could be serious.

Gamma radiation is produced by nuclei in a highly excited state. Gamma radiation is electromagnetic. The wavelength is extremely short (a small fraction of an ångstrom) and the photon of this radiation has a very high energy. Its range may be measured in light years as it streaks through the void of the universe from distant stars. The gamma radiation in this case is commonly called **cosmic** radiation. The penetrating power of these cosmic rays, without mass or change, is remarkable. Neither lead walls nor deep caves below the earth can fully block the effect of cosmic radiation.

In many cases, when a nucleus is left in an excited state as a result of change, gamma rays are emitted as the nucleus proceeds toward the ground state. The usual gamma rays of the nuclear decay of familiar isotopes are not as penetrating as cosmic radiation.

Gamma radiation is very hazardous. The photons have sufficient energy to destroy the molecules of living protoplasm by gross alteration or "burning."

The effects of high energy radiation may be measured in a variety of ways, using several kinds of units of radiation such as roentgens (R), REM (roentgen equivalent in man), RBE (relative biological effectiveness), and rads. A rad is equivalent to one erg of energy absorbed per gram of tissue.

Rads are usually chosen as a measure of radioactive dosage; some effects are described in the following chart.

Dose in Rads	Probable Effect on Man
0–50	Possible genetic mutation
50–100	Headache, dizziness, listlessness
100–200	Radiation sickness and hair loss
200–500	Severe bleeding and tissue destruction
Over 500*	Death

When humans are exposed to radiation doses in excess of 50 rads, they may expect to suffer from "radiation sickness." The intensity, duration, and

---

* This is known as a lethal dose (LD).

TABLE 13-1. A SUMMARY OF RADIATION TYPES

Unit	Common Symbol	Modern Symbol	Mass (amu)	Charge
alpha	α	$^{4}_{2}He$	4	+2
beta	β^-	$^{0}_{-1}e$	0	−1
positron	β^+	$^{0}_{+1}e$	0	+1
gamma	γ	$h\nu$	0	0

ultimate effect of radiation sickness depends on the size of the dose and the length of time of exposure. In addition to some symptoms already included in the chart, there may be nausea, vomiting, insomnia, and shock. Intestinal ulceration, bacterial infection, and circulatory failure often lead to death.

The other unit of ionizing radiation that should be described is the **roentgen** (r). This is a quantity of *exposure* resulting from x-ray or gamma radiation. Exposure to radiation may be described as r per hour. Numerically, r units are practically equal to rads. However, rads measure radiation absorbed while roentgens measure exposure. The roentgen is defined as 2.58×10^{-4} coulombs of charge per kilogram of air. This is the amount of radiation able to produce two billion ion pairs per cubic centimeter of air.

NUCLEAR EQUATIONS

The method for completing and balancing nuclear equations requires a periodic table and a familiarity with the charges and masses of nuclear particles. With these data in hand, the equations can be balanced in a manner consistent with the conservation laws.

Example I

Write the equation for $^{238}_{92}U$ emitting an alpha particle.
1. Write the equation, leaving a space for the product:

$$^{238}_{92}U \rightarrow\ ? + {}^{4}_{2}He$$

2. The product must have a mass which, added to the mass of the alpha particle (4), will equal 238. The charge on the product must be such that its addition to the charge on the alpha particle (+2) will equal 92.

$$^{238}_{92}U \rightarrow {}^{234}_{90}? + {}^{4}_{2}He \qquad 238 = 234 + 4$$
$$92 = 90 + 2$$

3. The element can be identified by its atomic number (90):

$$^{238}_{92}U \rightarrow {}^{234}_{90}Th + {}^{4}_{2}He$$

Example 2

What product is formed as a result of positron emission from iodine 121 decay?
1. Write the equation:

$$^{121}_{53}I \rightarrow\ ? + {}^{0}_{+1}e$$

2. Balance the charges and masses:

$$^{121}_{53}I \rightarrow {}^{121}_{52}? + {}^{0}_{+1}e \qquad 121 = 121 + 0$$
$$53 = 52 + 1$$

3. Element number 52 is tellurium:

$$^{121}_{53}I \rightarrow {}^{121}_{52}Te + {}^{0}_{+1}e$$

Example 3

What happens when nitrogen 15 gains a proton in the formation of carbon 12?

1. Write the equation:

$$^{15}_{7}N + {}^{1}_{1}H \rightarrow {}^{12}_{6}C + ?$$

2. Balancing the masses requires an addition of 4, and the charge must be a total of $+2$ for balancing. A single particle fitting that description is an alpha.

$$^{15}_{7}N + {}^{1}_{1}H \rightarrow {}^{12}_{6}C + {}^{4}_{2}He \qquad 15 + 1 = 12 + 4$$
$$7 + 1 = 6 + 2$$

Example 4

When a uranium 235 nucleus is split by a high energy neutron, barium 143 and krypton 90 are formed. What else must be involved in the nuclear radiation?

1. Write the equation:

$$^{235}_{92}U + {}^{1}_{0}n \rightarrow {}^{143}_{56}Ba + {}^{90}_{36}Kr + ?$$

2. Balancing the equation requires an addition of three mass units and no change in the charge. No single particle fits that description. The only nuclear particle having a mass of one and no charge is the neutron. Therefore, three neutrons must be emitted:

$$^{235}_{92}U + {}^{1}_{0}n \rightarrow {}^{143}_{56}Ba + {}^{90}_{36}Kr + 3{}^{1}_{0}n \qquad 235 + 1 = 143 + 90 + 3$$
$$92 + 0 = 56 + 36 + 0$$

RATE OF NUCLEAR DECAY

The rate at which a nucleus decays is a measure of its stability. There are instruments designed to measure such rates, and they will be discussed later in this chapter. The usual method of describing a decay rate is to correlate a radioactive substance with its **half-life**. The half-life of an element is the time it takes for one half of the observed mass to decay by some form of emission—α, β, or γ. In Chapter 2 the derivation of a constant was used as an example of finding the slope of a line. Related empirical data can be found that yield the following relationship.

$$\underset{\text{half-life}}{t_{1/2}} = \frac{0.693 \overset{\text{2.3} \times \log 2}{}}{\underset{\text{decay constant}}{k}}$$

Rearrange the equation:

$$k = \frac{0.693}{t_{1/2}}$$

Enlarging the equation in order to solve problems involving radioactive decay, the logarithm of the ratio of an original amount of decaying substance to a final amount after a passage of time is related in the following way:

$$\log \frac{N_0}{N} = \frac{kt}{2.3}$$

N_0 = Original amount of material or the original rate of nuclear particles being emitted.

N = Final amount of material or rate of particle emission after a period of time.

t = Time of counting particles.

k = Decay constant $\dfrac{0.693}{t_{1/2}}$

2.3 = A conversion factor for changing logarithms from the base e to the base 10.

Example 5

If the ratio of carbon 14 in a freshly cut piece of wood to the amount of carbon 14 in some ashes found in a cave is 1.00:0.42, calculate the age of the cave dwellers.

1. Write the nuclear equation:

$$^{14}_{6}C \rightarrow {}^{14}_{7}N + {}^{0}_{-1}e$$

2. Obtain the half-life from the table:

$$t_{1/2} = 5{,}760 \text{ years}$$

3. Write the equation:

$$\log \frac{N_0}{N} = \frac{kt}{2.3} = \frac{0.693t}{t_{1/2}(2.3)}, \quad \text{Since} \quad k = \frac{0.693}{t_{1/2}}$$

4. Organize the data:

$$\log \frac{N_0}{N} \left| \frac{1.00}{0.42} \right. = 2.38, \log 2.38 = 0.377$$

t	?
$t_{1/2}$	5,760 years

5. Develop the equation:

$$0.377 = \frac{0.693(t)}{5{,}760(2.3)}$$

$$t = \frac{(3.77 \times 10^{-1})(5.76 \times 10^{3})(2.3)}{6.93 \times 10^{-1}}$$

$$t = 7{,}207 \text{ year old cave dwellers}$$

TABLE 13-2. HALF-LIVES AND EMISSION TYPES OF SOME RADIOACTIVE ISOTOPES

Isotope	Half-life	Type of Emission
H 3 (tritium)	12.3 years	beta
C 14	5,760 years	beta
P 32	14.3 days	beta
Co 60	5.27 years	beta and gamma
Sr 90	28 years	beta
I 131	8.05 days	beta and gamma
Po 218	3.05 minutes	alpha

Example 6

A radioactive sample emitted particles at a rate of 1420 counts per minute at 1 p.m. If the rate was 1060 at 4 P.M. what is the half-life of the isotope?

1. Write the equations:

$$\log \frac{N_0}{N} = \frac{kt}{2.3} \quad \text{and} \quad t_{1/2} = \frac{0.693}{k}$$

2. Organize the data:

N_0	1420 counts
N	1060 counts
k	?
t	1 P.M. → 4 P.M. = 3 hours

$$\frac{N_0}{N} = \frac{1.42 \times 10^3}{1.06 \times 10^3} = 1.34$$

$$\log 1.34 = 0.126$$

3. Develop the equation:

$$k = \frac{(0.126)(2.3)}{3 \text{ hours}} = 0.0965$$

4. Final calculation

$$t_{1/2} = \frac{0.693}{k} = \frac{0.693}{0.0965} = 7.18 \text{ hours}$$

Example 7

If a 3.0 mg sample of iodine 131 is used as a beta emitter for 3 weeks, how much will remain?

1. Write the nuclear equation:

$$^{131}_{53}\text{I} \rightarrow ^{131}_{54}\text{Xe} + _{-1}^{0}e$$

2. Write the equations:

$$\log \frac{N_0}{N} = \frac{kt}{2.3} \quad k = \frac{0.693}{t_{1/2}}$$

3. Look up the half-life of iodine 131 and organize the data:

N_0	3.0 mg
N	?
$t_{1/2}$	8.05 days
t	3 weeks = 21 days
k	$\dfrac{0.693}{8.05} = 0.0861 = 8.61 \times 10^{-2}$

4. Develop the equation:

$$\log \frac{3.0}{N} = \frac{(8.61 \times 10^{-2})(21)}{2.3}$$

$$\log \frac{3.0}{N} = 0.786$$

5. Take the **antilog** of both sides of the equation:

$$\frac{3.0}{N} = 6.11$$

$$N = \frac{3.0}{6.1} = 0.49 \text{ mg remaining}$$

INSTRUMENTS FOR RADIOACTIVITY DETECTION

It was pointed out earlier that alpha and beta particles have the ability to cause ionization when they collide with electrons of atoms. The transfer of kinetic energy to the atom's electron enables the electron to move free of the atom in the process of ionization. Instruments designed to measure α and β decay rates are special electroscopes called ionization chambers, proportional counters, solid-state detectors, and Geiger-Müller counters.

The most common device, the **Geiger-Müller** counter (Fig. 13-2), is composed of a detecting tube and an amplifier for the production of audible "clicks," or recording on an electronic counter, called a "scaler." The Geiger-Müller counter is usually a metal tube containing a mica window. Since beta particles, and especially alpha particles, have very limited penetrating power, the window is necessary. The tube contains argon gas. The cylinder wall serves as the cathode in which a centrally located wire is the anode. A high voltage is established between the central wire and the cylinder wall so that any ions produced will be strongly and quickly attracted to the electrodes. The electron flow from the anode wire is amplified for counting. The amount of radioactive material in a sample is proportional to the current produced by the ionization of the gas.

Another type of detector suitable for nonionizing gamma rays and weak beta particles is the **scintillation counter**. There are two varieties of the scintillation counter—liquid and solid. Scintillation means the production of a flash of light. There are some substances, called **phosphors**, that produce this spark of light when a nuclear particle or a photon of gamma radiation transfer their energy to a phosphor molecule. As the excited phosphor

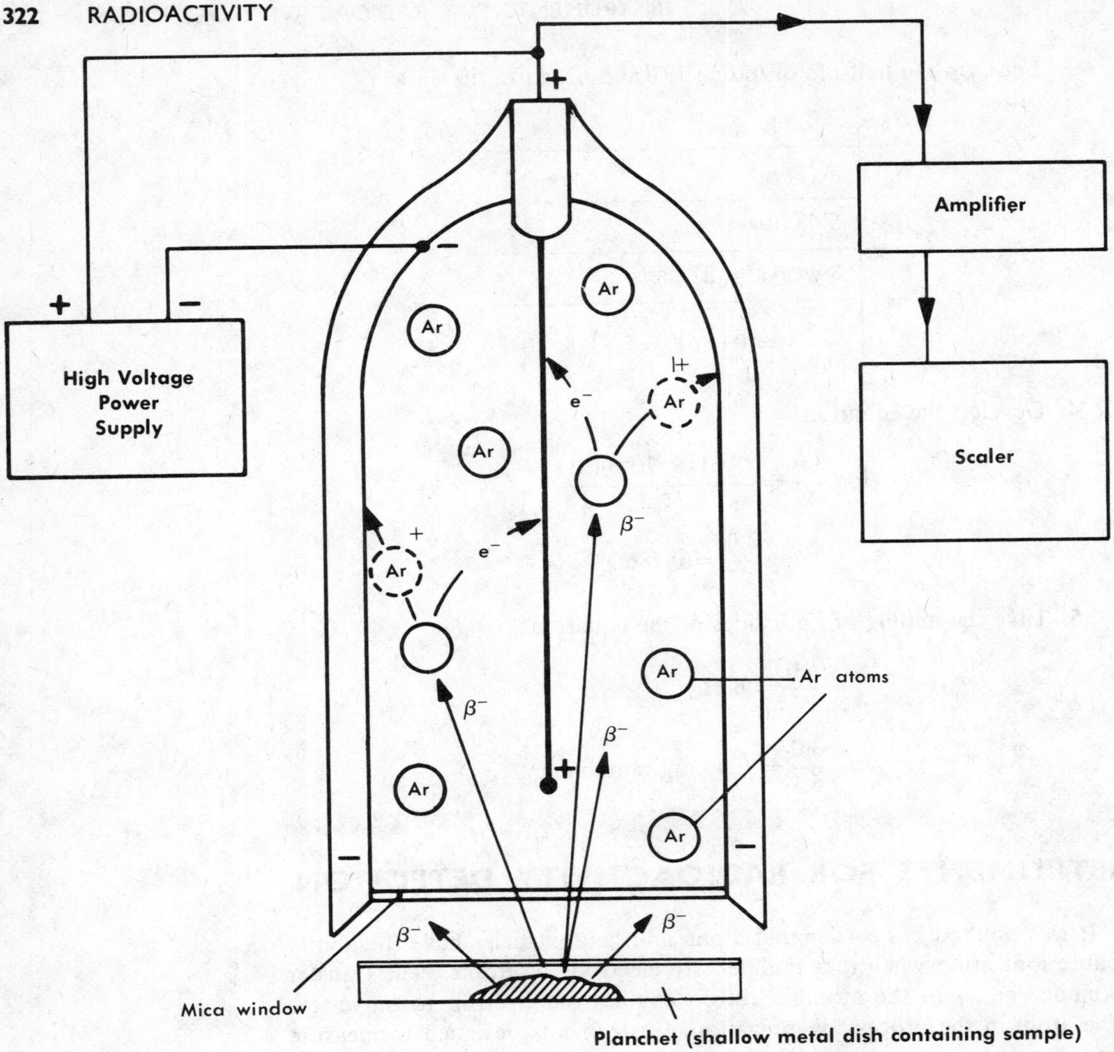

Geiger-Müller tube
Figure 13-2.

returns to its ground state, the flash of light is emitted. The intensity of the light flash is proportional to the energy of the radiation. The light flash is detected by a photomultiplier tube (see Chapter 2) and amplified for registration on the scaler.

Some phosphors are NaI crystals (solid) for gamma ray detection, anthracene crystals (solid) for beta counting, and a variety of aromatic compounds dissolved in organic solvents for liquid scintillation. Dissolving some sample in the scintillator solution is best for low energy beta particles. A diagrammatic scintillation counter is shown in Figure 13-3. In this diagram, the gamma radiation causes light emission from the phosphor. The light photon causes an electron to be ejected from the first **dynode**. This **photoelectric effect** is multiplied by the dynodes to a growing cascade of electrons that are finally collected by the final dynode where the current has been effectively multiplied thousands of times. There may be as many as fifteen dynodes in a photomultiplier. The final current is amplified for the

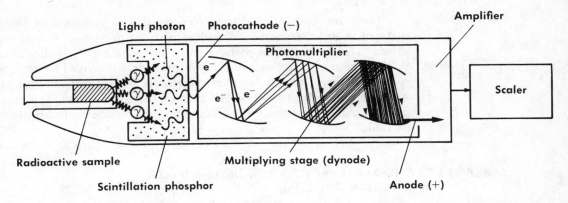

Scintillation counter
Figure 13-3.

scaler. The scaler readings must be adjusted so that counts due to the high voltage applied to the photomultiplier tube, called **dark current**, will not introduce a measurement error.

Other possible errors may be due to cosmic radiation, which produces a **background count**, and inefficient operation of the detection tubes. Background counts can be observed in the absence of samples, and efficiency can be estimated by using standardized samples. The effect of stray currents is always a concern when using electronic measuring instruments. The amplified stray current is detected as static, or **electronic noise**, and the amount of this noise is described as a **signal-to-noise ratio**. When the desired current, i.e., the current produced by the object being investigated is low, the signal-to-noise ratio approaches unity and has no value as a result.

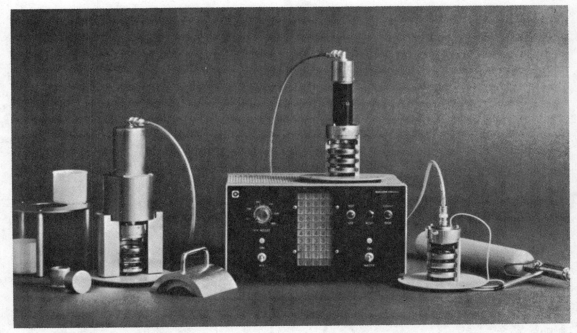

Figure 13-4. Alpha, beta, gamma laboratory system. It consists of Geiger, gasflow, and scintillation. (Courtesy of Nuclear-Chicago Corporation.)

There are scintillation detectors equipped with **multichannel analyzers** that can record the number of nuclear particles in each of several energy values. The energies of the particles are described in MeV's (million electron volts). The rate of radioactive decay is commonly described in terms of its **specific activity**. This is the number of **curies per gram** of radiactive substance. A curie (Ci) is defined as 3.7×10^{10} disintegrations per second. It is approximately equal to the activity of a gram of radium. Smaller units are described in Table 13-3.

TABLE 13-3. CONVERSION FACTORS FOR UNITS SMALLER THAN THE CURIE

Unit	Conversion Factor	dps (disintegrations per second)
Ci curie	basic unit	3.7×10^{10} dps
mCi millicurie	10^3 mCi Ci^{-1}	3.7×10^7 dps
μCi microcurie	10^6 μCi Ci^{-1}	3.7×10^4 dps
nCi nanocurie or mμCi millimicrocurie	10^9 nCi Ci^{-1}	37 dps
pCi picocurie or $\mu\mu$Ci	10^{12} pCi Ci^{-1}	3.7×10^{-2} dps

Radiation Safety

Laboratory personnel who work regularly with radiation or radioactive isotopes need to observe certain safety precautions in order to avoid an accumulating dosage. Laboratories and personnel can be monitored to assure safe levels of radioactivity. Photographic film badges, Geiger-Müller counters, and the use of standard labels can serve this purpose.

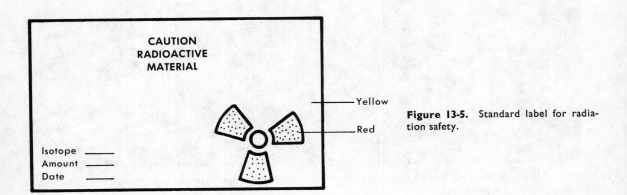

Figure 13-5. Standard label for radiation safety.

Specific procedures to follow when using radioactive materials are listed.
1. Wear protective coat and gloves.
2. Cover work area with disposable material.
3. Label all radioactive material.
4. Containers—rinse glassware with solutions similar to radioactive sample before using. This lowers specific activity.
5. Label all contaminated materials for discard into special receptacles.
6. Monitor hands, clothes, and work area before leaving.

PRACTICAL APPLICATIONS OF RADIOACTIVITY

Carbon 14 Dating

Example 5 is an illustration of the use of carbon 14 in the dating of historical objects and their related cultures. High energy cosmic radiation (gamma rays) is constantly bombarding our atmosphere. The molecules of nitrogen have neutrons ejected from their atomic nuclei as a result of the colliding radiation, which forms carbon 14.

$$^{14}_{7}N + ^{1}_{0}n \rightarrow ^{14}_{6}C + ^{1}_{1}H$$

The carbon 14 reacts with oxygen to form the carbon dioxide used by plants in the photosynthetic process. The plants and the animals that eat the plants metabolize the radioactive carbon into living tissue. When the plants and animals die, the amounts of carbon 14 taken in during their lives undergoes slow decay. The half-life of carbon 14 is about 5,760 years. It would take about seventy millennia for the radioactive carbon to practically disappear.

$$^{14}_{6}C \rightarrow ^{14}_{7}N + ^{0}_{-1}e$$

If the ratio of carbon 14 to carbon 12 in ancient plant and animal remains is compared to living tissue, the elapsed time—in years—can be calculated. An example of the usefulness of carbon 14 dating is the support given to the authenticity of the Dead Sea Scrolls (a wood product) which proved to be nearly 2000 years old.

Analytical Chemistry with Radioactivity

Radioactive isotopes can be used to quantitatively analyze molar fractions of ions in solutions that are too small to be measured by other methods. Very small molar fractions in the microgram range are described as **trace amounts**. For example, the concentration of a trace amount of chloride ion can be determined by adding silver nitrate solution in which the silver ion has been "tagged" or "labeled" by using silver 110, which is radioactive. The silver ion causes the precipitation of the chloride ion.

$$^{110}Ag^+ + Cl^- \rightarrow ^{110}AgCl(s)$$

(radioactive precipitate)

By measuring the radioactivity of the precipitate formed, it is possible to calculate the chloride ion concentration despite the possibility that it may be in the micromolar range.

When a particular substance in a mixture can be separated, but only in small amounts, a technique called **isotope dilution** may be used. For example, if the amount of an amino acid, alanine, were to be determined in a mixture of amino acids, some pure alanine would have to be isolated. A measured amount of radioactive alanine (commercially obtainable) is added to the isolated sample. Radioactive alanine is identical to normal alanine except that some molecules contain the radioisotope carbon 14. By measuring the specific activity (counts per minute per gram) of the radioactive sample, and diluting it with the isolated sample, the specific activity of the diluted compound can be measured.

$$\frac{\text{Specific activity}}{\text{of the diluted sample}} = \frac{\text{Specific activity of the radioactive sample}}{\text{dilution factor}}$$

From the specific activity information, the weight of alanine can be determined.

$$\text{Specific activity} = \text{Counts min}^{-1}\,\text{g}^{-1}$$

Another analytical method involves the exposure of trace amounts of materials to neutron bombardment. This converts elements to radioactive forms. The method is known as **activation analysis**. Microgram amounts of elements can be detected in this way. The value of this type of analysis is illustrated by the possible measurement of trace amounts of poisonous elements in food material. It may also be used in industry to check on contaminants, uniform dimensions of metal sheets, and corrosion.

Radioisotopes are used extensively in the medical and biochemical fields. The use of cobalt 60 in cancer treatment, and iodine 131 in the case of thyroid disease have been mentioned. In both examples, the radiation is localized so that very specific tissues (cancerous or diseased) are destroyed. Small amounts of radioactive material may be used to detect circulatory disorders as the flow of blood is followed by a counting device such as a Geiger-Müller tube.

The important field of environmental control explores the use of radiation and radioactive isotopes. The increased yield of better crops has emerged from radiation experiments performed on ungerminated seeds. Food spoilage has been sharply retarded by irradiation.

Animal gonads are particularly sensitive to all forms of high energy radiation, from x-rays to gamma rays. The effect of this radiation has been to cause genetic mutations by altering the nucleic acid geometry. While the results of some mutations may be hideous or worthless, others lead to the development of superior stock animals. The dangers of DDT for insect pest control are so well established that an alternative must be found. A great possibility is the use of radiation to effectively sterilize the insects or alter the reproductive cycle so that no live offspring result. Mankind may yet be free of disease bearing pests without poisoning himself in the process. The Atomic Energy Commission has made, and is making, significant strides in development of useful applications of radioactivity.

Biochemical research laboratories use radioactive elements as **tracers** in an attempt to understand the complexity of biochemical reactions. The use of labeled carbon dioxide (carbon 14) has increased man's understanding of the photosynthetic process of green plants. The growth of plants, and the formations of nucleic acids and amino acids can be studied with tritium (hydrogen 3) labeling. In this way, the separation of chromosomes in living cell nuclei can be observed, and protein formation can be measured.

Experiments designed to learn about the rates and complexity of chemical reactions occurring during the process of respiration make use of phosphorus 32. Phosphorus is a necessary part of the energy storing capability of living cells. The respiratory process involves a series of oxidation-reduction reactions that convert "energy poor" ADP (adenosine diphosphate) to "energy rich" ATP (adenosine triphosphate). The complex chain of reactions proceeds at varying rates of speed, depending on enzymes, substrates, activation energies, and metabolic pathways.

When ^{32}P is used as a tracer element, because of its incorporation into the ATP molecule, the chemical reactions can be followed. When the biochemical activity is abruptly stopped at controlled time intervals, the amounts of ^{32}P appearing at various points in the chain of reactions can be assayed. This type of investigation, known as **reaction kinetics**, has illuminated our understanding of the production of life energy.

PROBLEMS

Set A

1. Explain or define the following terms:

a. alpha particle
b. positron
c. rad
d. specific activity

e. roentgen (R) unit
f. half-life
g. phosphor
h. electronic noise

2. What is the relationship of neutron-to-proton ratio and isotope stability?

3. Qualitatively compare the energies liberated in chemical reaction to nuclear transformations.

4. Polonium 214 has a half-life of 164 microseconds. How useful is this isotope? Explain.

5. A safety rule dictates that a radioactive substance should be stored for seven half-lives before discarding. How many days should iodine 131 be stored?

6. Use a diagram to explain how a Geiger-Müller counter works.

7. Compare alpha and beta particles in the ability to form ion pairs.

8. Complete and balance the following equations:

a. Beryllium 9 gains an alpha particle in the formation of carbon 12.
b. Carbon 14 is transformed to nitrogen 14.
c. When chlorine 35 gains a proton, it proceeds to emit an alpha particle.

9. Balance the following equations.

a. $_{28}^{60}Ni + _{0}^{1}n \rightarrow _{27}^{60}Co + ?$
b. $_{92}^{235}U + _{0}^{1}n \rightarrow _{42}^{102}Mo + _{50}^{131}Sn + ?$
c. $_{7}^{14}N + ? \rightarrow _{8}^{17}O + _{1}^{1}H$

10. Calculate the age of an ancient pair of fiber sandals if the carbon 14 content of a live plant compared to the sandals is a ratio of 1.14.

11. The decay rate of an isotope was 2440 CPM on July 3rd. If the rate was 2160 CPM on July 7th, what is the half-life?

12. Calculate the decay constant of iodine 131 which has a half-life of 8.05 days.

13. How much of a 20 g sample of actinium 225 will be left after 3 days if the half-life is 10.0 days?

14. Convert 0.0035 mCi to nCi. How many disintegrations per second is this?

15. Describe three practical applications of radioactivity.

Set B

16. Explain or define the following terms:

a. beta (β^-) particle
b. gamma ray
c. curie unit (Ci)
d. lethal dose

e. scintillation
f. dark current
g. background count
h. activation analysis

17. What property would an element be likely to have if its atoms have a high n^0/p^+ ratio?

18. Why would you expect $_{17}^{35}Cl$ and $_{17}^{37}Cl$ to have the same chemical properties?

19. If carbon 14 in living plants has a specific activity of 15 counts $min^{-1} g^{-1}$ is the carbon 14 method reliable for dating objects 23,000 years old? Why?

20. How does a scintillation counter work?

21. What is the special danger encountered when alpha emitters are powders or gases?

22. Complete and balance the following equations:

a. The alpha emission of oxygen 17.
b. Carbon 10 loses a positron.
c. Potassium 40 gains a neutron in the transformation to chlorine 37.

23. Balance the following equations:

a. $^2_1H + {}^3_1H \rightarrow {}^4_2He + ?$
b. $^{35}_{16}S \rightarrow ? + {}^0_{-1}e$
c. $^{63}_{29}Cu + {}^2_1H \rightarrow {}^{64}_{29}Cu + ?$

24. If a radioactive isotope emits 1555 CPM at 4 P.M. and 960 CPM at 4.30 P.M., what is the half-life?

25. Sulfur 35 has a half-life of 87.1 days. How much of a 10.0 mg sample will remain after six months?

26. Calculate the decay constant for iron 59 which has a half-life of 45 days.

27. What is the essential difference between rads and roentgens?

28. The decay constant for platinum 197 is $3.85 \times 10^{-2}\ hr^{-1}$. How much of a 20 mg sample remains after 9 hours?

29. Describe three ways in which radioactivity can be used for environmental control.

SUPPLEMENTARY READING

Chase, G. D., and Rabinowitz, J. L.: *Principles of Radioisotope Methodology.* Burgess Publishing Co., Minneapolis, 1966.

Very useful for the technician.

Harper, D.: *Isotopes in Action.* Pergamon Press, New York, 1963.

Informative discussion of isotope application.

Choppin, G. R.: *Nuclei and Radioactivity.* W. A. Benjamin, Inc., New York, 1964.

A well-illustrated short text.

Casarett, A. P.: *Radiation Biology.* Prentice-Hall Co., Englewood Cliffs, New Jersey, 1968.

A fine text dealing with the human hazards and benefits of radioactivity.

14 Chemical Instrumentation (I)

There are many physical and chemical events in which the participating substances cannot be directly observed because of their small amounts, fast rates of reaction, or hazards in handling. The procedures of analysis, quantitative assay, and physical separation of the species of matter involved in such events must necessarily be done by sophisticated instruments.

Electronic instruments are designed to have the capacity to detect, measure, control, amplify, and record submicroscopic phenomena that exceed the human range of direct observation and manual dexterity. One of the greatest contributions of physics to our understanding and measurement of atomic and molecular activity comes from the fields of **optics**. Optics is concerned with the application of light energy to the behavior of matter. This relationship was emphasized in Chapter 3, which is recommended for review at this time. The instrumental applications of radiant energy control, detection, and measurement to chemistry is described by the general title **photometry**. Photometry (light measurement) is accomplished by **colorimeters**, **spectrophotometers**, and **fluorimeters**, which will be discussed in this chapter. The mention or photograph of any particular instrument is not to be interpreted as an endorsement of the superiority of that instrument. There are simply too many commercially competing manufacturers to permit any attempt toward an "equal time" recitation of their particular virtues. The technician must ultimately adapt himself to the brand found in his laboratory.

The intent of this discussion is to present the principles upon which all the various instruments operate. It is infinitely more profitable for a laboratory investigator to understand what the instrument is supposed to do and how it does it, rather than become preoccupied with a detailed list of operating procedures. Once an appreciation of the theory has been gained, it is relatively easy to learn what switches to throw, knobs to turn, and meters to read.

RADIANT ENERGY AND MATTER

Radiant energy is described in terms of the electromagnetic spectrum. The optical instruments commonly employed in the research laboratory today are restricted to the infrared, visible, and ultraviolet sections of the spectrum.

It should be remembered that radiation, while it is a wave phenomenon, is described as having a dual nature, since it also exhibits a particle behavior. These particles, or packets, of light energy are called **photons**, and their energies are directly related to their frequencies. The photons of radiant energy are produced by vibrating electrical charges which may be inner, middle, or valence electrons in atoms, or the vibrations of molecular atoms (Fig. 14-1).

Since electromagnetic radiation is produced when oscillating electrical charges emit energy, it is reasonable to assume that atoms and molecules absorbing this energy will exhibit a reverse process. However, quantum theory postulates that electrons are restricted to absorbing or emitting very specific amounts of energy, and molecules are limited to very definite vibrational frequencies. This means that it is possible to identify, both qualitatively and quantitatively, the particular atoms or molecules because of the wavelength of radiation they selectively absorb or emit.

In any discussion of spectra measurement, or **spectroscopy**, it becomes necessary to make a distinction between emission and absorption.

An emission spectrum is *produced by the object* under investigation. This emission of radiant energy is caused when the atom's electrons absorb some available form of energy as they move from lower to higher energy levels. The quantity of energy that was absorbed, is almost instantly emitted as radiant energy of a specific wavelength. The wavelength is related to the energy value emitted. Fluorescent lights (an example of a gas discharge tube) produce an emission spectrum. If a sodium salt, for example, is placed over a burner, it will produce a line of yellow light when viewed through a spectroscope. The energy of the emitted yellow light is equal to the energy absorbed. More important, from a spectrophotometric point of view, is the identification of sodium made possible by its characteristic emission of specific lines of color. More than 50 elements and 700 substances can be identified by this process called **flame photometry**.

When some of the energy absorbed by molecules has a higher frequency (shorter wavelength) than the light emitted, it may be assumed that some of the energy was radiated as heat when the molecule's vibrational frequency dropped, while the remaining energy was emitted as light, having a longer wavelength. An example of this **fluorescence** phenomenon is the ability of a variety of compounds to absorb ultraviolet light and emit light in the visible spectrum. The zinc sulfide coating, called a **phosphor**, on a television tube absorbs the energy of fast moving electrons, and emits the visible light composing the picture. When there is a time lag between the absorption and emission of radiant energy, a long lasting "glow" may persist. This is known as **phosphorescence**. The fluorescent effect that is achieved by molecules

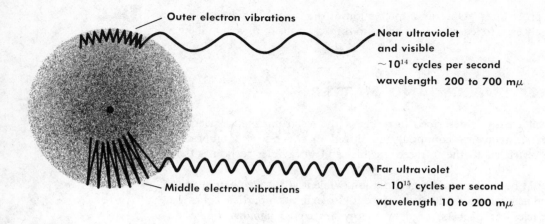

Outer electron vibrations

Near ultraviolet
and visible
~ 10^{14} cycles per second
wavelength 200 to 700 mμ

Far ultraviolet
~ 10^{15} cycles per second
wavelength 10 to 200 mμ

Middle electron vibrations

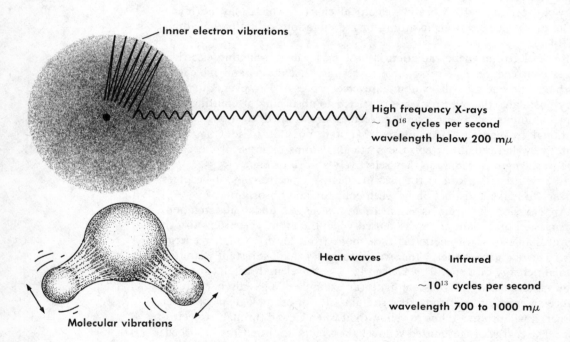

Inner electron vibrations

High frequency X-rays
~ 10^{16} cycles per second
wavelength below 200 mμ

Heat waves   Infrared
~ 10^{13} cycles per second
wavelength 700 to 1000 mμ

Molecular vibrations

absorbing some energy in shifting to an excited state so that the emerging photons have a lower frequency is called the **raman effect**. Raman spectroscopy supplements infrared spectroscopy in the analysis of organic compounds.

Absorption spectroscopy is concerned with the particular wavelengths of light that are absorbed by molecules placed in the light path. The important distinction between an emission spectrum and an absorption spectrum is that the absorption spectrum is produced by an external light source from which specific wavelengths are absorbed by the sample being investigated. The resulting spectrum has dark lines (an absence of color) instead of a few lines of color on an otherwise dark observation screen (Fig. 14-2).

Sources of radiant energy vary from good heat emitters, such as a nichrome wire coil for infrared radiation, through tungsten filament light

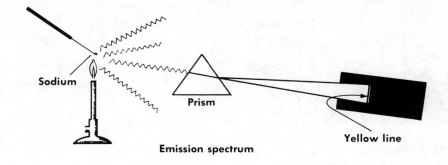

Emission spectrum

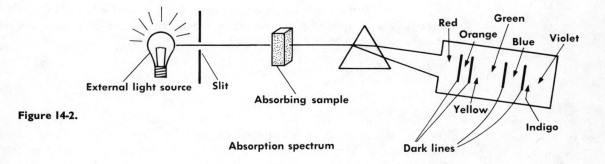

Figure 14-2.

Absorption spectrum

bulbs for visible radiation, to xenon, hydrogen, or mercury gas discharge tubes for ultraviolet rays.

The specific questions of sample preparation, wavelength selection, and treatment of data will be taken up in the individual discussions of the colorimeter, uv-vis (ultraviolet and visible) spectrophotometer, IR (infrared) spectrophotometer, atomic absorption spectrometer, and the spectrofluorimeter.

The Colorimeter

Colorimeters are designed to operate within the visible spectrum. An ordinary tungsten filament light bulb is the light source and the wavelength selection is performed by the selective use of a variety of color filters. The test tubes used are carefully standardized and their positions are fixed in order to increase the precision of the measurements.

The principle on which the colorimeter is based is that the amount of light absorbed by a colored solution is directly proportional to its concentration. This is known as Beer's Law. The unabsorbed light strikes a photoelectric cell in which the electric current produced is proportional to the light intensity. The current is then measured by a galvanometer. The galvanometer is calibrated to read the light absorbance by the solution or, conversely, the per cent of light transmitted through the solution. The **absorbance** is the log of the reciprocal of the transmittance.

$$A = \log \frac{1}{T}$$

absorbance   transmittance

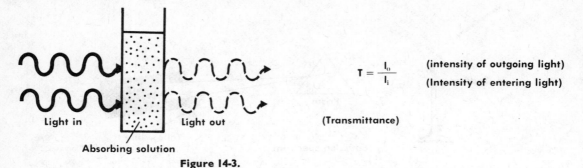

$$T = \frac{I_o}{I_i}$$

(intensity of outgoing light)

(Intensity of entering light)

(Transmittance)

Light in   Light out

Absorbing solution

Figure 14-3.

The **transmittance** is the ratio of the light intensity coming into the sample compared to the light intensity emerging from the sample (Fig. 14-3). For example, if the meter indicated 80 per cent of the light was transmitted through the sample, the transmittance would be 80 per cent, and the absorbance would be

$$80 \text{ per cent} = 0.80$$

$$A = \log \frac{1}{0.80} = \log 1.25$$

$$A = 0.096$$

It is usually possible to obtain a convenient conversion table for transmittance and absorbance from a reference text. Such a conversion table is found in Table A-11 in Appendix A.

The method for accurate reading of the absorbance of a solution is to first correct the meter reading for zero absorbance (100 per cent transmittance) by using a water blank. If the solvent is not water, use whatever the solvent is and then adjust the reading to zero. In this way, the absorbance due to the test solution alone is assured.

Filters are used to increase the efficiency of the colorimeter by blocking portions of the spectrum that are unlike the color of the test solution. If a red solution is being measured, the blue portion of the spectrum can add nothing but possible interference. Therefore, a blue filter would be indicated. Table 14-1 suggests filters that are most appropriate for different colored solutions.

The physical structure of the colorimeter is relatively simple. The diagram in Figure 14-4 points out its essential features.

The current produced by the photo cell is proportional to the light intensity striking it. The galvanometer is calibrated in per cent of trans-

TABLE 14-1. VARIOUS TEST SOLUTIONS AND APPROPRIATE FILTERS.

Colors of Test Solution	Filter Spectral Range (mμ)	Filter Color
Red, orange, yellow, green and cloudy solutions (turbid)	400 to 460	blue
Red, orange, yellow and blue, indigo, purple	500 to 575	green
Purple, indigo, blue, green, yellow	630 to 700	red

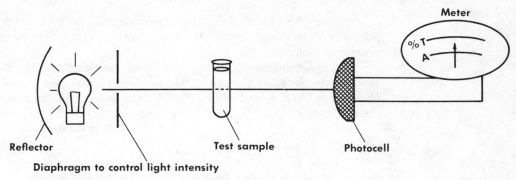

Figure 14-4. Diagrammatic illustration of the essential parts of colorimeter.

mittance and absorbance units. A photograph of a typical colorimeter is illustrated in Figure 14-5.

When turbid (cloudy) solutions are being measured for their absorbance properties, the colorimeter may still be used effectively. A blue filter usually functions most satisfactorily. The photometric measurements of turbidity is called **nephelometry**.

The Spectrophotometer

Spectrophotometers are designed to provide both qualitative and quantitative information about test solutions. As the sample selectively

Figure 14-5. The Klett-Summerson colorimeter. A, Scale knob; B, meter; C, pointer; D, pointer (the two pointers must be aligned); E, test sample; F, light switch; G, zero adjustment; H, photo cell—meter switch. (Courtesy of Klett-Summerson Manufacturing Co.)

absorbs wavelengths of light, the degree of absorbance is related to the *thickness* of the sample container, called a **cuvet**, and the *concentration* of the absorbing molecules (Fig. 14-6).

The absorbance of light at a particular wavelength is directly proportional to the thickness and concentration. Another commonly used name for absorbance (A) is **optical density** (O.D.). The proportionality is expressed in the equation

$$A = \varepsilon \, c \, l$$

absorbance

sample thickness (cm)

concentration (molar)

proportionality constant is the **absorptivity** or the **molar extinction coefficient.**

The molar extinction coefficient is always constant for a particular substance at a given wavelength. It can usually be found in the literature, or it can be easily determined by preparing a standard solution of known concentration and reading the absorbance on the spectrophotometer.

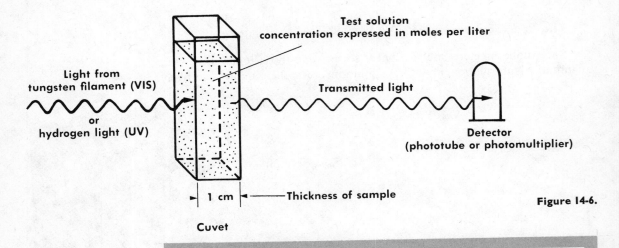

Light from tungsten filament (VIS) or hydrogen light (UV)

Test solution concentration expressed in moles per liter

Transmitted light

Detector (phototube or photomultiplier)

|← 1 cm →|←———— Thickness of sample

Cuvet

Figure 14-6.

Example 1

Calculate the molar extinction coefficient (absorptivity) of NADH (nicotinamide adenine dinucleotide, reduced) where a 3.63 millimolar solution has been diluted sixtyfold.

1. Read the absorbance (optical density) at 340 mμ:

$$A = 0.378$$

2. Develop the equation:

a. $A = \varepsilon \, c \, l$

b. $\varepsilon = \dfrac{A}{c \, l} \times$ dilution factor

c. $\varepsilon = \dfrac{0.378}{(3.63 \times 10^{-3} \text{ M})(1 \text{ cm})} \times 60$

d. $\varepsilon = 6.25 \times 10^3$

Example 2

A solution of NADH has an absorbance reading of 0.448 when it has been diluted to ten times its original volume. Calculate the molar concentration.

1. Obtain the molar extinction coefficient:

$$6.25 \times 10^3$$

2. Develop the equation:

a. $A = \varepsilon c \, l$

b. $C = \dfrac{A}{\varepsilon \, l} \times \text{dilution}$

c. $C = \dfrac{0.448}{(6.25 \times 10^3)(1 \text{ cm})} \times 10$

d. $C = 7.17 \times 10^{-4} \text{ M}$

The spectrophotometer has several advantages over the colorimeter. Three significant features found in uv-visible spectrophotometry that are not available in colorimetry are: (1) the reproductibility of absorption curves, (2) the selection of specific wavelengths, and (3) the ultraviolet light source as an option.

The part of the spectrophotometer designed to select specific wavelengths is called the **monochromator**. A **prism** or **diffraction grating** is calibrated to reflect a specific wavelength through a sample. The diffraction grating spreads light out into a "fan-like" spread of wavelengths as a prism does (Fig. 14-7). By rotating the prism or grating, the selected wavelength can be directed, or reflected, through the test sample which stands between the light source and the light detector (Fig. 14-8).

The detection system in the spectrophotometer may be a phototube or a photomultiplier. The current from the detector tube is directly proportional to the light intensity after the light has passed through the test sample. The galvanometer is designed to be read as transmittance and absorbance rather than amps. If the sample absorbs most of the light at a particular wavelength, the absorbance reading is high while the per cent of transmittance is low. Before a sample is "read," a **solvent blank** is used to adjust the absorb-

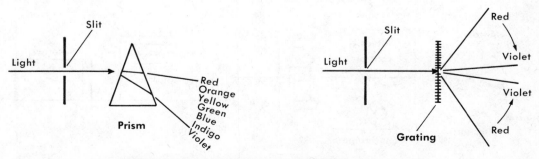

Figure 14-7. Prism and grating light diffraction.

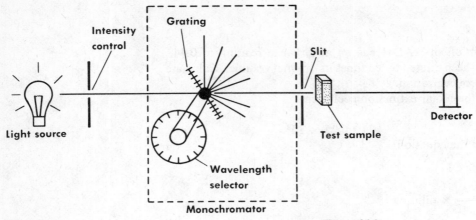

Figure 14-8.

ance reading to zero. Great care must be taken in the ultraviolet region since ordinary glass cuvets absorb ultraviolet light. Very expensive (*handle with care!*) optically matched quartz cuvets must be used. The solvent in which the test sample is dissolved (water, alcohol, or various organic solvents) must be used as the blank so that its absorbance will not be confused with the absorbance characteristics of the compound under investigation.

The technician has to exercise care in adjusting the sensitivity, solvent absorption, dark current, and transmittance so that test samples will be accurately reflected in terms of their true absorbance properties. The sensitivity control is especially critical when the operator switches from the phototube detector to the photomultiplier since the photomultiplier increases the detection sensitivity about tenfold. This is an essential procedure when very low concentrations of sample are used.

The basic arrangement of the components of a uv-visible spectrophotometer are diagrammed in Figure 14-9. The Zeiss spectrophotometer is shown in Figure 14-10.

A spectrum obtained by scanning a portion of the wavelength range on the spectrophotometer often provides a means of identifying a compound. Each substance has its own peculiar maximum absorbance points in the spectrum. In effect, these peak absorbance characteristics provide a "fingerprint" of the compound. The chlorophyll spectrum is illustrated in Figure 14-11 as an example. Other compounds, even though they may be similar to chlorophyll in molecular structure, will not absorb a maximum amount of light at the same wavelengths as chlorophyll.

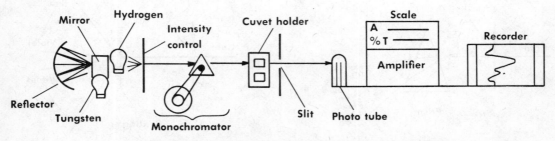

Figure 14-9.   Basic arrangement of spectrophotometer components.

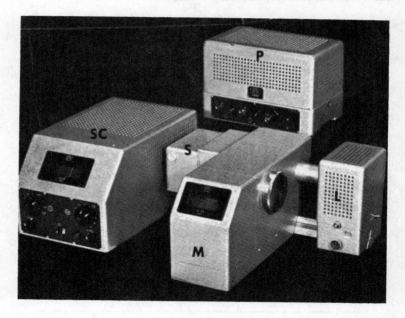

Figure 14-10. Zeiss spectrometer. L = light source; M = monochromator; S = sample compartment; P = power supply; SC = sensitivity control.

An analytical approach, known as finding a **difference spectrum** provides another method for determining the concentration of an absorbing substance. This is done by measuring the absorbances of the oxidized and reduced forms of a sample and using the difference between the optical densities. For example, when the respiratory cofactor, coenzyme Q, is dissolved in isopropanol and diluted fivefold, an absorbance peak at 264 mμ is obtained. If the coenzyme Q is reduced by adding sodium borohydride, there is no appreciable absorbance at that wavelength. This difference in absorbance yields a quantitative difference in the amount of coenzyme Q present (Fig. 14-12).

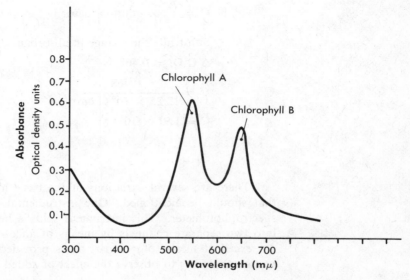

Figure 14-11. Absorbance spectrum of chlorophyll.

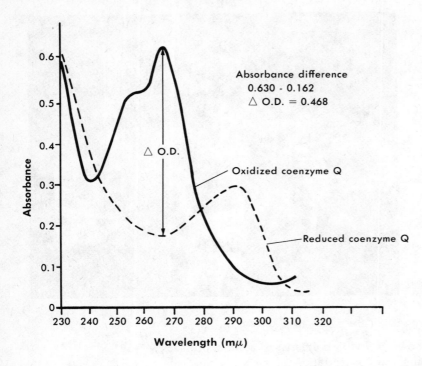

Absorbance difference
0.630 - 0.162
△ O.D. = 0.468

△ O.D.

Oxidized coenzyme Q

Reduced coenzyme Q

Figure 14-12. The coenzyme Q difference spectrum.

Example 3

Use the spectrophotometric data from the difference spectrum to calculate the molar concentration of coenzyme Q.

1. Obtain the molar extinction coefficient from the literature:

$$\varepsilon = 12.25 \times 10^3$$

2. Develop the equation:
   a. $A = \varepsilon \, c \, 1$

   b. $C = \dfrac{A}{\varepsilon \, 1} \times$ dilution factor

   c. Substitute the change in absorbance (Δ O.D.) for A:

   Δ O.D. $= 0.468$

   $$C = \frac{0.468}{12.25 \times 10^3 \, (1 \text{ cm})} \times 5$$

   $C = 1.91 \times 10^{-4} \text{ M}$

There are several variations of the basic uv-visible spectrophotometer that should be mentioned. One instrumental variation is the **split beam** spectrophotometer. This instrument splits a beam of a single wavelength into two separate channels by means of an oscillating mirror attachment. An excellent means of comparison is provided in this way when the investigator wishes to observe the effect of added substances on the structure

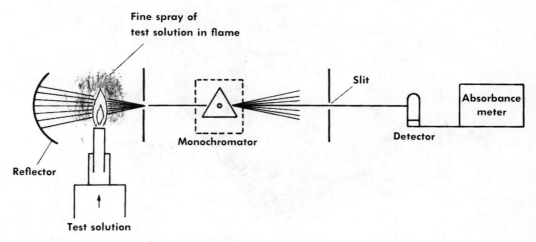

Figure 14-13.

of the molecule in question. As the molecular structure is altered, the absorbance property is changed also. This is called a **photometric titration.**

Another variation is the **double beam** spectrophotometer. This instrument allows the simultaneous passage of two different wavelengths through the sample. However, the two beams may be alternated by using a "chopper."* The double-beam spectrophotometer is especially suited for the study of the rates and complexities of biochemical reactions. It also permits absorption studies on low concentrations where turbidity may interfere.

A special attachment may be used with some instruments for the purpose of **flame photometry.** Instead of a visible or ultraviolet light source, a hot flame such as burning hydrogen is used to excite atomic electrons in a fine spray of test solution. This produces the emission spectrum characteristics of the atoms being investigated. The monochromator is used to find the specific wavelengths being emitted by the excited electrons. This method of flame photometry provides very rapid analyses. A diagrammatic representation of the flame photometry adaptation is illustrated in Figure 14-13.

An instrument related to the spectrophotometer with a flame photometry attachment is the **atomic absorption spectrophotometer.** Both methods of analysis may be included under the general heading of **flame spectroscopy.** Atomic absorption differs from flame photometry since the atomic absorption spectrophotometer measures light absorption by the atoms while the flame photometer measures light emitted by atoms. The atomic absorption spectrophotometer has the advantage of greater sensitivity than the flame photometer.

The atomic absorption spectrophotometer (Fig. 14-14) has a light source provided by a special cathode lamp which produces brilliant, sharp lines. The atomic absorption spectrophotometer is equipped with a "chopper," which is synchronized with the amplifier so that light emission from the light source can be distinguished from radiation due to the flame.

In the atomic absorption spectrophotometer the light from the cathode lamp is directed through flame in which the sprayed test solution absorbs

---

* A "chopper" is usually a disc with an opening. As the chopper revolves at a selected rate, the beam of light is alternately shut off and then permitted to pass through the opening.

Figure 14-14. Atomic absorption spectro-photometer.

wavelengths characteristic of its components. The monochromator sorts out the wavelengths so that the phototube detector can respond to the varying intensities at different wavelengths. The intensity of the emitted light is proportional to the concentrations of the elements in the test sample (Fig. 14-15).

Atomic absorption spectrophotometry is ideally suited for the quantitative analysis of trace metal atoms in oil, food, and the products of biomedical research laboratories.

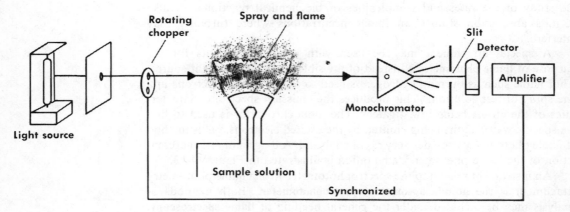

Figure 14-15. Diagrammatic illustration of atomic absorption spectrophotometer.

Fluorimetry

The fluorescence phenomenon in which a portion of the energy absorbed by molecules is degraded to heat, while the remainder is radiated as lower frequency photons produced by molecules returning to their vibrational ground states, is another optical method of analysis.

Many compounds absorb light in the ultraviolet and emit visible light in the process of fluorescence. A **spectrofluorimeter** is an instrument designed to measure the intensity of fluorescence, which is proportional to the concentration of the test compound.

In principle, the spectrofluorimeter is similar to the spectrophotometer. In fact, some spectrophotometers can be adapted for fluorescence measurements. There are critical factors, however, affecting fluorescence measurements that have to be taken into account because of their effect on the fluorescence of the compound being investigated. Some of these critical factors are pH, temperature, and fluorescent contaminants that water can extract from rubber and plastic laboratory apparatus.

Some of the most important fluorometric analyses are performed in the study of biochemical reactions. An example of how a reaction can be studied by fluorometric analysis is the case of nicotinamide adenine dinucleotide (NAD) fluorescence change by rat liver mitochondria, as the mitochondria change the NAD to a nonfluorescing compound in the presence of an appropriate substrate and enzyme. The rate of NAD change is regulated by varying the concentrations of the interacting substances. The change is observed as a reduction of the light being fluoresced. Finally, when all the NAD has been changed, the fluorescence is "quenched." Experiments of this type are performed by using a monochromator to select the wavelengths of ultraviolet light that the selected components will absorb in the process of fluorescence. Quantitatively, the amount of fluorescing substance that is present in a test sample is determined by comparing the absorbance of the sample to the absorbance of a standard reference where the concentration is known.

Infrared Spectrophotometry

Infrared spectrophotometry is an invaluable method of analysis in organic chemistry. The infrared spectrophotometer differs from the uv-visible instruments in several ways:

1. The radiation source produces "heat" waves, where the wavelengths emitted from a glowing wire or tube are in the micron range.
2. The test sample may be a thin solid wafer as well as a liquid or a gas.
3. The detector is a **thermocouple** rather than a phototube or photomultiplier. A thermocouple is made of two strips of different metals joined at the ends. The amount of electric current generated by the thermocouple is proportional to the amount of heat absorbed at one end of the junction.

The absorption of infrared radiation depends primarily on the different vibrational characteristics of the covalent bonds in organic molecules. When a sufficiently energetic photon of infrared radiation permits a bond to be *stretched* or *bent* to a degree that is greater than its normal vibrations, that photon is absorbed. The wavelength of that photon is then blocked from the thermocouple detector which results in the recording of absorbance. The effect of bending and stretching is to increase the polarity of the molecule by separating the centers of positive and negative charge more than usual (Fig. 14-16).

The stretching vibrations are described as **symmetric** or **asymmetric** (Fig. 14-17).

One other type of stretching vibrations is called **in-plane**, or **scissoring** and **rocking**. The other type is **out-of-plane**, or **wagging** and **twisting** (Fig. 14-18).

A test sample may be identified by its own "fingerprint" since its absorbance regions will depend on the types and numbers of its bonds. Double bonds require higher energies for vibrations, and triple bonds

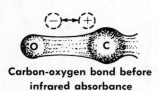

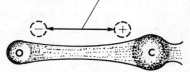

Greater separation of centers of charge

Figure 14-16.

Carbon-oxygen bond before
infrared absorbance

Stretched carbon-oxygen
bond after infrared absorbance

Slight polarity

Greater polarity

require still more energy. These energies are associated with specific wavelengths that can be selected by the instrument.

The stretching and bending of different types of bonds—C—C, C═C, C≡C, C—H, C═O, O—H, and others—have been associated with specific regions of the infrared spectrum as a result of massive amounts of experimental evidence. When infrared spectroscopic data are used in conjunction with other data such as mass spectrographic molecular weight measurements, light refraction measurements, magnetic spectroscopic measurements, and physical property observations, a compound can be qualitatively and quantitatively described.

The conventional method of plotting infrared spectra is different from the absorbance vs. wavelength format used in uv-visible spectrophotometry. The IR plot, whether hand drawn or automatically recorded, plots the per cent of transmittance against wavelength and wave number. The wave

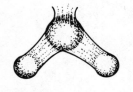

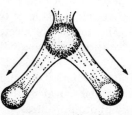

Before energy absorbance

After energy absorbance

Symmetrical stretching

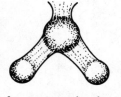

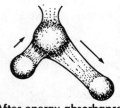

Before energy absorbance

After energy absorbance

Figure 14-17.

Asymmetrical stretching

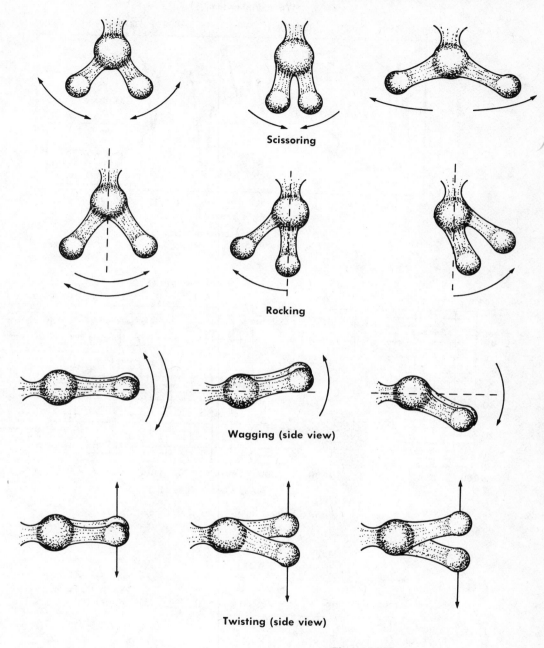

Scissoring

Rocking

Wagging (side view)

Twisting (side view)

Figure 14-18.

number is the reciprocal of the wavelength and therefore is written as cm^{-1}. This is sometimes erroneously called frequency.

Typical IR spectra of ethanol and acetone are illustrated in Figures 14-19 and 14-20.

Once the "fingerprint" spectrum of an unknown substance is obtained, it is possible to begin the identification process by matching the functional groups with the wave numbers or wavelengths at which absorption occurs. Table 14-2 hints at some characteristic absorptions that may be found in much greater detail in standard reference texts.

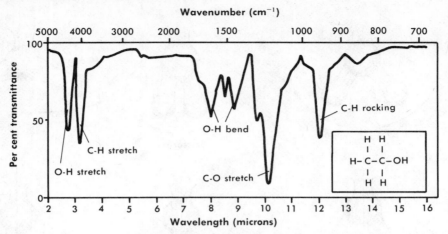

Figure 14-19.   IR spectrum of ethanol.

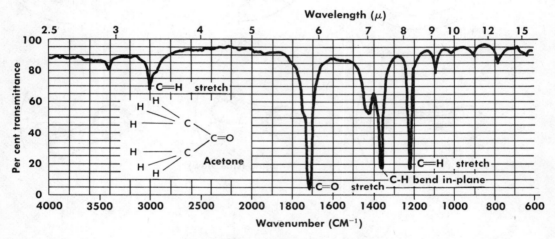

Figure 14-20.   IR spectrum of acetone.

TABLE 14-2

Functional Group	Wavelength Range (mμ)	Wave Number Range (cm⁻¹)
alcohols and phenols strong O—H stretching	2.7–3.3	3700–3000
carboxylic acids strong —C=O stretching	5.4–6.1	1900–1600
medium O—H bending	6.9–8.3	1500–1200
ketones and aldehydes strong C=O stretching	5.4–6.1	1900–1600
hydrocarbons strong —C—H stretching strong C—H bending	3.2–3.3 6.7–13.3	3000–3050 1500–750

Another method of identification takes advantage of published tables of standard spectra. Such publications are produced by the Sadtler Research Laboratories, Inc., of Philadelphia and Varian Associates of California. These organizations provide thousands of indexed spectra in print or on microfilm. The spectra are organized alphabetically, by chemical class, by molecular formula, and numerically according to the format devised by *Chemical Abstracts* (see Chapter 6). The Sadtler Corporation also publishes a coded system of identification which labels the strong absorbance bands in wavelength sequence to be matched with known compounds having the same characteristics. An example of an IR spectrophotometer is illustrated in Figure 14-21.

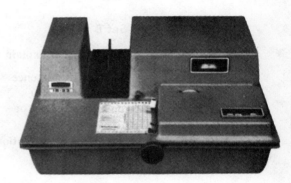

Figure 14-21. Infrared spectrophotometer. (Courtesy Beckman Instruments, Inc.)

PROBLEMS

Set A

1. Define or explain the following terms:

a. photometry
b. fluorescence
c. phosphor

d. Raman effect
e. optical density

2. What is emission spectroscopy? Name an instrument that demonstrates the use of this phenomenon.

3. If a solution shows a 40 per cent light transmittance in a colorimeter, what is the absorbance?

4. What are the three principal physical factors that affect the absorbance of molecules in a spectrophotometer?

5. Calculate the molar extinction coefficient of a substance if 0.02 molar solution, diluted tenfold, is observed to have an absorbance of 0.125.

6. A protein solution has an absorbance of 0.462 after a fivefold dilution. If the molar extinction coefficient is 3.2×10^2, calculate the molar concentration.

7. Find the molar concentration of a substance from its difference spectrum if the oxidized form has an optical density of 0.830 compared to the O.D. for the reduced form of 0.280. The dilution factor is 10, and the molar extinction coefficient is 2.4×10^4.

8. What is the essential difference between the atomic absorption spectrophotometer and the flame photometer?

9. What types of molecular vibrations can be detected by infrared spectrophotometers?

10. What is the function of filters in colorimetry?

Set B

11. Define or explain the following terms:

a. phosphorescence
b. nephelometry
c. monochromator

d. flame photometry
e. wave number

12. What is absorption spectroscopy? Name an instrument that uses this phenomenon.

13. What special function does the spectrofluorimeter perform?

14. Name three advantages that a uv-vis spectrophotometer has over a colorimeter.

15. What is the molar extinction coefficient of a substance where a fiftyfold dilution of a 6.5×10^{-3} molar solution has an optical density of 0.482?

16. Calculate the molar concentration of a vitamin extract if a twofold dilution has a transmittance of 15 per cent? The molar extinction coefficient is 2.2×10^3.

17. What is the absorbance value of a test solution which shows a 60 per cent transmittance?

18. A sample is placed in a special cuvet that has a light path of 1.0 mm. The difference spectrum shows a Δ O.D. of 0.326 with a fivefold dilution. If the molar extinction coefficient is 1.8×10^4, calculate the molar concentration.

19. What special type of experiment is possible with a split beam spectrophotometer?

20. What instrument is ideally suited for the identification of trace metal atoms in food?

SUPPLEMENTARY READING

Bauman, R. P.: *Absorption Spectroscopy.* J. Wiley and Sons, Inc., New York, 1962.

Contains a great deal of useful and very interesting information on spectroscopic theory, instruments, and usage.

Laser and Light. Reading from *Scientific American.* W. H. Freeman and Co., San Francisco, 1969.

A most informative and interesting collection of articles from Scientific American *journals.*

Hercules, D. M. (ed.): *Fluorescence and Phosphoresence Analysis.* Interscience Publishing Co., New York, 1967.

Excellent contributions by a number of scientists to explain the principles and applications described in the title.

Connors, K. A.: *A Textbook of Pharmaceutical Analysis.* J. Wiley and Sons, Inc., New York, 1967.

Very good chapters on instrumental analyses related to the pharmaceutical industry.

15 Chemical Instrumentation (II)

NUCLEAR MAGNETIC RESONANCE (NMR)

NMR spectroscopy is an invaluable supplement to the infrared method of analysis. NMR uses the magnetic properties of an atomic nucleus to identify the types and arrangements of functional groups in complex molecules. The most commonly investigated atomic nucleus is that of hydrogen. This is especially fortuitous because of the great importance of hydrogen in organic chemistry. Since the hydrogen nucleus is composed of one proton, the terms hydrogen ion and proton are synonymous.

The identification of the positions and linkages of protons in complex molecules can be determined by the response of the proton to an external magnetic force. The hydrogen nucleus will respond differently, depending on such structural factors such as:

1. The number of hydrogens attached to a carbon atom.

$$
\begin{array}{ccc}
\text{H} & \text{H} & \text{H} \\
| & | & | \\
\text{H--C--H} & \text{--C--} & \text{--C--Cl} \\
| & | & | \\
& \text{H} & \text{Cl} \\
\text{3 hydrogens} & \text{2 hydrogens} & \text{1 hydrogen}
\end{array}
$$

2. Hydrogen attached to an oxygen atom.

$$—O—H$$

3. A proton functioning as a hydrogen bond.

4. The effective shielding of a hydrogen by surrounding groups of atoms whose nuclei and electrons have their own magnetic influence.

The property of **nuclear spin** is the phenomenon that gives rise to the analytical capacity of NMR spectroscopy. The proton behaves as though it were a ball of electrical charge spinning about on an axis. A rotating unit of charge generates a magnetic field since it acts like a miniature magnet (Fig. 15-1).

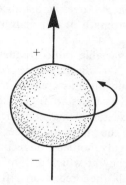
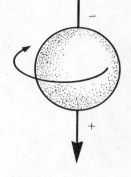

Figure 15-1. Idealized model representing nuclear spin.

When the spinning nucleus is in an external magnetic field, its axis shifts so that it is aligned to become nearly parallel or nearly antiparallel to the external magnetic field. The slight shifting of the axis of rotation causes a circular path to be described by the axis. An analogy can be found in a toy gyroscope as its axis makes a circle when it is not perpendicular to the ground. This circular path is called a **precessional orbit** (Fig. 15-2).

The strength of the external magnetic field is determined by the frequency of an oscillating frequency generator which operates between zero and several million cycles per second. A million cycles per second is called a

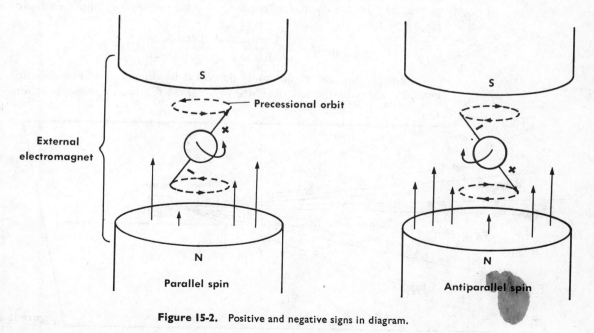

Figure 15-2. Positive and negative signs in diagram.

megahertz (MHz), and frequencies in this order of magnitude are in the radio wave portion of the electromagnetic spectrum.

Some widely used NMR instruments that are primarily designed for proton observations, operate at a radio frequency of 60 MHz and a corresponding magnetic field strength of 1.4×10^4 gauss (G). A **gauss** is a unit of magnetic force. The gauss may be usefully defined by analogy in which case one gauss is roughly equivalent to the force exerted by 100 large ants as they drag a gram weight over a distance of 10 centimeters (Fig. 15-3). The magnetic field strength of our spinning planet is rated at approximately at $\frac{1}{2}$G.

When the variable radio frequency generator is adjusted to a specific frequency, which is directly proportional to the strength of the magnetic field the spinning nucleus creates, the lower energy nucleus (parallel spin) "flips" over and assumes an antiparallel orientation. This "flip" reduces the strength of the external magnetic field as a result of the energy absorbed. The change in the magnetic force causes a flow of current that can be measured by a galvanometer or recorded as a peak on a chart.

The frequency at which a "flip" is induced occurs when the rotational frequency equals the precessional frequency. This is called the **resonance** frequency. The energy required to produce the nuclear "flip" is different for a particular type of nucleus, depending on its physical position in a complex molecule. The difference between the observed resonance frequency causing a "flip" or recorded "peak" in the NMR spectrogram, measured against the position of the peak produced by the protons of a standard compound, tetramethylsilane (TMS), is called a **chemical shift**. In this way, an increase in current at a specific frequency produces the NMR spectrum that identifies a functional group.

The NMR spectrogram pictures the peaks of signal intensity separated by a distance representing the chemical shift. The frequency at which these peaks occur is measured in convenient arbitrary units called ppm (parts per million). The ppm units are obtained by dividing the chemical shift (cycles per second) by the oscillator frequency (cycles per second) and then multiplying by 10^6.

$$\text{ppm} = \frac{\text{chemical shift (cps)}}{\text{oscillator frequency (cps)}} \times 10^6$$

The spectrogram is commonly described by directions of **downfield** for chemical shifts to the left of TMS, and **upfield** for the opposite direction.

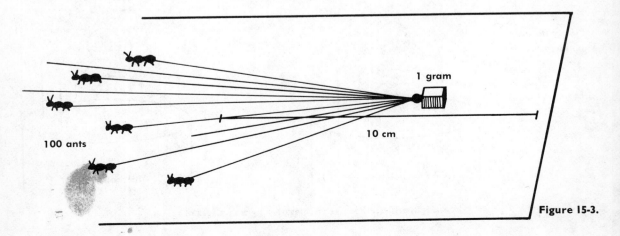

100 ants

1 gram

10 cm

Figure 15-3.

However, it should be noted that the upfield and downfield designations are more accurately applied to the description of decreasing and increasing oscillator frequency. There are very few compounds that absorb to the right of TMS. A sample spectrum of ethanol is shown in Figure 15-4 to illustrate the labeling and interpretation of NMR data.

One other line on the spectrogram which contributes to the full interpretation of the NMR spectrum is the **integration**. This is a single line drawn from the downfield peak toward the upfield TMS standard. It is called integration because its height is proportional to the areas under the individual peaks. The integration of the ethanol peaks is emphasized in the illustration in Figure 15-5. The millimeter distances on the integration curve for

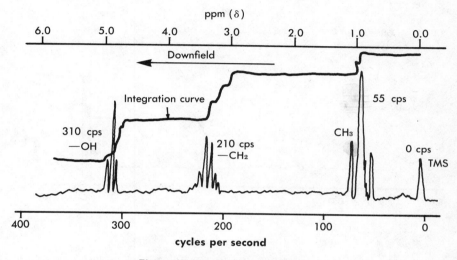

Figure 15-4. NMR spectrum of ethanol.

x, y, and z are in the ratio of $1:2:3$. This ratio corresponds to the number of protons in the ethanol molecules.

$$\left.\begin{array}{l} x = 1 = \underline{\text{—OH}} \\ y = 2 = \underline{\text{—CH}_2\text{—}} \\ z = 3 = \underline{\text{—CH}_3} \end{array}\right\} \begin{array}{l} \text{CH}_3\text{CH}_2\text{OH} \\ \\ \text{ethanol} \end{array}$$

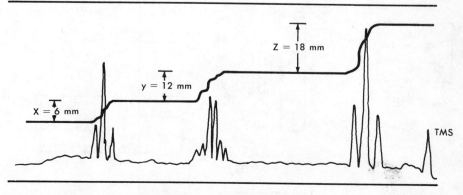

Figure 15-5. Ethanol integration curve.

The calculation of the chemical shift (ppm) measurement is easily done according to the equation

$$\text{ppm} = \frac{\text{chemical shift}}{\text{oscillator frequency}} \times 10^6$$

For the CH_3 peak when the oscillator frequency is 50 MHz (megahertz or megacycles), substitute:

$$\text{ppm} = \frac{55 \, \text{sec}^{-1}}{50 \times 10^6 \, \text{sec}^{-1}} \times 10^6$$

$$\text{ppm} \doteq 1.1$$

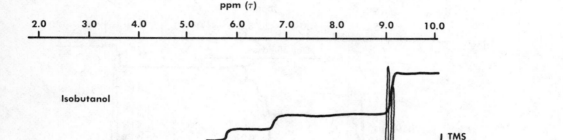

Figure 15-6

While the chemical shift is always measured in ppm, as described above, there is one other commonly used unit, called τ (the Greek letter tau). This scale is not based on the oscillator frequency. The TMS standard peak is simply defined as 10 tau. The scale based on the oscillator frequency is symbolized as δ (the Greek letter delta) for contrast. The τ value is obtained simply by subtracting the δ value from 10. For example, if δ equals 2.2 ppm, the τ value is $10 - 2.2 = 7.8$. On the spectrogram, the scale relationship would appear as shown in the illustration in Figure 15-6.

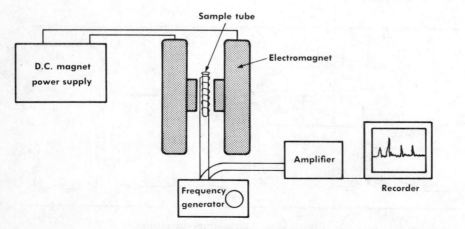

Figure 15-7.

Simplified diagram of NMR apparatus

NMR spectroscopy has the advantages of working with small samples of less than a milliliter, and concentrations as low as 10 per cent by weight. The best solvents are those having no single proton atoms, such as carbon tetrachloride. If the sample is not soluble in CCl_4, other solvents may be used so long as their proton resonances do not occur at the same absorption frequencies as the sample. A simplified diagram of NMR apparatus is shown in Fig. 15-7.

ELECTRON SPIN RESONANCE SPECTROSCOPY (ESR)

ESR, which is also known as **electron paramagnetic resonance** (EPR), has a direct application to the study of **free radical** structure. Free radicals may be single atoms of elements that exist normally as diatomic molecules, such as hydrogen, oxygen, or chlorine. Other examples are sulfates, nitrates, carbonates, and transition metal complex ions. Biochemical investigations are especially concerned with metal ions in enzyme systems and the role of free radicals in the formation of cancerous cells. An important generalization that can be made regarding free radicals is that they tend to be very chemically active and, as a result, very short-lived as free, electrically charged particles.

ESR instruments as so designed that free radicals may be observed as they are generated by various forms of high energy radiation or electrochemical redox reactions. By observing the free radicals in the process of formation, they may be "caught" before they have the short time necessary for chemical bonding and the resulting elimination of the species in question from the free radical state.

In addition to free radicals, ESR is used to assay trace amounts of metal ions in biological media. If the metal ion has an unpaired electron in its valence shell, it will respond to an external magnetic field. This phenomenon is known as **paramagnetism**, a term generally used to describe a relatively weak magnetic interaction. When lone (unpaired) electrons absorb quantum amounts of energy from an electromagnetic wave, the electrons will interact with the electrons of other atoms in the test sample. The resonance effect is due to the "flip" of the spinning electron as it pairs (couples) with the other electrons. This resonance is similar to, but more complex than, the resonance effect described for spinning nuclei.

ESR instruments operate at very high frequencies—usually about 9.5×10^3 megacycles per second—and a corresponding magnetic field strength of 3.4×10^3 gauss. The more modern notation for the frequency is 9.5 GHz, which stands for 9.5 gigahertz. The prefix giga- is equivalent to a billion. Although the efficiency of energy transfer from the magnet to the sample is usually less than 30 per cent, it is still adequate for detecting electron spin resonance in a small sample containing on the order of a trillionth of a mole of free radicals.

The graphic recording of ESR spectra also involves the use of standard reference substances. Two common examples are the free radical 1,1-diphenyl-2-picrylhydrazyl (Fig. 15-8) and a portion of a ruby crystal which contains a trace amount of chromium (III) ion. The spectrograms indicate the number of groups of atoms and the number of atoms in each group. The number of groups is related to the number of "spikes," and the number of atoms in each group is represented by the intensity of each "spike." An

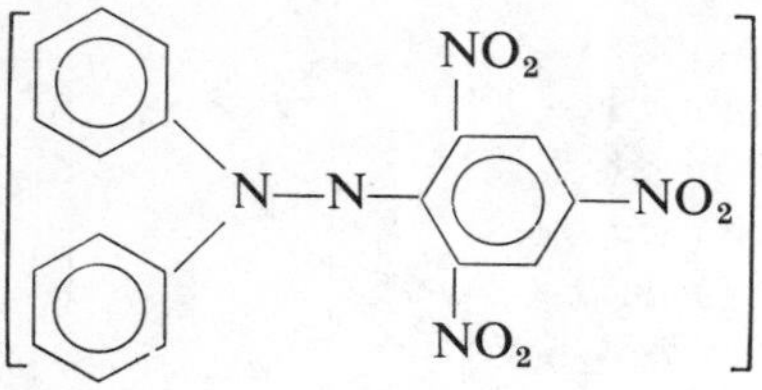

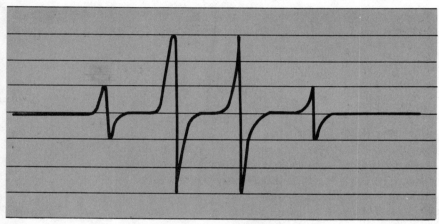

Figure 15-8.

idealized ESR spectrogram is presented in Figure 15-9 to illustrate an imaginary free radical formation due to the gamma radiation of a small amount of organic material. There are four lines having intensity ratios of 1:3:3:1.

Figure 15-9

ESR spectrogram

GAS CHROMATOGRAPHY

Gas chromatography is a method of separating and qualitatively or quantitatively analyzing vapors or materials in the gas state. A long tube is filled with any one of a variety of commercially obtainable adsorbents that is coated with a selected solvent. A relatively inert gas such as helium is used to conduct the mixture of vapor through the column. The conducting gas is known as the **carrier**. The gases in the mixture being analyzed emerge at the far end of the column at different times, depending on how strongly they are adsorbed by the material used to pack the column. One common method of detecting the emerging gases is by measuring the change in the electrical conductance of a thermistor as the movement of the gas causes a change in the resistance by removing heat. A diagrammatic representation of a gas chromatograph is shown in Figure 15-10.

Gas chromatographs (GC) have two types of packing material used in the column. The packing or **stationary phase** may be a solid, such as silica, charcoal, or alumina, or it could be a nonadsorbing solid coated with a liquid. The nonadsorbing solids commonly used are made of small uniform grains of porous materials such as dry clay (diatomaceous earth), plastics, or glass. The liquids are selected with regard to their degree of polarity and the temperature range, which still permits them to remain in the liquid state.

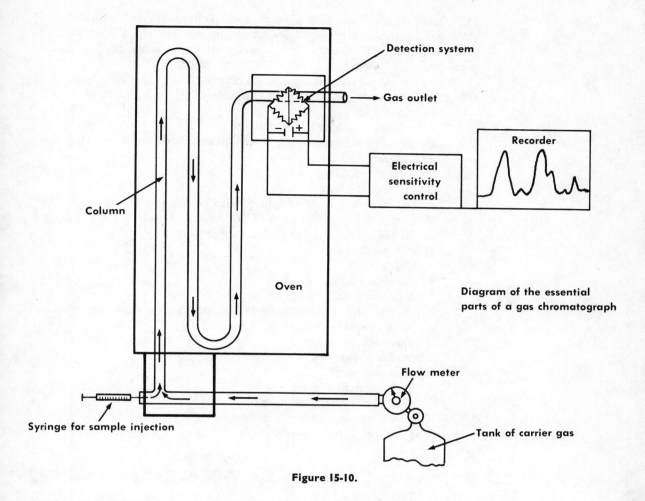

Figure 15-10.

Nonpolar gases would require a nonpolar liquid phase while polar gases would require appropriately polar liquids in the column.

One of the great advantages of GC is the capability of the instrument to analyze volumes of organic compound mixtures in the microliter range. Special syringe type sample injectors are commercially available. They may be calibrated in 0.1 μl units. The GC apparatus is equipped with a rubber port which leads into the base of column where the injected sample can be conducted upwards by the carrier gas.

The various compounds in the sample mixture will move through the stationary phase at different rates of speed depending on the tenacity of the adsorbance. For example, a mixture of simple nonpolar hydrocarbons will move through a nonpolar liquid in about five minutes and be separated very nicely. However, if a polar liquid phase were used, all the hydrocarbons would move through in less than a minute, resulting in a poor separation.

There are two major types of detectors in use. One type, called a **thermal conductivity** (TC) detector, will respond to differing concentrations of gases. The TC detector is usually a microthermistor. When a gas has passed through the adsorbing column, it will lower the controlled temperature of a microthermistor and thereby reduce the applied voltage. When gases to be analyzed have high thermal conductivities as a physical property, the

TC method is excellent. The change in the voltage can be electronically amplified or attenuated (reduced) depending on its value and the sensitivity of the recorder. TC detectors have the advantages of simplicity and high precision.

Some typical gas chromatograms are illustrated in Examples 1, 2, and 3. Information added to a chromatogram should be:

1. Instrument used
2. Length and diameter of the column
3. Commercial names of stationary phase, both solid and liquid. Examples:
   a. Squalene ($C_{30}H_{62}$)
   b. Apiezon-L grease
   c. DC-550, Dow-Corning (phenyl silicone oil)
   d. Ucon Oil LB-550-X, Union Carbide (polyalkalene-glycol)
   e. DEGS (diethylene glycol succinate polyester)
   f. Porapak-S
   g. Chromosorb
   h. Carbowax 1540
   i. Anakrom ABS
4. Size of particles in the stationary phase, described in standard "mesh" units. Example: 60–80 mesh
5. Temperature of the oven
6. Flow rate of carrier gas in ml per minute
7. Type of detector
   a. TC (thermal conductivity)
   b. FID (flame ionization detector)
8. Degree of attenuation or amplification
9. The order of compounds as they are eluted

Example 1

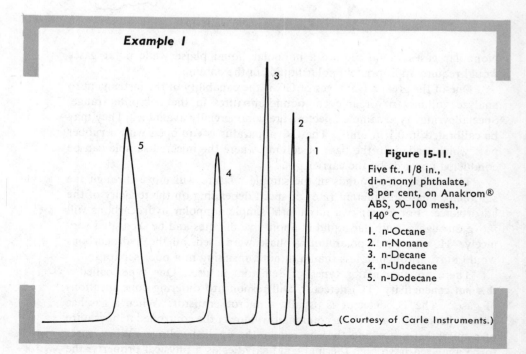

Figure 15-11.

Five ft., 1/8 in., di-n-nonyl phthalate, 8 per cent, on Anakrom® ABS, 90–100 mesh, 140° C.

1. n-Octane
2. n-Nonane
3. n-Decane
4. n-Undecane
5. n-Dodecane

(Courtesy of Carle Instruments.)

Example 2

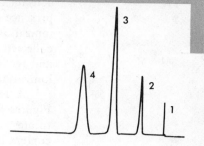

Figure 15–12. Eight foot, ⅛ in., G. E. Silicone SF-96. 15 percent, on—

Eight ft., 1/8 in.,
G. E. Silicone SF–96,
15 per cent on
Chromosorb® P, 60–80
mesh, 25° C.

1. Air
2. Methylene chloride
3. Chloroform
4. Carbon tetra-
   chloride

Example 3

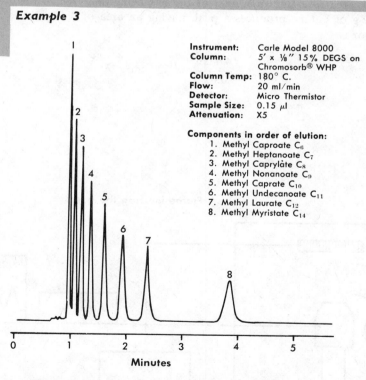

Instrument:	Carle Model 8000
Column:	5' x ⅛" 15% DEGS on Chromosorb® WHP
Column Temp:	180° C.
Flow:	20 ml/min
Detector:	Micro Thermistor
Sample Size:	0.15 μl
Attenuation:	X5

Components in order of elution:
1. Methyl Caproate C_6
2. Methyl Heptanoate C_7
3. Methyl Caprylate C_8
4. Methyl Nonanoate C_9
5. Methyl Caprate C_{10}
6. Methyl Undecanoate C_{11}
7. Methyl Laurate C_{12}
8. Methyl Myristate C_{14}

Figure 15-13. Short chain fatty acid esters. Excellent peak symmetry and exceptionally short analysis time for the lower molecular weight fatty acid esters are easily achieved on a 5 foot, ⅛-inch column. The C_6 through C_{14} esters are separated isothermally in less than four minutes. The benefits of on-column sample injection and the rapid response of the small column detector are essential for this level of performance.

The other common type of detector is the FID (**flame ionization detector**). The principal advantage of the FID is that it can signal the presence of specific gases. The FID method ionizes gases by supplying the ionization energy as intense heat. The ions, being electrically charged, are collected at oppositely charged electrodes. The resulting current is amplified and recorded. A simplified diagrammatic illustration of a hydrogen flame ionization detector is shown in Figure 15-14.

A few examples of gas chromatograms using FID are illustrated in Figures 15-15 and 15-16. The first one (Fig. 15-15) is an analysis of barbiturates in blood, and the second (Fig. 15-16) is an analysis of the alcohol content in blood.

Whenever gas is left in the column, the detectors function inefficiently. A method of removing the volume of gas in the apparatus (called the **holdup volume**) is to flush the detector directly with carrier gas.

Quantitative analysis of a mixture of gases is based on the fact that the area under the peak recorded on the chromatogram is proportional to the number of moles of gas causing the response. However, different compounds affect the detector system to varying degrees. In other words, a mole of methane will produce a peak whose area may be larger or smaller than a mole of propane. This problem is eliminated by determining a **response factor** for each gas being quantitatively analyzed. The response factor is the ratio of the peak area to a known weight of the gas being analyzed. For example, if 0.02 g of butane produces a peak having an area of 4.4 cm², the response factor is

$$\frac{4.4 \text{ cm}^2}{0.02 \text{ g}} = 220 \text{ cm}^2\text{g}^{-1}$$

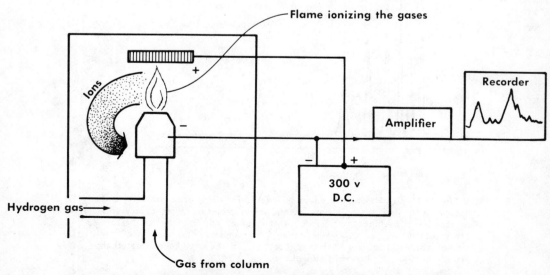

Figure 15-14. Diagrammatic illustration of hydrogen FID system.

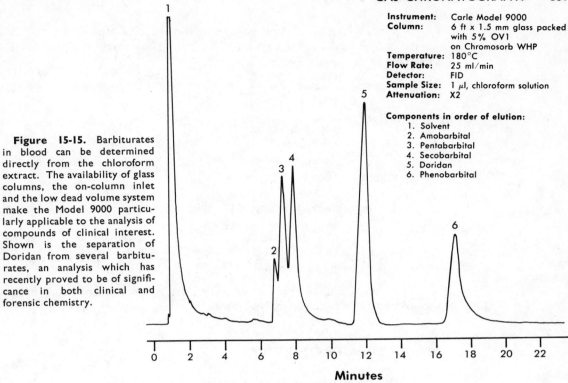

Instrument: Carle Model 9000
Column: 6 ft x 1.5 mm glass packed
with 5% OV1
on Chromosorb WHP
Temperature: 180°C
Flow Rate: 25 ml/min
Detector: FID
Sample Size: 1 μl, chloroform solution
Attenuation: X2

Components in order of elution:
1. Solvent
2. Amobarbital
3. Pentabarbital
4. Secobarbital
5. Doridan
6. Phenobarbital

Figure 15-15. Barbiturates in blood can be determined directly from the chloroform extract. The availability of glass columns, the on-column inlet and the low dead volume system make the Model 9000 particularly applicable to the analysis of compounds of clinical interest. Shown is the separation of Doridan from several barbiturates, an analysis which has recently proved to be of significance in both clinical and forensic chemistry.

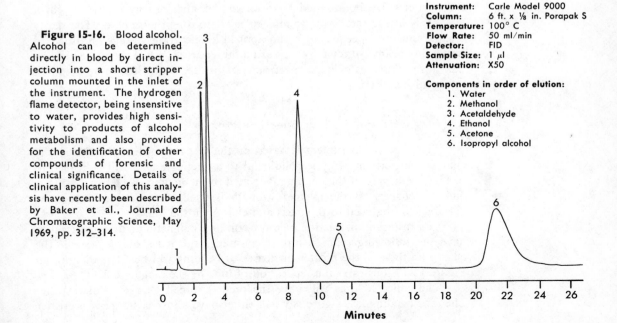

Instrument: Carle Model 9000
Column: 6 ft. x ⅛ in. Porapak S
Temperature: 100° C
Flow Rate: 50 ml/min
Detector: FID
Sample Size: 1 μl
Attenuation: X50

Components in order of elution:
1. Water
2. Methanol
3. Acetaldehyde
4. Ethanol
5. Acetone
6. Isopropyl alcohol

Figure 15-16. Blood alcohol. Alcohol can be determined directly in blood by direct injection into a short stripper column mounted in the inlet of the instrument. The hydrogen flame detector, being insensitive to water, provides high sensitivity to products of alcohol metabolism and also provides for the identification of other compounds of forensic and clinical significance. Details of clinical application of this analysis have recently been described by Baker et al., Journal of Chromatographic Science, May 1969, pp. 312–314.

Suppose 0.52 g of a sample containing some butane produced a characteristic butane peak having an area of 3.2 cm². How would the percentage of butane in the sample be calculated?

1. If response factor $= \dfrac{\text{area}}{\text{weight}}$ then

2. weight $= \dfrac{\text{area}}{\text{response factor}}$

weight $= \dfrac{3.2 \text{ cm}^2}{220 \text{ cm}^2 \text{ g}^{-1}}$

weight $= 0.015$ g

3. The percentage of butane in the sample is

$$\frac{0.015 \text{ g}}{0.52 \text{ g}} \simeq 3 \text{ per cent}$$

The use of GC in conjunction with infrared spectroscopy and mass spectrometry is a valuable analytical tool in modern organic chemistry.

MASS SPECTROMETRY

A mass spectrometer is an instrument that can determine molecular weights and atomic masses with great precision. It has very important applications to radiochemistry and gas chromatographic methods of analysis because of the additional capability of sorting out molecules and isotopes by taking advantage of mass differences.

Basically, a mass spectrometer is a tube in which a vacuum can be created so that the path of rapidly moving positive ions will be without obstructions. As the gas to be analyzed or separated is admitted to the tube, the particles are bombarded by a stream of high energy electrons which effectively ionize the atoms or molecules composing the sample. The newly created ions are speeded up by an applied accelerating voltage and focused into a fine beam as they move toward a target area. While there are about a half dozen varieties of mass spectrometers in use, two of the more common types will be discussed.

Magnetic Field Mass Spectrometer

This type of instrument is based on the fact that a beam of electrically charged particles moving perpendicular to a magnetic field will be forced to bend. The degree of the bending is related to the magnet strength, ion mass, and ion charge. If the magnet strength is fixed at a constant value, the variables are reduced to the mass and charge on the ion.

The resulting interaction of ion beam and magnetic field is observed to have the following relationship: the smaller the mass or the greater the charge on the ion, the greater its degree of bending will be. The diagram in Figure 15-17 illustrates the separation of ions by the magnetic field.

The collected ions will have the predetermined charge to mass ratio. The ions bending less vigorously will either have a lower charge or a greater

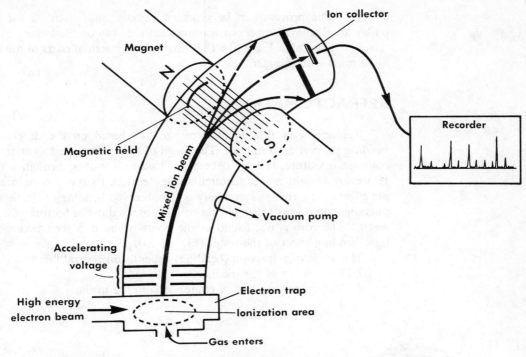

Figure 15-17. Separation of ions in a magnetic field mass spectrometer.

mass or both. The ions bending more sharply downward will have the opposite properties.

The Time-of-Flight Mass Spectrometer

This type of instrument is based on the principle that a heavier ion moves more slowly than a light one. The time-of-flight instrument is sufficiently long so that the ions of varying masses will arrive at a detecting plate at distinctly different time intervals. As the ions of a particular mass strike the detector, electrons are knocked out and conducted to an electron multiplying

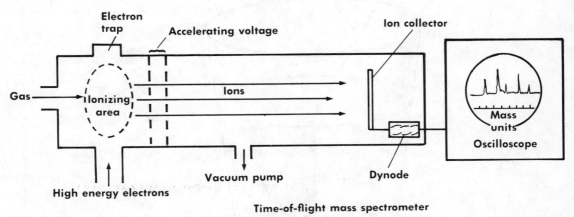

Time-of-flight mass spectrometer
Figure 15-18.

dynode. The process can be repeated rapidly and often as the current pulses are displayed on a synchronized cathode ray oscilloscope.

The diagram in Figure 15-18 illustrates the essential parts of the time-of-flight mass spectrometer.

REFRACTOMETRY

Refractometry is a method of analysis based on the different light-bending properties of matter. The speed of light, often used as an important constant in nature, is 3×10^{10} cm sec^{-1} when it moves through a vacuum. However, as light moves through some medium, its speed is reduced. The net effect of the change in velocity is the observed bending of the light. This phenomenon is illustrated when a coin is observed in the bottom of a glass of water. The coin is not found at the point where it is seen because of the light-bending effect of the water (Fig. 15-19).

The angle of refraction (bending) depends on several factors:
1. The density of the media
2. Molecular structure of compounds in the media

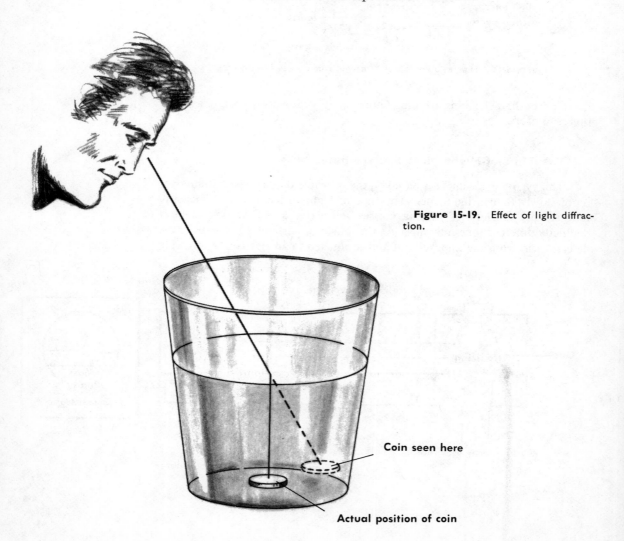

Figure 15-19. Effect of light diffraction.

Coin seen here

Actual position of coin

3. Media temperatures

4. Wavelength of light being diffracted

The material under investigation may be identified when the temperature is controlled while the media and wavelength are selected to conform with standard tables for purposes of comparison.

The measurement that is actually used for identification is called the **index of refraction,** which is constant for a particular medium. The index of refraction (n) is the ratio of the sine of the angle for light entering a medium from air, to the sine of the angle resulting from the bending of light in a medium. The sine of an angle is nothing more than the result of dividing the length of the side of a triangle opposite the angle by the hypotenuse (Fig. 15-20).

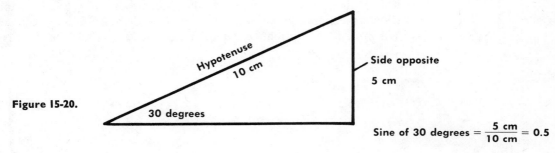

Figure 15-20.

Side opposite
5 cm

$$\text{Sine of 30 degrees} = \frac{5 \text{ cm}}{10 \text{ cm}} = 0.5$$

For practical purposes, the sine of an angle may be obtained from tables of natural trigonometric functions. Using the coin in a glass of water example, the index of refraction for water is illustrated in Figure 15-21.

$$n = \frac{\text{sine air angle}}{\text{sine water angle}} = \frac{\text{sine } 20° \, 10'}{\text{sine } 15° \, 00'}$$

$$n = \frac{0.345}{0.259} = 1.33$$

The index of refraction for water is 1.33.

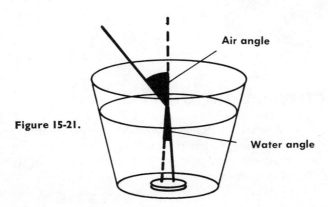

Figure 15-21.

Air angle

Water angle

The refractometer is an instrument capable of selecting a single wavelength of light. This is a more precise method for constructing comparative tables of indices of refraction. The most common wavelength is 589 mμ, which is produced by a fixed prism in the instrument. The 589 mμ wavelength corresponds to an emission line of sodium, called the D line. The reported index of refraction uses the symbol D with the temperature as a superscript. For example, the index of refraction of acetone at 20° would be noted as

$$n_D^{20°} = 1.36$$

Since the presence of impurities in a medium alters the index of refraction, the degree of contamination can be assayed by comparing the refractive indices of the sample and the established value found in a reference table. If a pure medium is unknown, the observed index of refraction can serve to aid in its identification. A typical refractometer is illustrated in Figure 15-22.

Figure 15-22. Refractometer. (Courtesy of Bausch and Lomb.)

CENTRIFUGATION

A laboratory technician often has the problem of separating particles of varying size and weight from a suspension of mixed particles. One of the most practical methods of performing this separation is described as **differential centrifugation**.

Differential centrifugation takes advantage of the commonly observed

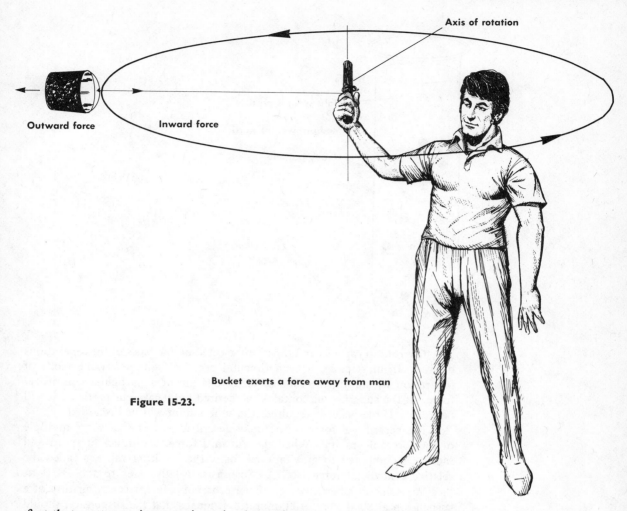

Axis of rotation

Outward force

Inward force

Bucket exerts a force away from man

Figure 15-23.

fact that matter, as it moves in a circular path, exerts a force away from the axis of rotation (Fig. 15-23).

The differential characteristic arises from the additional observation that the outward force of the whirling object depends on the following factors:

1. The *speed* of rotation
2. The *radius* of the circle
3. The *mass* of the object

If particles of varying size in a suspension need to be separated, the tube can be spun in the centrifuge for a period of time at a speed sufficient to cause the heavier particles to move to the bottom of the tube before the lighter particles. When a suspension is moving at a fixed speed at a definite distance from the axis of rotation (radius), the only variable is the particle mass. The most massive particles exert the greatest outward force and consequently move outward in the shortest time.

The illustrations in Figures 15-24, 15-25, and 15-26 demonstrate the separation made possible by selecting the speed and time of rotation. The lighter particles can be poured off (**decanted**), which effectively separates them from the heavier particles. The heaviest particles could have been separated by using a speed and time adequate for their sedimentation but inadequate for sedimentation of the lighter particles (Fig. 15-27).

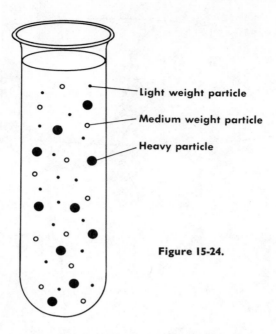

Light weight particle

Medium weight particle

Heavy particle

Figure 15-24.

The rotation speeds and times are selected on the basis of the separations desired. In most cases, an experimental protocol will specify the mode of separation by stating the time of centrifugation at a particular centrifugal force. If the supernatant contains the desired lighter weight particles, it will be saved. If the pellet is required, the supernatant will be discarded.

The centrifugal force is routinely described in terms of some multiple of the force of gravity. When the outward force, a product of speed and radius, is compared to gravitational force, the resulting value is called the **relative centrifugal force** (RCF). Scientists usually refer to the RCF as "g's" or some number times g. If the protocol calls for centrifugation of a suspension at 8000 × g for 10 minutes, it means that the suspension should

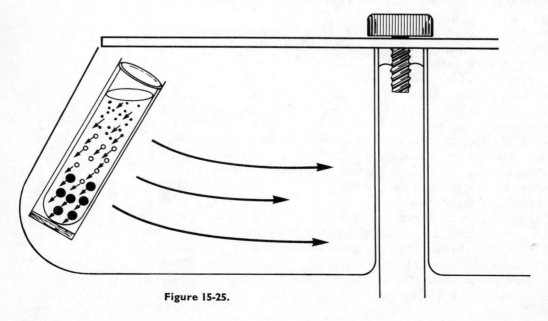

Figure 15-25.

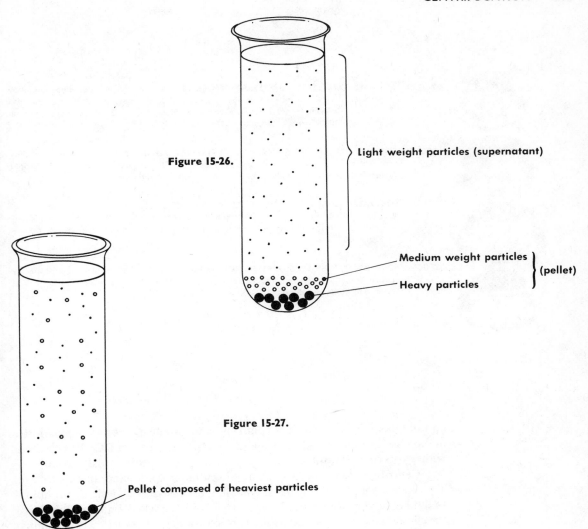

Figure 15-26.

Light weight particles (supernatant)

Medium weight particles ⎫
⎬ (pellet)
Heavy particles ⎭

Figure 15-27.

Pellet composed of heaviest particles

be "spun" at a speed which will produce an RCF equivalent to 8000 times the force of gravity. The specific speed is related to the radius of the tube holder, called the **rotor head** or simply the **rotor**. Rotor heads vary in size, weight, places for tubes, and radial distance from the motor's extended rod (rotor). High speeds of the order of 10,000 to 50,000 revolutions per minute (rpm) require light rotor heads machined for small plastic tubes. Examples of typical rotor heads are illustrated in Figure 15-28. The suitability of various types of plastic tubes for centrifugation is reviewed in Appendix A.

The centrifuges used in today's modern laboratories are often equipped with controls for regulating the speed, time, and temperature. Refrigerated centrifuges are extremely important when biological material is being used. The manufacturer usually supplies a chart which allows the technician to quickly find the speed required to produce a g-value (RCF) for a specific rotor head corresponding to the protocol specifications. If a chart is not available, a nomogram (Dole and Cortzias, 1951) may be used to calculate centrifugal force or the rpm necessary to achieve a certain centrifugal force. By placing a straightedge across the chart in order to line up two factors, such as radius

28020 SORVALL TYPE SS-34 SUPERSPEED CENTRIFUGE ROTOR, 34° angle, aluminum-alloy construction, with 8 numbered compartments for 50 ml tubes, complete with air-tight Cover and Carrying Handle [Szent-Gyorgyi & Blum 8-Tube Continuous Flow System adaptable] (17,000 rpm — 34,800 x G)

29017 SORVALL TYPE SM-24 SUPERSPEED CENTRIFUGE ROTOR, 28° angle, aluminum-alloy construction, with 24 numbered compartments for 15 ml tubes, complete with airtight Cover and Carrying Handle (16,000 rpm — 31,550 x G)

8136 SORVALL TYPE GSA HIGH-SPEED CENTRIFUGE ROTOR, 28° angle, aluminum-alloy construction, with 6 numbered compartments for 250 ml plastic bottles (or 315 ml stainless steel bottles), complete with air-tight Cover and Carrying Handle (9,500 rpm — 14,600 x G)

27004 SORVALL TYPE SE-12 SUPERSPEED CENTRIFUGE ROTOR, 40° angle, aluminum-alloy construction, with 12 numbered compartments for 15 ml tubes, complete with air-tight Cover and Carrying Handle (20,000 rpm — 41,545 x G)

Figure 15-28.

and number of g's, the proper rpm can be determined. For example, in the nomogram in Figure 15-29 a straightedge indicates that an RCF of 8000 × g with a rotor head radius of 9 cm requires a speed setting of 8,250 rpm.

Centrifuges may be classified as clinical, high speed, and ultracentrifuge. The **clinical centrifuge** is usually a small, inexpensive type designed to hold 4 to 8 tubes having volumes as large as 15 ml. The small size and portability of the clinical centrifuge permit its use in a coldroom if necessary. The maximum speed of a typical clinical centrifuge is about 1800 rpm or 2000 × g, although some models exceed these limits. The tubes are often glass (which would be preferable for some routine operations such as blood analysis), the principal advantage being the easier sterilization of glass as compared to some plastics. An example of a clinical centrifuge is shown in Figure 15-30.

The high speed centrifuge is too large for bench top use and portability. They are usually refrigerated because of biological requirements and because of the heat generated by friction with the air. A typical high speed centrifuge may produce nearly 50,000 × g with speeds up to 20,000 rpm. This instrument is ideally suited for the differential centrifugation of cellular particles. Some units are equipped with inlet and outlet ports which permit them to be *continuous flow* instruments. The clear supernatant is continuously siphoned off while the sediment builds up in the bottom of tubes while the centrifuge is in operation. An example of a high speed centrifuge is presented in Figure 15-31.

The **ultracentrifuge** is an extremely high speed instrument capable of RCF values up to nearly 500,000 × g and speeds of nearly 70,000 rpm. The ultracentrifuge may be used for the separation of very light cellular fragments, or it may be used as an analytical instrument by measuring rates

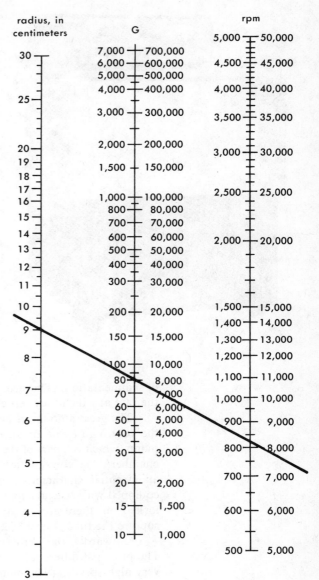

radius, in
centimeters

G

rpm

Figure 15-29.   Centrifugation nomogram.

Figure 15-30.   Clinical centrifuge.

Figure 15-31.   Refrigerated high speed centrifuge.

of sedimentation.   The size, shape, and *density*, of particles may be related to the rate at which they settle to the bottom of a tube.

The great amount of heat that would be generated by air friction due to the very high speed of an ultracentrifuge is eliminated by using a vacuum pump (which is a part of the instrument) to create a vacuum in the rotor head chamber.   The combination of powerful motor, vacuum, and refrigeration is an essential characteristic of the ultracentrifuge.   The sample tubes are equipped with special caps.   When the tube is filled to the brim, and the cap locked on, there are no air bubbles present to expand in the vacuum and rupture the tube (Fig. 15-32).

Outwardly, the ultracentrifuge is similar to any high speed instrument. The principal difference is the vacuum condition necessary for operation at very high speeds.   New models continue to be developed with the hope that eventually small molecules may be separated by differential centrifugation. The proper use of any centrifuge may be governed by a list of important steps that can be called standard operating procedure.

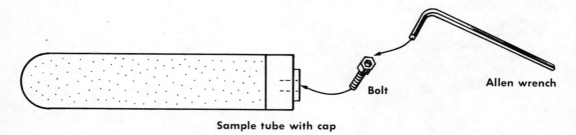

Sample tube with cap

Figure 15-32.   Ultracentrifuge tube.

Figure 15-33.

1. The tubes to be spun must be arranged in balanced pairs placed directly opposite each other in the rotor head. If an attempt is made to spin a single tube or two tubes of unequal weight, the centrifuge will vibrate violently. This can result in extensive damage to the instrument and danger to persons nearby. A good method of balancing tubes is to place them in standardized cups on either side of a double pan balance. Add or subtract sample suspension until the balance indicates zero. (Fig. 15-33). If there is enough suspension for only one tube, it should be balanced by another tube containing water. The tubes are considered to be balanced when their *weights* are equal, not necessarily their volumes.

2. If sudden vibration and noise occur, the centrifuge should be shut off immediately. Examination may show a broken tube or spillage that has upset the balance.

3. If the tube holders contain rubber pads, they should be used. This eases the stress on glass tubes and thereby reduces breakage.

4. Only tubes appropriate to the instrument should be used. Overly long or faulty tubes make breakage probable.

5. The centrifuge cover must be in place during operation. The hazards of glass bits moving at high speeds is obvious.

6. The centrifuge should never be stopped by hands or objects. The instrument must be allowed to come to a stop by itself. Magnetic braking accomplishes this in a reasonable length of time.

PROBLEMS

Set A

1. Define or explain the following terms:

a. precessional orbit
b. parallel spin
c. MHz
d. gauss

2. What are some factors affecting the NMR spectrum of hydrogen atoms?

3. What device is used to promote the chemical shift in NMR apparatus?

4. Define or explain the following terms:

a. carrier gas   c. index of refraction
b. stationary phase   d. RCF

5. Calculate the chemical shift in ppm when a 60 megacycle frequency generator produces a peak at 110 cps.

6. If an NMR spectrum shows a peak at 6.4 ppm for the tau (τ) value, what is the corresponding delta (δ) value?

7. What physical property of a gas is necessary to permit its detection by a microthermistor in a gas chromatograph?

8. Upon what fundamental principle is the magnetic field mass spectrometer based?

9. What happens to light as it passes through media having different densities?

10. Use the nomogram on page 371 to find the speed necessary to achieve a centrifugal force of 500 g's when the radius of the rotor head is 12.0 cm.

Set B

11. Define or explain the following terms:

a. resonance frequency   c. downfield
b. chemical shift   d. free radical

12. What use can be made of the integration curve on an NMR spectrogram?

13. If an infrared spectrum is used to identify an organic compound as an alcohol, how would the NMR spectrum add to an investigator's knowledge?

14. Define or explain the following terms:

a. paramagnetic   c. FID
b. GHz   d. centrifugal force

15. With a frequency generator operating at 50 MHz, a peak at 240 Hz downfield is observed. What is the chemical shift?

16. Assume the chemical shift in Problem 15 was 4.2 ppm δ. What is the tau (τ) value?

17. What special advantage does the flame ionization detector have over the thermal conductivity detector in gas chromatography?

18. What is the basis of operation for the time-of-flight mass spectrometer?

19. A centrifuge tube contains 10 ml of suspension weighing 12.5 grams. Should a tube containing 10 ml or 12.5 ml of water be used for balancing?

20. Use the nomogram on page 371 to find the RCF produced when a 15 cm radius rotor head is spun at 8000 rpm.

SUPPLEMENTARY READING

Wagner, J. J.: Nuclear Magnetic Resonance Spectroscopy—An Outline. *Chemistry*, March, 1970, p. 13.

A short and clear account of what NMR is and what it can do.

Ewing, G. W.: *Instrumental Methods of Chemical Analysis*, 3rd ed. McGraw-Hill Book Co., New York, 1969.

A superb reference text that is completely up to date.

Robinson, J. W.: *Undergraduate Instrumental Analysis*. Marcel Dekker, Inc., New York, 1970.

A very readable text including many fine illustrations.

Roberts, J. D.: *Nuclear Magnetic Resonance*. McGraw-Hill Book Co., Inc., New York, 1959.

Excellent chapters on the NMR phenomenon and the chemical shift. Clear illustrations.

Van Norman, R. W.: *Experimental Biology*. Prentice-Hall, Inc., Englewood Cliffs, New Jersey, 1963.

Contains a short, concise chapter on centrifugation.

Appendix A

TABLE A-I. ATOMIC WEIGHTS
Based on Carbon 12

Element	Symbol	Atomic Number	Atomic Weight	Element	Symbol	Atomic Number	Atomic Weight
Actinium	Ac	89	227	Mercury	Hg	80	200.59
Aluminum	Al	13	26.9815	Molybdenum	Mo	42	95.94
Americium	Am	95	[243]*	Neodymium	Nd	60	144.24
Antimony	Sb	51	121.75	Neon	Ne	10	20.183
Argon	Ar	18	39.948	Neptunium	Np	93	[237]
Arsenic	As	33	74.9216	Nickel	Ni	28	58.71
Astatine	At	85	[210]	Niobium	Nb	41	92.906
Barium	Ba	56	137.34	Nitrogen	N	7	14.0067
Berkelium	Bk	97	[249]	Nobelium	No	102	[253]
Beryllium	Be	4	9.0122	Osmium	Os	76	190.2
Bismuth	Bi	83	208.980	Oxygen	O	8	15.9994
Boron	B	5	10.811	Palladium	Pd	46	106.4
Bromine	Br	35	79.909	Phosphorus	P	15	30.9738
Cadmium	Cd	48	112.40	Platinum	Pt	78	195.09
Calcium	Ca	20	40.08	Plutonium	Pu	94	[242]
Californium	Cf	98	[251]	Polonium	Po	84	210
Carbon	C	6	12.01115	Potassium	K	19	39.102
Cerium	Ce	58	140.12	Praseodymium	Pr	59	140.907
Cesium	Cs	55	132.905	Promethium	Pm	61	[145]
Chlorine	Cl	17	35.453	Protactinium	Pa	91	231
Chromium	Cr	24	51.996	Radium	Ra	88	226.05
Cobalt	Co	27	58.9332	Radon	Rn	86	222
Copper	Cu	29	63.54	Rhenium	Re	75	186.2
Curium	Cm	96	[247]	Rhodium	Rh	45	102.905
Dysprosium	Dy	66	162.50	Rubidium	Rb	37	85.47
Einsteinium	Es	99	[254]	Ruthenium	Ru	44	101.07
Erbium	Er	68	167.26	Samarium	Sm	62	150.35
Europium	Eu	63	151.96	Scandium	Sc	21	44.956
Fermium	Fm	100	[253]	Selenium	Se	34	78.96
Fluorine	F	9	18.9984	Silicon	Si	14	28.086
Francium	Fr	87	[223]	Silver	Ag	47	107.870
Gadolinium	Gd	64	157.25	Sodium	Na	11	22.9898
Gallium	Ga	31	69.72	Strontium	Sr	38	87.62
Germanium	Ge	32	72.59	Sulfur	S	16	32.064
Gold	Au	79	196.967	Tantalum	Ta	73	180.948
Hafnium	Hf	72	178.49	Technetium	Tc	43	[99]
Helium	He	2	4.0026	Tellurium	Te	52	127.60
Holmium	Ho	67	164.930	Terbium	Tb	65	158.924
Hydrogen	H	1	1.00797	Thallium	Tl	81	204.37
Indium	In	49	114.82	Thorium	Th	90	232.038
Iodine	I	53	126.9044	Thulium	Tm	69	168.934
Iridium	Ir	77	192.2	Tin	Sn	50	118.69
Iron	Fe	26	55.847	Titanium	Ti	22	47.90
Krypton	Kr	36	83.80	Tungsten	W	74	183.85
Lanthanum	La	57	138.91	Uranium	U	92	238.03
Lawrencium	Lw	103	[257]	Vanadium	V	23	50.942
Lead	Pb	82	207.19	Xenon	Xe	54	131.30
Lithium	Li	3	6.939	Ytterbium	Yb	70	173.04
Lutetium	Lu	71	174.97	Yttrium	Y	39	88.905
Magnesium	Mg	12	24.312	Zinc	Zn	30	65.37
Manganese	Mn	25	54.9380	Zirconium	Zr	40	91.22
Mendelevium	Md	101	[256]				

* A value given in brackets denotes the mass number of the longest-lived or best-known isotope.

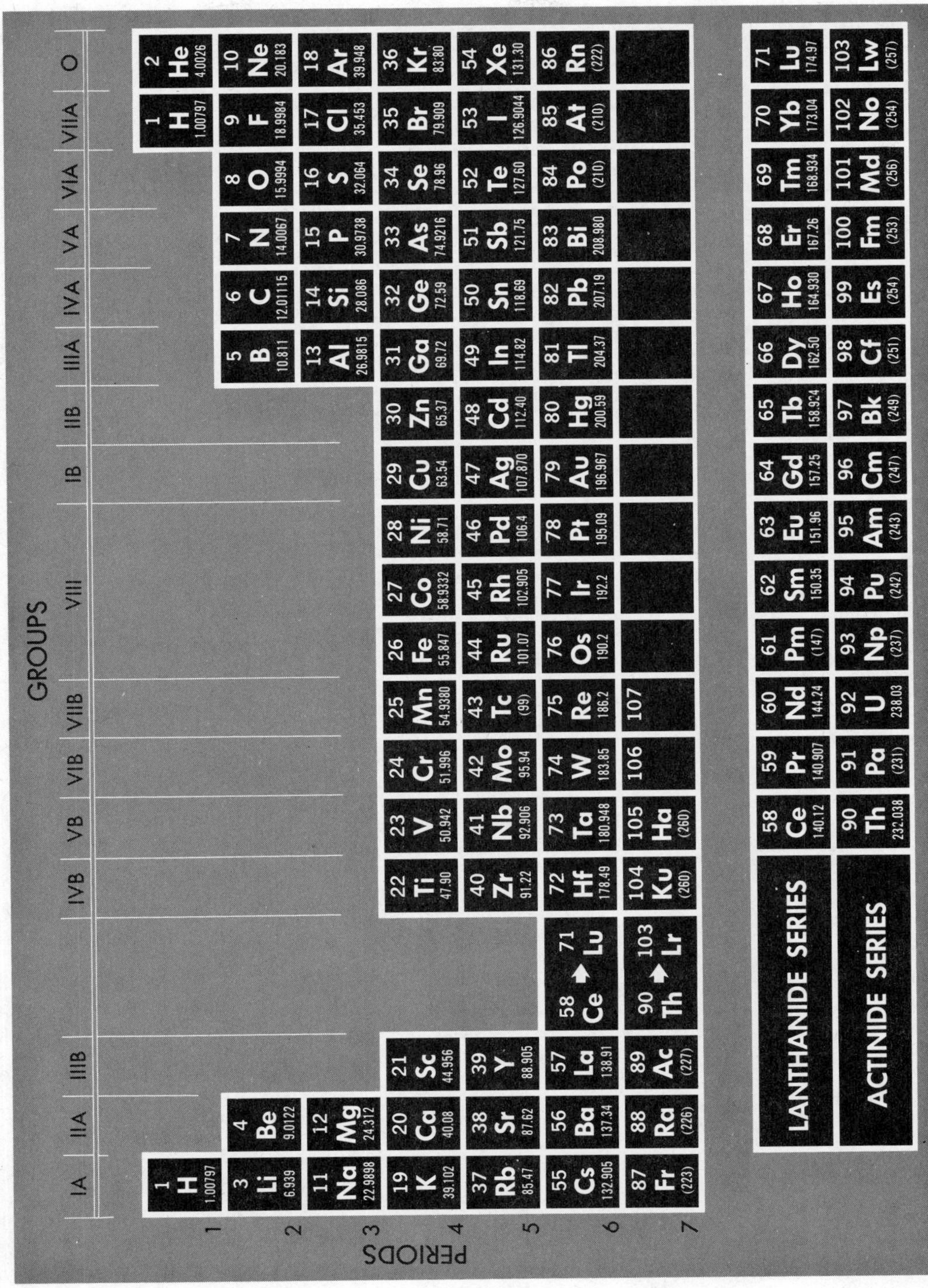

Figure A-1. A periodic chart of the elements.

TABLE A-2. SOLUBILITY OF COMMON IONIC SOLIDS IN COLD WATER

Anions	Cations	Solubility
acetate chlorate nitrate	most	soluble
chloride bromide iodide	lead(II), silver, mercury(I)	insoluble
	most others	soluble
sulfate	lead(II), mercury(I), barium, calcium	insoluble
	most others	soluble
carbonate phosphate chromate	Group I metals, ammonium	soluble
	most others	insoluble
sulfide	Group I metals, ammonium, barium, calcium, magnesium	soluble
	most others	insoluble
hydroxide	Group I metals, barium	soluble
	most others	insoluble

TABLE A-3. ACID-BASE INDICATORS

Indicator	Color Change with Increasing pH	pH Range
Thymol blue	red to yellow	1.2–2.8
Bromphenol blue	yellow to blue	3.0–4.6
Methyl orange	red to yellow	2.8–4.0
Bromcresol green	yellow to blue	3.8–5.4
Methyl red	red to yellow	4.2–6.2
Litmus	red to blue	4.5–8.3
Bromthymol blue	yellow to blue	6.0–7.6
Phenol red	yellow to red	6.8–8.4
Phenolphthalein	colorless to red	8.3–10.0
Alizarin yellow R	yellow to violet	10.1–12.0
Indigo carmine	blue to yellow	11.6–13.0
1,3,5-Trinitrobenzene	colorless to orange	12.0–14.0

TABLE A-4. INDICES OF REFRACTION $n_D^{20°}$

Material	Index	Material	Index
acetone	1.36	glycerin	1.47
aluminum oxide	1.76	ice	1.31
carbon disulfide	1.63	isopropyl alcohol	1.38
carbon tetrachloride	1.46	methyl alcohol	1.33
diamond	2.42	oleic acid	1.46
ethyl alcohol	1.36	palmitic acid	1.43
ethyl ether	1.35	propionic acid	1.39
fluorite	1.43	quartz,	
glass, zinc crown	1.52	ordinary crystal	1.54
light flint	1.57	sodium chloride	1.54
heavy flint	1.65	stearic acid	1.43
heaviest flint	1.89	water	1.33

TABLE A-5. BOILING POINTS OF COMMONLY USED SOLVENTS

Name	Additional Names	Molecular Weight	Boiling Point*
Acetic acid	Ethanoic acid	60.05	118.5^{760}
Acetoacetic acid	3-Oxobutanoic acid	102.09	$\leq 100°d†$
Acetone	2-Propanone	58.08	56.2
Aniline	Aminobenzene	93.13	184.3^{760}
Benzene		78.11	80.1
n-Butanol	1-Butanol	74.12	117.5^{760}
Carbon disulfide		76.14	45^{760}
Carbon tetrachloride	Tetrachloromethane	153.82	76.8^{760}
Chloroform	Trichloromethane	119.38	61.2^{760}
Ethanol	Ethyl alcohol	46.07	78.5
Ethyl acetate	Acetic acid ethyl ester	88.11	77.1^{760}
Ethyl ether	Diethyl ether	74.12	34.6
Ethylene dichloride	1,2-Dichloroethane	98.96	84^{760}
Heptane		100.21	98.4
Isoamyl acetate	Acetic acid 3-methylbutyl ester	130.2	142
Isoamyl alcohol	3-Methyl-1-butanol	88.15	131^{760}
Isobutyl alcohol	2-Methyl-1-propanol	74.12	108.4
Isopropyl alcohol	2-Propanol	60.09	82.4
Methanol	Carbinol; Methyl alcohol	32.04	65.0^{760}
Methyl isobutyl ketone	4-Methyl-2-propanone	100.16	116.9
Methylene chloride	Dichloromethane	84.93	40
Nitrobenzene		123.11	210.8^{760}
Petroleum ether		Varies with fraction	Appr. 40 to 120. Varies with fraction
Pyridine		79.10	115.5
Toluene	Methylbenzene	92.13	110.6
p-Xylene	1,4-Dimethylbenzene	106.16	138
m-Xylene	1,3-Dimethylbenzene	106.16	139
o-Xylene	1,2-Dimethylbenzene	106.16	144

* Superscript indicates the barometric pressure at which the boiling point was measured. If no figure is given, the barometric pressure was measured at approximately 1 atmosphere.

† d = decomposes.

TABLE A-6. VAPOR PRESSURE OF LIQUID WATER

°C	torrs	°C	torrs	°C	torrs
0	4.6	35	42.2	70	233.7
5	6.5	40	55.3	75	289.1
10	9.2	45	71.9	80	355.1
15	12.8	50	92.5	85	433.6
20	17.5	55	118.0	90	525.8
25	23.8	60	149.3	95	633.0
30	31.8	65	187.5	100	760.0

TABLE A-7. STRENGTHS OF CONCENTRATED ACIDS AND BASES

Compound	Molecular Weight	Density g ml^{-1}	Molarity
HCl	36.47	1.19	12.2
HNO_3	63.02	1.41	15.7
H_2SO_4	98.08	1.84	17.8
H_3PO_4	98.04	1.70	14.9
CH_3COOH	60.03	1.05	17.4
$NH_3(aq)$	17.03	0.90	14.8
NaOH	40.01	1.53	19.1
KOH	56.11	1.54	14.3
C_2H_5OH (absolute alcohol)	46.05	0.80	17.1

GREEK ALPHABET

A, α Alpha	H, η Eta	N, ν Nu	T, τ Tau
B, β Beta	Θ, θ Theta	Ξ, ξ Xi	Υ, υ Upsilon
Γ, γ Gamma	I, ι Iota	O, o Omicron	Φ, φ Phi
Δ, δ Delta	K, κ Kappa	Π, π Pi	X, χ Chi
E, ε Epsilon	Λ, λ Lambda	P, ρ Rho	Ψ, ψ Psi
Z, ζ Zeta	M, μ Mu	Σ, σ, ς Sigma	Ω, ω Omega

Figure A-2.

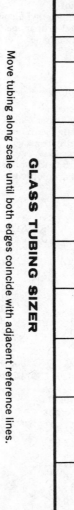

Move tubing along scale until both edges coincide with adjacent reference lines.

GLASS TUBING SIZER

Figure A-2. (Courtesy of Doerr Glass Company, Vineland, New Jersey.)

Figure A-3. (Courtesy of Doerr Glass Company, Vineland, New Jersey.)

GMCA CAP SIZER

Place the cap with the top down on the vertical line. Position the top of the cap so that it is tangent to the horizontal index line. The arc matching the bottom of the cap indicates the size.

Size refers to plastic caps

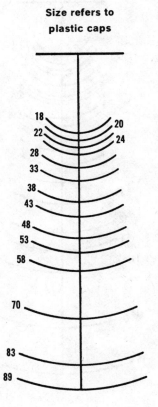

RUBBER TUBING SIZER
(Medium Wall)

Match the end of the tubing with one of the circles. Read off the dimensions as shown in the Bore Thickness and Wall Thickness columns.

Wall Thickness	Bore, Inches	
3/64	1/8	◎
3/64	5/32	◎
3/64	3/16	◎
1/16	1/4	◎
1/16	5/16	◎
1/16	3/8	◎
3/32	1/2	◎

Figure A-4. (Courtesy of Doerr Glass Company, Vineland, New Jersey.)

RUBBER STOPPER SIZER

Place the stopper with the top down on the vertical line. Position the top of the stopper so that it is tangent to the horizontal index line. The arc matching the bottom of the stopper indicates the size.

Size refers to top diameters

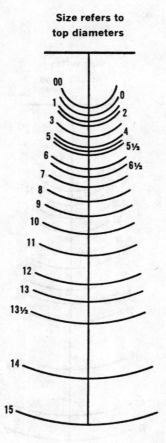

Figure A-5. (Courtesy of Doerr Glass Company, Vineland, New Jersey.)

CORK SIZER

Place the cork with the top down on the vertical line. Position the top of the cork so that it is tangent to the horizontal index line. The arc matching the bottom of the cork indicates the size.

Figure A-6. (Courtesy of Doerr Glass Company, Vineland, New Jersey.)

Size refers to top diameters

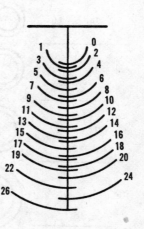

A GUIDE TO THE PHYSICAL AND CHEMICAL PROPERTIES OF THE RESINS USED IN PLASTIC LABWARE[1]

Polyolefins

This group of resins includes conventional and linear polyethylene, polyalomer, and polypropylene. All are unbreakable, nontoxic, non-contaminating. These are the only plastics lighter than water. They will easily withstand exposure to nearly all chemicals at room temperatures for up to 24 hr. Strong oxidizing agents will eventually cause embrittlement. All polyolefins can be damaged by long exposure to ultraviolet light.

Conventional polyethylene is translucent and flexible—ideally suited for squeeze-type wash bottles. A few surface-active agents, which normally do not attack conventional polyethylene, will cause cracking at points of physical stress.

Linear polyethylene is opaque, more rigid, and somewhat harder than conventional polyethylene. It possesses virtually the same excellent chemical resistance.

Polypropylene is translucent, and the most rigid polyolefin. It can be autoclaved repeatedly, and has no known solvent at room temperatures. Polypropylene is slightly more susceptible to strong oxidizing agents than conventional polyethylene. Because its coefficient of thermal expansion approximates that of water near room temperatures, polypropylene is excellent for graduated cylinders, and other calibrated ware.

Polyallomer is translucent, rigid, and the lightest of all commercially available thermoplastics. It displays the same excellent chemical resistance of polyethylene and polypropylene, is even more abrasion resistant than polypropylene, and has higher impact strength. Because its surface is slightly more resilient than polypropylene, it provides an excellent friction fit as in the Büchner funnel, and the standard taper stoppers. Polyalomer and polypropylene both have excellent dimensional stability.

Teflon

Both TFE and FEP have phenomenal chemical resistance, excellent adhesion resistance, are autoclavable, and unbreakable.

Teflon TFE is opaque, white, and has the lowest coefficient of friction of any solid. It makes a superior stopcock plug and separatory funnel plug because of its low friction and tight seal.

Teflon FEP is flexible, transparent, has a slight bluish cast, and a heavy feel because of its higher density. It resists all known chemicals except molten alkali metals, elemental fluorine, and fluorine precursors at elevated temperatures. FEP withstands temperatures from −270°C to +205°C. It has virtually zero moisture absorption, and may be sterilized repeatedly by all known chemical and thermal methods. Nonwetting surface is ideal for beakers and bottles because liquids drain completely.

---

[1] Courtesy of the Nalge Company, Division of Sybron Corporation.

TABLE A-8. PHYSICAL PROPERTIES OF PLASTICS

	Conventional Polyethylene	Linear Polyethylene	Polyallomer	Polypropylene
Temperature limit, °C	80	120	130	135
Specific gravity	0.92	0.95	0.90	0.90
Tensile strength, psi	2000	4000	2900	5000
Brittleness temperature, °C	−100	−100	−40	0
Water absorption, %	<0.01	<0.01	<0.02	<0.02
Flexibility	excellent	rigid	slight	rigid
Transparency	translucent	opaque	translucent	translucent
Relative O_2 permeability (Arbitrary scale for comparison purposes only)	0.40	0.08	0.20	0.11
Autoclavable	no	with caution	yes	yes

Styrene and Copolymer

Both high-impact polystyrene, and the copolymer, styrene-acrylonitrile, are rigid and have excellent dimensional stability. They are highly transparent except where pigment has been added.

Polystyrene has good chemical resistance to aqueous solutions. It can be easily molded into precision shapes, and is relatively inexpensive.

Styrene-acrylonitrile is a copolymer and has higher impact strength than polystyrene.

Polycarbonate

This is window-clear, amazingly strong, and rigid. It is autoclavable, nontoxic, unbreakable, and the toughest of all thermoplastics. Its strength and dimensional stability make it ideal for high-speed centrifuge ware. Polycarbonate is resistant to aliphatic hydrocarbons, alcohols, vegetable oils, ethers, neutral and acid salts, oxidizing and reducing agents. It should not be exposed to bases, halogenated hydrocarbons, aromatic hydrocarbons and their derivatives, esters, ethers, ketones, or amines.

Polyvinyl Chloride

This is transparent and has a slight bluish cast. The new narrow mouth bottles are relatively thin-walled and can be flexed slightly. PVC has good

TABLE A-8. (*Continued*)

	Teflon FEP	Polycarbonate	General Purpose Polystyrene	Styrene-Acrylonitrile	Polyvinyl Chloride	Nalgon Vinyl Tubing (PVC)
Temperature limit, °C	205	135	70	95	70	105
Specific gravity	2.15	1.20	1.07	1.07	1.34	1.20
Tensile strength, psi	3000	8000	6000	11,000	6500	1800
Brittleness temperature, °C	−270	−135	*	−25	−30	−45
Water absorption, %	<0.01	0.35	0.05	0.23	0.06	0.10
Flexibility	excellent	rigid	rigid	rigid	rigid	excellent
Transparency	transparent	clear	clear	clear	clear	clear
Relative O_2 permeability	0.59	0.15	0.11	0.03	0.01	0.01
Autoclavable	yes	yes	no	no	no	no

* Normally somewhat brittle at room temperatures.
(Courtesy of the Nalge Company, Division of Sybron Corporation)

chemical resistance at room temperatures. Exceptions are ketones, aromatic hydrocarbons, and esters. It also has extremely good resistance to oils and very low permeability to most gases. PVC bottles are well-suited for shipping oil or water samples.

INTERPRETATION OF CHEMICAL RESISTANCE

Table A-9 should be taken as a general guide only. It states that, for most compounds of the class indicated, the specific material is rated either Excellent, Good, Fair, or Not Recommended.

Because so many factors can affect the chemical resistance of a given product, we recommend you test under your own conditions if any doubt exists. The combination of compounds of two or more classes may cause an undesirable chemical effect. Other factors affecting chemical resistance include temperature, pressure and other stresses, length of exposure, and concentration of the chemical. As the maximum useful temperature of the plastic is approached, resistance to attack decreases. (Courtesy the Nalge Company, Division of Sybron Corporation.)

TABLE A-9. CHEMICAL RESISTANCE OF PLASTICS

Class of Substances	Conventional Poly-ethylene	Linear Poly-ethylene	Poly-allomer	Poly-propylene	Teflon FEP	Poly-carbonate	General Purpose Poly-styrene	Styrene Acrylo-nitrile	Poly-vinyl Chloride
Acids, inorganic	E	E	E	E	E	E	N	G	G
Acids, organic	G	E	E	E	E	F	G	G	G
Alcohols	E	E	E	E	E	G	G	G	G
Aldehydes	G	G	G	G	E	F	N	N	N
Amines	G	G	G	G	E	N	G	G	N
Bases	E	E	E	E	E	N	G	G	G
Dimethyl sulfoxide (DMSO)	E	E	E	E	E	N	N	N	N
Esters	G	G	G	G	E	N	N	N	N
Ethers	G	G	G	G	E	N	F	G	N
Foods	E	E	E	E	E	E	E	G	G
Glycols	G	E	E	E	E	G	G	G	F
Hydrocarbons, aliphatic	G	G	G	G	E	F	N	F	G

TABLE A-9. *(Continued)*

Class of Substances	Conventional Polyethylene	Linear Polyethylene	Polyallomer	Polypropylene	Teflon FEP	Polycarbonate	General Purpose Polystyrene	Styrene Acrylonitrile	Polyvinyl Chloride
Hydrocarbons, aromatic	G	G	G	G	E	N	N	N	N
Hydrocarbons, halogenated	G	G	G	G	F				N
Ketones	G	G	G						N
Mineral oil	E	E	E						E
Oils, essential	G	G	G	G					N
Oils, lubricating	G	E	E	E					E
Oils, vegetable	E	E	E	E					E
Proteins, unhydrolyzed	E	E	E	E					G
Salts	E	E	E	E					E
Silicones	G	E	E	E					G
Water	E	E	E	E					

E Excellent. Long exposures (up to 1 yr) at room temperature
G Good. Short exposures (less than 24 hr) at room temperature
F Fair. Short exposures at room temperature cause little or no
N Not Recommended. Short exposures may cause permanent
(Courtesy of the Nalge Company, Division of Sybron Corporation

TABLE A-I0. CHEMICAL RESISTANCE OF PLASTICS UNDER STRESS

Reagent	Poly-ethylene	Poly-propylene	Poly-carbonate	Cellulose nitrate
Acetaldehyde (100%)	M	S	U	U
Acetic acid (5%)	S	S	S	S
Acetic acid (60%)	M	U	U	U
Acetic acid (Glacial)	U	U	U	U
Acetone	S	S	U	U
Acid cleaning solution	M	M	U	U
Allyl alcohol	S	S	U	U
Alum concentrated	S	S	—	—
Aluminum chloride	S	S	S	S
Aluminum fluoride	S	S	U	U
Ammonium acetate	S	S	S	S
Ammonium carbonate	S	S	U	—
Ammonium hydroxide (10%)	S	S	U	S
Ammonium hydroxide (conc.)	S	S	U	U
Ammonium persulfate (sat'd)	S	—	—	—
Ammonium sulfide	S	S	U	U
Amyl alcohol	S	S	S	—
Aniline	M	—	U	—
Aqua regia	U	U	U	U
Benzene	U	U	U	U
Benzyl alcohol	U	U	U	U
Boric acid	S	S	S	S
Brine	S	S	S	S
N-Butyl alcohol	S	S	M	S
Calcium chloride	S	S	S	S
Calcium hypochlorite	S	S	S	—

Reagent	Poly-ethylene	Poly-propylene	Poly-carbonate	Cellulose nitrate
Ethyl acetate	S	S	S	S
Ethyl alcohol (50%)	S	S	U	U
Ethyl alcohol (95%)	S	U	U	S
Ethylene dichloride	U	S	S	—
Ethylene glycol	S	S	—	—
Ferric chloride	S	S	—	—
Fluoboric acid	S	M	U	U
Formaldehyde (40%)	S	S	—	—
Formic acid (100%)	S	U	S	—
Fuel oil	M	M	—	U
Gallic acid	S	U	U	U
Gasoline (refined)	U	—	—	—
Gasoline (sour)	U	—	—	—
2-Heptyl	S	S	S	S
Hydroformic acid (100%)	S	S	U	U
Hydrochloric acid (10%)	S	S	S	S
Hydrochloric acid (50%)	S	S	M	M
Hydrochloric acid (Conc.)	S	M	M	M
Hydrofluoric acid (10%)	S	S	U	U
Hydrofluoric acid (100%)	S	S	U	U
Hydrogen peroxide (3%)	S	S	S	S
Hydrogen peroxide (100%)	S	S	—	—
Hydroquinone	S	—	—	—
Isobutyl alcohol	S	S	U	U
Isopropyl alcohol	S	S	S	—
Lactic acid (20%)	S	S	—	—
Lactic acid (100))	S	S	—	—
Lauryl alcohol %	S	S	U	U
Lead acetate	S	S	S	S

Reagent	Poly-ethylene	Poly-propylene	Poly-carbonate	Cellulose nitrate
Phenyl ethyl alcohol	S	S	S	S
Phosphoric acid (10%)	S	S	S	M
Phosphoric acid (conc.)	S	S	U	—
Phosphorus trichloride	S	M	M	—
Potassium acetate	S	S	—	S
Potassium carbonate	S	S	S	S
Potassium chlorate	S	S	S	S
Potassium chloride	S	S	U	U
Potassium hydroxide (5%)	S	S	U	U
Potassium hydroxide (conc.)	S	—	—	—
Potassium permanganate	S	U	—	—
Silicic acid	M	—	—	—
Silicone fluids	S	S	—	—
Silver cyanide	S	S	++	++
Soap solutions	S	S	S	S
Sodium bisulfate	S	S	S	S
Sodium borate	S	S	S	S
Sodium carbonate	S	S	S	S
Sodium chloride (10%)	S	—	—	—
Sodium chloride (sat'd)	S	—	—	—
Sodium dichromate	S	—	S	S
Sodium hydroxide (less than 1%)	S	U	U	U
Sodium hydroxide (10%)	S	U	U	U
Sodium hydroxide (conc.)	S	S	U	S
Sodium hypochlorite	S	S	U	—
Sodium nitrate (10%)	S	S	—	—
Sodium peroxide	S	U	—	S

Chemical				
Carbon tetrachloride	U	U	S	S
Cetyl alcohol	S	—	S	U
Chlorine water	S	S	S	S
Chlorobenzene	U	U	U	S
Chloroform	U	U	U	U
Chromic acid (10%)	S	U	M	S
Chromic acid (50%)	S	S	U	S
Citric acid (10%)	S	U	U	M
Cresol	M	S	S	U
Cyclohexyl alcohol	S	—	S	U
Diacetone	S	—	S	U
Diazo salts	S	U	U	S
Diethyl ketone	M	U	U	M
Dimethylformamide	S	S	S	S
Dioxane	M	U	U	M
Distilled water	S	U	—	S
Ether diethyl	M	U	U	U

Chemical				
Linseed oil	S	—	—	S
Maleic acid	U		—	
Manganese salts	S	S	S	U
Magnesium hydroxide	U	U	S	U
Mercury	U	U	U	U
Methyl alcohol	S	U	U	U
Methyl ethyl ketone	S	U	U	U
Methyl salicylate	S	S	S	U
Methylene chloride	U	S	S	S
Nickel salts	S	S	S	S
Nitric acid (10%)	S	M	M	M
Nitric acid (50%)	U	U	U	U
Nitric acid (95%)	U	S	S	S
Oleic acid	S	S	S	S
Oxalic acid	U	—	S	—
Perchloric acid (10%)	U	—	—	—
Phenol	S	U	U	U

Chemical				
Sodium sulfide	S	S	S	—
Sodium thiosulfate	S	S	S	S
Sulfuric acid (10%)	S	S	U	U
Sulfuric acid (50%)	S	S	S	S
Sulfuric acid (75%)	U	S	U	U
Sulfuric acid (conc.)	S	S	S	U
Tannic acid	S	—	U	S†
Toluene	—	U	U	—
Trichlorethylene	S	U	M	—
Trisodium phosphate	S	M	U	U
Turpentine	M	U	S	S
Urea	S	S	S	S
Urine	S	S	U	S†
Xylene	U	U	U	S
Zinc chloride	S	S	S	S

KEY:

S Satisfactory.

M Moderate attack, may be satisfactory for use in a centrifuge depending on length of exposure, speed involved, etc.; suggest testing under actual conditions of use.

U Unsatisfactory, not recommended.

— Performance unknown; suggest testing, using a blank to avoid loss of valuable material.

† If pure.

‡ Because of the many types of soap solutions, it is impossible to predict performance.

(Courtesy of the Nalge Company, Division of Sybron Corporation)

Note: This chemical-resistance chart is the product of several years' experience in researching plastics and their application to centrifuge ware. A plastic material may offer satisfactory chemical resistance in static use but be unsatisfactory for use under centrifugal force.

Because mechanical stress exists in a centrifuge, the ware can fail when exposed to both chemical and physical stress. For example, a chemical may not cause stress cracking when simply stored in a bottle or tube. But, centrifuge the chemical in that bottle or tube, and stress cracking may appear in a short time. Since no organized chemical resistance data under centrifugal conditions exist, these recommendations are conservative and are intended to serve as a guide.

TABLE A-II. TRANSMITTANCE TO ABSORBANCE CONVERSION TABLE

Per cent Transmittance to Absorbance

	0.0	0.1	0.2	0.3	0.4	0.5	0.6	0.7	0.8	0.9
100	.000	.000	.000	.000	.000	.000	.000	.000	.000	.000
99	.004	.004	.003	.003	.003	.002	.002	.001	.001	.000
98	.008	.008	.008	.007	.007	.006	.006	.006	.005	.005
97	.013	.013	.012	.012	.011	.011	.011	.010	.010	.009
96	.017	.017	.017	.016	.016	.015	.015	.014	.014	.014
95	.022	.022	.021	.021	.020	.020	.019	.019	.018	.018
94	.026	.026	.026	.025	.025	.024	.024	.023	.023	.022
93	.031	.031	.030	.030	.029	.029	.028	.028	.027	.027
92	.036	.035	.035	.034	.034	.034	.033	.033	.032	.032
91	.041	.040	.040	.039	.039	.038	.038	.037	.037	.036
90	.045	.045	.044	.044	.043	.043	.043	.042	.041	.041
89	.050	.050	.049	.049	.048	.048	.047	.047	.046	.046
88	.055	.055	.054	.054	.053	.053	.052	.052	.051	.051
87	.060	.059	.059	.059	.058	.058	.057	.057	.056	.056
86	.065	.065	.064	.064	.063	.063	.062	.062	.061	.061
85	.070	.070	.069	.069	.068	.068	.067	.067	.066	.066
84	.075	.075	.074	.074	.073	.073	.072	.072	.071	.071
83	.080	.080	.080	.079	.078	.078	.077	.077	.076	.076
82	.086	.085	.085	.084	.084	.083	.083	.082	.081	.081
81	.091	.091	.090	.096	.089	.088	.088	.087	.087	.086
80	.096	.096	.095	.095	.094	.094	.093	.093	.092	.092
79	.102	.102	.101	.101	.100	.100	.099	.098	.098	.097
78	.108	.107	.107	.106	.106	.105	.104	.104	.103	.103
77	.113	.113	.112	.112	.111	.111	.110	.110	.109	.108
76	.119	.119	.118	.117	.117	.116	.116	.115	.115	.114
75	.125	.124	.124	.123	.122	.122	.121	.121	.120	.120
74	.130	.130	.129	.129	.128	.128	.127	.127	.126	.126
73	.136	.136	.135	.135	.134	.133	.133	.132	.132	.131
72	.142	.142	.141	.140	.140	.139	.139	.138	.137	.137
71	.148	.148	.147	.147	.146	.145	.145	.144	.143	.143
70	.155	.154	.153	.153	.152	.151	.151	.150	.150	.149
69	.161	.160	.160	.159	.158	.158	.157	.156	.156	.155
68	.167	.167	.166	.165	.164	.164	.163	.163	.162	.162
67	.174	.173	.172	.172	.171	.170	.169	.169	.168	.168
66	.180	.179	.179	.178	.177	.177	.176	.175	.175	.174
65	.187	.186	.185	.185	.184	.183	.183	.182	.181	.181
64	.193	.193	.192	.192	.191	.190	.189	.189	.188	.187
63	.200	.200	.199	.198	.198	.197	.196	.196	.195	.194
62	.207	.207	.206	.205	.205	.204	.203	.202	.202	.201
61	.214	.214	.213	.212	.211	.211	.210	.209	.209	.208
60	.221	.221	.220	.219	.218	.218	.217	.216	.216	.215
59	.229	.228	.227	.227	.226	.225	.225	.224	.223	.222
58	.236	.235	.235	.234	.234	.233	.232	.231	.231	.230
57	.244	.243	.243	.242	.241	.240	.239	.239	.238	.237
56	.252	.251	.250	.250	.249	.248	.247	.246	.246	.245
55	.260	.259	.258	.257	.257	.256	.255	.254	.253	.253
54	.267	.266	.266	.265	.264	.263	.263	.262	.261	.260
53	.275	.275	.274	.273	.272	.271	.271	.270	.269	.268
52	.284	.283	.282	.281	.280	.279	.279	.278	.277	.276
51	.292	.291	.290	.290	.289	.288	.287	.286	.285	.284
50	.301	.300	.299	.298	.298	.297	.296	.295	.294	.293
49	.310	.309	.308	.307	.306	.306	.305	.304	.303	.302
48	.319	.318	.317	.316	.315	.314	.313	.313	.312	.311
47	.328	.327	.326	.325	.324	.323	.322	.322	.321	.320
46	.337	.336	.335	.331	.334	.333	.332	.331	.330	.329
45	.347	.346	.345	.344	.343	.342	.341	.340	.339	.338
44	.357	.356	.355	.354	.353	.352	.351	.350	.349	.348
43	.367	.366	.365	.364	.363	.362	.361	.360	.359	.358
42	.377	.376	.375	.374	.373	.372	.371	.370	.369	.368
41	.387	.386	.385	.384	.383	.382	.381	.380	.379	.378
40	.398	.397	.396	.395	.394	.393	.392	.390	.389	.388
39	.409	.408	.407	.406	.405	.404	.402	.401	.400	.399
38	.420	.419	.418	.417	.416	.415	.414	.412	.411	.410

TABLE A-II. *(Continued)*

Per cent Transmittance to Absorbance

	0.0	0.1	0.2	0.3	0.4	0.5	0.6	0.7	0.8	0.9
37	.432	.431	.430	.428	.427	.426	.425	.424	.423	.424
36	.444	.443	.441	.440	.439	.438	.437	.435	.434	.433
35	.456	.455	.454	.452	.451	.450	.449	.447	.446	.445
34	.469	.467	.466	.465	.463	.462	.461	.460	.459	.457
33	.481	.480	.479	.477	.476	.475	.474	.472	.471	.470
32	.495	.494	.492	.491	.490	.488	.487	.485	.484	.483
31	.509	.507	.506	.504	.503	.502	.500	.499	.498	.496
30	.522	.521	.520	.518	.517	.516	.514	.513	.511	.510
29	.537	.536	.534	.533	.531	.530	.528	.527	.526	.524
28	.553	.551	.550	.548	.546	.545	.543	.542	.540	.539
27	.568	.567	.565	.563	.562	.560	.559	.557	.556	.554
26	.585	.583	.581	.580	.578	.577	.575	.573	.572	.570
25	.602	.600	.598	.597	.595	.593	.592	.590	.588	.586
24	.620	.618	.616	.614	.612	.610	.609	.607	.605	.604
23	.638	.636	.634	.632	.631	.629	.627	.625	.623	.621
22	.657	.655	.653	.651	.650	.648	.646	.644	.642	.640
21	.678	.676	.673	.671	.669	.667	.665	.663	.661	.659
20	.699	.697	.695	.692	.690	.688	.686	.684	.682	.680
19	.721	.719	.717	.715	.712	.710	.708	.706	.703	.701
18	.745	.742	.740	.738	.735	.733	.731	.728	.726	.724
17	.770	.767	.764	.762	.760	.757	.755	.752	.750	.747
16	.796	.793	.791	.788	.785	.783	.780	.777	.775	.772
15	.824	.821	.818	.815	.812	.810	.807	.804	.801	.799
14	.854	.851	.848	.845	.842	.839	.836	.833	.830	.827
13	.886	.883	.880	.876	.873	.870	.867	.863	.860	.857
12	.921	.917	.914	.910	.907	.903	.900	.896	.893	.890
11	.959	.955	.951	.947	.943	.939	.936	.932	.928	.925
10	1.000	.996	.991	.987	.983	.979	.975	.971	.967	.963
9	1.046	1.041	1.036	1.031	1.027	1.022	1.017	1.013	1.009	1.004
8	1.097	1.092	1.086	1.081	1.076	1.071	1.066	1.060	1.056	1.051
7	1.155	1.149	1.143	1.137	1.131	1.125	1.119	1.114	1.108	1.102
6	1.222	1.215	1.208	1.201	1.194	1.187	1.181	1.174	1.168	1.162
5	1.301	1.292	1.284	1.276	1.268	1.260	1.252	1.244	1.237	1.229
4	1.398	1.387	1.377	1.367	1.357	1.347	1.337	1.328	1.319	1.310
3	1.522	1.509	1.495	1.481	1.469	1.456	1.444	1.432	1.420	1.409
2	1.699	1.678	1.657	1.638	1.620	1.602	1.585	1.568	1.553	1.537
1	2.000	1.959	1.921	1.886	1.854	1.824	1.798	1.770	1.745	1.721
0		3.000	2.699	2.522	2.398	2.301	2.222	2.155	2.097	2.046

(Courtesy Beckman Instruments, Inc.)

GLASSWARE CLEANING METHODS

The list of cleaning solutions and methods for specific contaminants presented here should cover most of the glassware cleaning operations a technician may have to perform. If an automatic dishwasher is used, glassware should still be given a thorough distilled water rinse before drying.

I. *Cleaning Solutions*

    A. *Chromic Acid Mixture.* This combination of sulfuric acid and sodium dichromate may be purchased under any one of a variety of trademarks from chemical supply houses, or it may be prepared cautiously as follows:

        1. Dissolve about 60 g of potassium dichromate in hot water.

        2. Slowly (to avoid splattering) add enough concentrated sulfuric acid to make a liter.

    B. *Basic Permanganate Mixture.* Dissolve a combination of 20 g of potassium permanganate and 50 g of sodium hydroxide in a liter of water.

    C. *Hot Detergent Mixture.* Laboratory detergents are commercially available in powder or tablet form. The detergents should be used sparingly to avoid excess sudsing.

    D. *Fuming Sulfuric Acid*

    E. *Acetone*

    F. *Nitric Acid*

II. *Cleaning Methods*

    A. *Stopcock Grease* (petroleum base)

        1. Dissolve grease in acetone.

        2. Wash with detergent.

        3. Rinse with tap water four times.

        4. Rinse with distilled water three times.

    B. *Stopcock Grease* (silicone base)

        1. Soak for one half to two hours in fuming sulfuric acid or deca hydronaphthalene (decalin)

        2. Rinse with acetone.

        3. Wash with detergent.

        4. Rinse with tap water four times.

        5. Rinse with distilled water three times.

    C. *Bacteriological Contamination*

        1. Soak in chromic acid mixture.

        2. Rinse with tap water four times.

        3. Rinse with distilled water three times.

    D. *Fat and Oil Contamination*

        1. Soak in basic permanganate mixture.

        2. Rinse with tap water four times

        3. Rinse with distilled water three times.

    E. *Blood, body fluids, and albuminous "crusts"*

        1. Soak in chromic acid mixture.

        2. Rinse with tap water four times.

        3. Rinse with distilled water three times.

    F. *Iron Stains*

        1. Rinse with about 10 molar hydrochloric acid.

2. Rinse with tap water four times.
3. Rinse with distilled water three times.

G. *Permanganate Stains*
  1. Swirl a small volume of acidified iron (II) sulfate solution in the stained glass.
  2. Wash with detergent.
  3. Rinse with tap water four times.
  4. Rinse with distilled water three times.

H. *Enzyme Contamination*
  1. Rinse with dilute nitric acid.
    Note: Chromic acid should *not* be used since any trace amount of chromium ion may inhibit enzyme activity.
  2. Wash with detergent.
  3. Rinse with tap water four times.
  4. Rinse with distilled water three times.

The rinsing operation must always be carried out thoroughly.* Trace amounts of metal ions that remain due to carelessness may seriously affect organism growth and testing procedures.

When a piece of glassware must be dried quickly after it has been cleaned, a high grade of acetone may be used.

The cleaning operation is usually simplified if the pipets, beakers, graduated cylinders, test tubes, and flasks are immediately placed in a detergent solution after use.

Delicate (and expensive) spectrophotometer cuvets must be handled with extreme care and never exposed to the harsher cleaning agents.

---

* The test for truly clean glassware in all cases is the absence of water droplets that adhere to glass when contaminants are still present.

Appendix B
Answers to Selected
Numerical Problems

Chapter 1

1.1   a. 3.20×10^2 d. 6.07×10^7 g. 1.92
     b. 1.60×10^4 e. 5×10^{-2} h. 1.49×10^{-7}
     c. 9.124×10^3 f. 2.38×10^{-4}

1.2   a. 183 c. 516 e. 378
     b. 181 d. 422 f. 735

1.3   a. 1.83×10^4 c. 0.516 e. 3.78×10^3
     b. 1.81×10^8 d. 4.22×10^2 f. 7.35×10^{-5}

1.4   a. 102.40 c. 3.60×10^{-6}
     b. 4.02 d. 10.6

1.5   a. 2.5×10^3 m   c. 1.25×10^{-5} cm   e. 2.5×10^{-2} mm
     b. 3.2×10^2 mm   d. 2.8×10^2 mμ f. 6.2×10^2 mμ

1.6   a. 3.2×10^{-2} kg   c. 4×10^3 μg   e. 4×10^{-4} g
     b. 2.2×10^3 mg   d. 0.5 mg   f. 1.5×10^3 μg

1.7   a. 0.23ℓ c. 30 ml   e. 90 ml
     b. 1.50 ml   d. 85 ml   f. 5λ

1.8   a. $94.8°F$ c. $-15.7°C$
     b. $26.8°C$ d. $-328°F$

1.9   a. $278°K$ c. $127°C$
     b. $253°K$ d. $0°C$

1.10 3×10^8 m sec^{-1}

Chapter 2

2.4 0.6 g ml^{-1}
2.5 0.19 ml
2.6 8.0 g
2.7 1.84
2.9 2×10^3 ohms
2.10 114 volts

Chapter 3

3.3   2.8 kcal
3.4 2.8×10^3 cm sec^{-1}
3.5 3×10^2 ergs
3.6 2.6×10^{-17} kcal
3.7 9.4×10^{14} sec^{-1}
3.8 3.3×10^{-11} erg
3.9   32.2

Chapter 4

4.6   60 g
4.7 3.01×10^{23} atoms, 6.6×10^{-23} g
4.8   448 ml
4.9   a. C = 28.3 per cent, O = 71.7 per cent
      b. Na = 43.4 per cent, C = 11.3 per cent, O = 45.3 per cent
      c. C = 52.2 per cent, H = 13.0 per cent, O = 34.8 per cent
4.12 0.32 g

Chapter 5

5.3   27.3 mg ml^{-1}
5.4   28.4 mg ml^{-1}
5.5   3.9 per cent error
5.6   a. 0.60 ml
      b. 1.36 ml
5.7   0.20 ml
5.8 108 ± 2 mg

Chapter 7

7.1   a. 11.9 g solute plus water to 35 ml
      b. 25.8 g solute plus water to 160 ml
      c. 0.76 g solute plus water to 450 ml
7.2   b. 2.66 g solute plus water to 120 ml
      c. 52.5 g solute plus water to 2.5ℓ
7.3   0.8 g
7.4   0.89 M
7.5   a. dilute 29.2 ml stock to 500 ml
      b. dilute 7.0 ml stock to 350 ml
      c. dilute 6.7 ml stock to 50 ml
7.6   a. 10.3 M, 10.3 N
      b. 17.6 M, 35.2 N
      c. 3.55 M, 3.55 N
7.7   0.143 M $CaCl_2$, 0.143 M NaCl
7.8 1.65×10^{20} molecules per ml
7.9   20 per cent
7.10 0.2 M at 20°C, 1.5 M at 100°C

Chapter 8

8.4   3.18 N, 3.18 M
8.5   0.25 ml
8.6   2.44
8.7   8.11
8.8 3.8×10^{-4} M
8.10 0.75

Chapter 9

9.3   a. 5+   b. 4+   c. 5+   d. 5+   e. 6+
9.6   8.2 mg
9.7   447 ml
9.8   2.78 M
9.9   0.38 N
9.10 0.27 N, 0.053 M

Chapter 10

10.2 4.82×10^4 coulombs
10.3   80.5 hours
10.5   100.1 grams
10.7   0.01 mho
10.8   c. 0.12 volts
10.10 a. 0.48 volts
       b. 0.46 volts
10.12 a. -18.5 kcal, feasible
       b. $+58.8$ kcal, not feasible

Chapter 13

13.5   56.3 days
13.10 1,080 years
13.11 22.8 days
13.12 0.086
13.13 16.3 grams
13.14 3.5×10^3 nCi, 1.3×10^5 dps

Chapter 14

14.3   0.398
14.5   62.5
14.6 7.2×10^{-3} M
14.7 2.5×10^{-4} M

Chapter 15

15.5   1.83 ppm
15.6   3.6
15.10 2,140 rpm

Index